T0184444

Lecture Notes in Mathematics

Edited by J.-M. Morel, F. Takens and B. Teissier

Editorial Policy for Multi-Author Publications: Summer Schools / Intensive Courses

1. GENERAL. Lecture Notes aim to report new developments in all areas of mathematics and their applications – quickly, informally and at a high level. Manuscripts should be reasonably self-contained and rounded off. Thus they may, and often will, present not only results of the author but also related work by other people. They should provide sufficient motivation, examples and applications. There should also be an introduction making the text comprehensible to a wider audience. This clearly distinguishes Lecture Notes from journal articles or technical reports which normally are very concise. Articles intended for a journal but too long to be accepted by most journals, usually do not have this "lecture notes" character.

2. In general SUMMER SCHOOL and other similar INTENSIVE COURSES are held to present mathematical topics that are close to the frontiers of recent research to an audience at the beginning or intermediate graduate level, who may want to continue with this area of work, for a thesis or later. This makes demands on the didactic aspects of the presentation. Because the subjects of such schools are advanced, there often exists no textbook, and so ideally, the publication resulting from such a school could be a first approximation to such a textbook. Usually several authors are involved in the writing, so it is not always simple to obtain a unified approach to the presentation.

   For prospective publication in LNM, the resulting manuscript should not be just a collection of course notes, each of which has been developed by an individual author with little or no co-ordination with the others, and with little or no common concept. The subject matter should dictate the structure of the book, and the authorship of each part or chapter should take secondary importance. Of course the choice of authors is crucial to the quality of the material at the school and in the book, and the intention here is not to belittle their impact, but simply to say that the book should be planned to be written by these authors jointly, and not just assembled as a result of what these authors happen to submit.

   This represents considerable preparatory work (as it is imperative to ensure that the authors know these criteria before they invest work on a manuscript), and also considerable editing work afterwards, to get the book into final shape. Still it is the form that holds the most promise of a successful book that will be used by its intended audience, rather than yet another volume of proceedings for the library shelf.

3. Manuscripts should be submitted (preferably in duplicate) either to one of the series editors or to Springer-Verlag, Heidelberg. Volume editors are expected to arrange for the refereeing, to the usual scientific standards, of the individual contributions. If the resulting reports can be forwarded to us (series editors or Springer) this is very helpful. If no reports are forwarded or if other questions remain unclear in respect of homogeneity etc, the series editors may wish to consult external referees for an overall evaluation of the volume. A final decision to publish can be made only on the basis of the complete manuscript, however a preliminary decision can be based on a pre-final or incomplete manuscript. The strict minimum amount of material that will be considered should include a detailed outline describing the planned contents of each chapter.

   Volume editors and authors should be aware that incomplete or insufficiently close to final manuscripts almost always result in longer evaluation times. They should also be aware that parallel submission of their manuscript to another publisher while under consideration for LNM will in general lead to immediate rejection.

Continued on inside back-cover

Lecture Notes in Mathematics 1816

Editors:
J.-M. Morel, Cachan
F. Takens, Groningen
B. Teissier, Paris

Springer
Berlin
Heidelberg
New York
Hong Kong
London
Milan
Paris
Tokyo

S. Albeverio W. Schachermayer M. Talagrand

Lectures on Probability Theory and Statistics

Ecole d'Eté de Probabilités
de Saint-Flour XXX - 2000

Editor: Pierre Bernard

Springer

Authors

Sergio Albeverio
Institute for Applied Mathematics,
Probability Theory and Statistics
University of Bonn
Wegelerstr. 6
53115 Bonn, Germany

e-mail: *albeverio@uni-bonn.de*

Walter Schachermayer
Department of Financial and
Actuarial Mathematics
Vienna University of Technology
Wiedner Hauptstraße 8–10/105
1040 Vienna, Austria

e-mail: *wschach@fam.tuwien.ac.at*

Michel Talagrand
Equipe d'Analyse
Université Paris VI
4 Place Jussieu
75230 Paris Cedex 05
France

e-mail: *mit@ccr.jussieu.fr*

Editor

Pierre Bernard
Laboratoire de Mathématiques Appliquées
UMR CNRS 6620, Université Blaise Pascal
Clermont-Ferrand, 63177 Aubière Cedex
France

e-mail: *pierre.bernard@math.univ-bpclermont.fr*

Cover: Blaise Pascal (1623-1662)

Cataloging-in-Publication Data applied for
Bibliographic information published by Die Deutsche Bibliothek

Die Deutsche Bibliothek lists this publication in the Deutsche Nationalbibliografie;
detailed bibliographic data is available in the Internet at http://dnb.ddb.de

Mathematics Subject Classification (2000):
60-01, 60-06, 60G05, 60G60, 60J35, 60J45, 60J60, 70-01, 81-06, 81T08, 82-01, 82B44, 82D30, 90-01, 90A09

ISSN 0075-8434 Lecture Notes in Mathematics
ISSN 0721-5363 Ecole d'Eté des Probabilités de St. Flour
ISBN 3-540-40335-3 Springer-Verlag Berlin Heidelberg New York

This work is subject to copyright. All rights are reserved, whether the whole or part of the material is
concerned, specifically the rights of translation, reprinting, reuse of illustrations, recitation, broadcasting,
reproduction on microfilm or in any other way, and storage in data banks. Duplication of this publication
or parts thereof is permitted only under the provisions of the German Copyright Law of September 9, 1965,
in its current version, and permission for use must always be obtained from Springer-Verlag. Violations are
liable for prosecution under the German Copyright Law.

Springer-Verlag Berlin Heidelberg New York a member of BertelsmannSpringer
Science + Business Media GmbH

http://www.springer.de

© Springer-Verlag Berlin Heidelberg 2003
Printed in Germany

The use of general descriptive names, registered names, trademarks, etc. in this publication does not imply,
even in the absence of a specific statement, that such names are exempt from the relevant protective laws
and regulations and therefore free for general use.

Typesetting: Camera-ready TeX output by the authors

SPIN: 10931677   41/3142/du - 543210 - Printed on acid-free paper

INTRODUCTION

This volume contains lectures given at the Saint-Flour Summer School of Probability Theory during the period August 17th - September 3d, 2000. This school was Summer School 2000 of the European Mathematical Society.

We thank the authors for all the hard work they accomplished. Their lectures are a work of reference in their domain.

The School brought together 90 participants, 39 of whom gave a lecture concerning their research work.

At the end of this volume you will find the list of participants and their papers.

Thanks. We thank the European Math Society, the European Commission DG12, Blaise Pascal University, the CNRS, the UNESCO, the city of Saint-Flour, the department of Cantal, the Region of Auvergne for their helps and sponsoring.

Finally, to facilitate research concerning previous schools we give here the number of the volume of "Lecture Notes" where they can be found:

Lecture Notes in Mathematics

1971 : n° 307 – 1973 : n° 390 – 1974 : n° 480 – 1975 : n° 539 –
1976 : n° 598 – 1977 : n° 678 – 1978 : n° 774 – 1979 : n° 876 –
1980 : n° 929 – 1981 : n° 976 – 1982 : n° 1097 – 1983 : n° 117 –
1984 : n° 1180 – 1985–1986 et 1987 : n° 1362   – 1988 : n° 1427 –
1989 : n° 1464 – 1990 : n° 1527 – 1991 : n° 1541 – 1992 : n° 1581 –
1993 : n° 1608 – 1994 : n° 1648 – 1995 : n° 1690 – 1996 : n° 1665 –
1997 : n° 1717 – 1998 : n° 1738 – 1999 : n° 1781 – 2000 : n° 1816

Lecture Notes in Statistics

1986 : n° 50

Table of Contents

Part III Michel Talagrand: Mean field models for spin
glasses: a first course

Part I

Sergio Albeverio: Theory of Dirichlet forms
and applications

Table of Contents

Summary. The theory of Dirichlet forms, Markov semigroups and associated processes on finite and infinite dimensional spaces is reviewed in an unified way. Applications are given including stochastic (partial) differential equations, stochastic dynamics of lattice or continuous classical and quantum systems, quantum fields and the geometry of loop spaces.

0 Introduction

The theory of Dirichlet forms is situated in a vast interdisciplinary area which includes analysis, probability theory and geometry.

Historically its roots are in the interplay between ideas of analysis (calculus of variations, boundary value problems, potential theory) and probability theory (Brownian motion, stochastic processes, martingale theory).

First, let us shortly mention the connection between the "phenomenon" of Brownian motion, and the probability and analysis which goes with it. As well known the phenomenon of Brownian motion has been described by a botanist, R. Brown (1827), as well as by a statistician, in connection with astronomical observations, T.N. Thiele (1870), by an economist, L. Bachelier (1900), (cf. [455]), and by physicists, A. Einstein (1905) and M. Smoluchowski (1906), before N. Wiener gave a precise mathematical framework for its description (1921-1923), inventing the prototype of interesting probability measures on infinite dimensional spaces (Wiener measure). See, e.g., [394] for the fascinating history of the discovery of Brownian motion (see also [241], [16] for subsequent developments).

This went parallel to the development of infinite dimensional analysis (calculus of variation, differential calculus in infinite dimensions, functional analysis, Lebesgue, Fréchet, Gâteaux, P. Lévy...) and of potential theory.

Although some intimate connections between the heat equation and Brownian motion were already implicit in the work of Bachelier, Einstein and Smoluchowski, it was only in the 30's (Kolmogorov, Schrödinger) and the 40's that the strong connection between analytic problems of potential theory and fine properties of Brownian motion (and more generally stochastic processes) became clear, by the work of Kakutani. The connection between analysis and probability (involving the use of Wiener measure to solve certain analytic problems) as further developed in the late 40's and the 50's, together with the application of methods of semigroup theory in the study of partial differential equations (Cameron, Doob, Dynkin, Feller, Hille, Hunt, Martin, ...).

The theory of stochastic differential equations has its origins already in work by P. Langevin (1911), N. Bernstein (30's), I. Gikhman and K. Ito (in the 40's), but further great developments were achieved in connection with the above mentioned advances in analysis, on one hand, and martingale theory, on the other hand.

By this the well known relations between Markov semigroups, their generators and Markov processes were developed, see, e.g. [162], [160], [207], [208], [209], [276], [463].

This theory is largely concerned with processes with "relatively nice characteristics" and with "finite dimensional state space" E (in fact locally compact state spaces are usually assumed). From many areas, however, there is a demand of extending the theory in two directions:

1) "more general characteristics", e.g. allowing for singular terms in the generators
2) infinite dimensional (and nonlinear) state spaces.

As far as 1) is concerned let us mention the needs of handling Schrödinger operators and associated processes in the case of non smooth potentials, see [70].

As far as 2) is concerned let us mention the theory of partial differential equations with stochastic terms (e.g. "noises"), see, e.g. [201], [28], [37], [38], [129], [127] the description of processes arising in quantum field theory (work by Friedrichs, Gelfand, Gross, Minlos, Nelson, Segal...) or in statistical mechanics, see, e.g. [16], [15], [344], [242]. Other areas which require infinite dimensional processes are the study of variational problems (e.g. Dirichlet problem in infinite dimensions) [278], the study of certain infinite dimensional stochastic equations of biology, e.g. [474], the representation theory of infinite dimensional groups, e.g. [68], the study of loop groups, e.g. [30], [12], the study of the development of interest rates in mathematical finance, e.g. [416], [337], [502].

The theory of Dirichlet forms is an appropriate tool for these extensions. In fact it is central for it to work with reference measures μ which are neither necessarily "flat" nor smooth and in replacing the Markov semigroups on continuous functions of the "classical theory" by Markov semigroups on

$L^2(\mu)$-spaces (thus making extensive use of "Hilbert space methods" [211]). The theory of Dirichlet forms was first developed by Feller in the 1-dimensional case, then extended to the locally compact case with symmetric generators by Beurling and Deny (1958-1959), Silverstein (1974), Ancona (1976), Fukushima (1971-1980) and others (see, e.g., [244], [258]).(Extensions to non symmetric generators were given by J. Elliott, S. Carrillo-Menendez (1975), Y. Lejan (1977-1982), a.a., see, e.g. [367]).

The case of infinite dimensional state spaces has been investigated by S. Albeverio and R. Høegh-Krohn (1975-1977), who were stimulated by previous analytic work by L. Gross (1974) and used the framework of rigged Hilbert spaces (along similar lines is also the work of P. Paclet (1978)). These studies were successively considerably extended by Yu. Kondratiev (1982-1987), S. Kusuoka (1984), E. Dynkin (1982), S.Albeverio and M.Röckner (1989-1991), N. Bouleau and F. Hirsch (1986-1991), see [39], [147], [278], [367], [230], [172], [465], [234], [235], [236], [237], [238], [239], [256].

An important tool to unify the finite and infinite dimensional theory was provided by a theory developed in 1991, by S. Albeverio, Z.M. Ma and M. Röckner, by which the analytic property of quasi regularity for Dirichlet forms has been shown in "maximal generality" to be equivalent with nice properties of the corresponding processes.

The main aim of these lectures is to present some of the basic tools to understand the theory of Dirichlet forms, including the forefront of the present research. Some parts of the theory are developed in more details, some are only sketched, but we made an effort to provide suitable references for further study.

The references should also be understood as suggestions in the latter sense, in particular, with a few exceptions, whenever a review paper or book is available we would quote it rather than an original reference. We apologize for this "distortion", which corresponds to an attempt of keeping the reference list into some reasonable bounds - we hope however the references we give will also help the interested reader to reconstruct historical developments.

For the same reason, all references of the form "see [X]" should be understood as "see [X] and references therein".

1 Functional analytic background: semigroups, generators, resolvents

1.1 Semigroups, Generators

The natural setting used in these lectures is the one of normed linear spaces B over the closed algebraic field $\mathbb{K} = \mathbb{R}$ or $\mathbb{C}$. Some of the results are however depending on the additional structure of completeness, therefore we shall assume most of the time that B is a Banach space.

We are interested in describing operators like the Laplacian Δ and the associated semigroup (heat semigroup), and vast generalizations of them.

Let $L \equiv (L, D(L))$ be a linear operator on a normed space B over $\mathbb{K}$, defined on a linear subset $D(L)$ of B, the definition domain of L.

We say that two such operators $L_i, i = 1, 2$ are equal if $D(L_1) = D(L_2)$ and $L_1 u = L_2 u, \forall u \in D(L_1)$.

L is said to be bounded if $\exists C \geq 0$ s.t. $\|Lu\| \leq C\|u\|, \forall u \in D(L) = B$.

We then have, setting $\|L\| \equiv \sup_{u \in B, \|u\| \leq 1} \|Lu\| \in [0, +\infty]$

$$L \text{ bounded} \Leftrightarrow \|L\| < +\infty.$$

L is said to be continuous at 0 ($\in D(L)!$) if $u_n \to 0, u_n \in D(L)$ implies $Lu_n \to 0, n \to \infty$.

L is said to be continuous if $u_n \to u, u_n \in D(L)$ implies $u \in D(L)$ and $Lu_n \to Lu, n \to \infty$.

One easily shows

$$L \text{ bounded} \Leftrightarrow L \text{ continuous at } 0 \Leftrightarrow L \text{ continuous.}$$

We define $L = \alpha_1 L_1 + \alpha_2 L_2, \alpha_i \in \mathbb{K}, i = 1, 2$, by
$D(L) = D(L_1) \cap D(L_2), Lu = \alpha_1 L_1 u + \alpha_2 L_2 u, \forall u \in D(L).$
Moreover we define for L_1, L_2
$L_1 L_2 u \equiv L_1(L_2 u), \forall u \in D(L_1 L_2) \equiv L_1 D(L_2) \equiv \{u \in B | L_2 u \in D(L_1)\}$

Definition 1. *A linear bounded operator A on a normed linear space B is a contraction if $\|A\| \leq 1$. A family $T = (T_t)_{t \geq 0}$ of linear bounded operators on B is said to be a strongly continuous semigroup or C_0-semigroup if*

  i) *$T_0 = 1$ (the identity on B)*
  ii) *$\lim_{t \downarrow 0} T_t u = u, \forall u \in B$ (strong continuity)*
  iii) *$(T_t)_{t \geq 0}$ is a semigroup i.e.*
    $T_t T_s = T_s T_t = T_{s+t}, \forall t, s > 0.$
    $(T_t)_{t \geq 0}$ is said to be a C_0-semigroup of contractions or a

    C_0-contraction semigroup if, in addition,
  iv) *T_t is a contraction for all $t \geq 0$.*

Exercise 1. Show that i),ii),iv) imply that $t \to T_t u$ is continuous, for all $t \geq 0, \forall u \in B$.

Definition 2. *Let $T \equiv (T_t)_{t \geq 0}$ be a C_0-contraction semigroup on B. The linear operator L is said to be generator of T if:*

i) $D(L) \equiv \left\{ u \in B \mid \lim_{t \downarrow 0} \frac{1}{t}(T_t u - u) \text{ exists in } B \right\}$

ii) $Lu = \lim_{t \downarrow 0} \frac{1}{t}(T_t u - u) \, \forall u \in D(L)$

Exercise 2. Show that the "strong derivative" $\frac{d}{dt} T_t u \equiv \lim_{h \downarrow 0} \frac{(T_{t+h} - T_t)u}{h}$ exists in B, for all $u \in D(L)$ and $\frac{d}{dt} T_t u = L T_t u = T_t L u \, \forall t \geq 0, \forall u \in D(L)$. In particular $Lu = \frac{d}{dt} T_t u|_{t=0}, \forall u \in D(L)$.

It is easy to convince oneself that even simple operators like the Laplacian Δ are not bounded, e.g. in $B = L^2(\mathbb{R}^d)$. For this reason it is useful to introduce the concept of a closed operator.

Definition 3. *A linear operator L in B is called <u>closed</u> if $u_n \in D(L), u_n \to u$ as $n \to \infty$, $L u_n$ convergent as $n \to \infty$, in B, imply that $u \in D(L)$, and $L u_n \to L u$.*

Exercise 3. Show that L closed $\Leftrightarrow G(L)$ closed in $B \times B$, where $G(L) \equiv \{\{u, Lu\}, u \in D(L)\}$ is the graph of L.

Proposition 1. *Let $T = (T_t)_{t \geq 0}$ be a C_0-contraction semigroup on a Banach space B, with generator L. Then $T_t u = u + \int_0^t T_s L u \, ds, u \in D(L)$ where the integral on the r.h.s is to be understood in the natural sense of strong integrals on Banach spaces (Bochner integral [1]).*

Proof. This follows immediately from Exercise 2, via integration.   □

Proposition 2. *The generator L of a C_0-contraction semigroup $T = (T_t)_{t \geq 0}$ on a Banach space is a closed operator.*

Proof. This easily follows from Proposition 1, the strong continuity (Exercise 1), the fact that for $u_n \to u$, $L u_n$ convergent to v, $\|T_s L u_n\| \leq \|L u_n\| \leq C$, for some $C \geq 0$, independent of n, as $L u_n$ converges, and dominated convergence.   □

Proposition 3. *The generator L of a C_0-contraction semigroup $T = (T_t)_{t \geq 0}$ on a Banach space is densely defined.*

---

[1] See, e.g. [506], p.132

Proof. One easily shows that for any $u \in B$, with $v_t \equiv \int_0^t T_s u \, ds$:

$$\frac{1}{r}[v_{t+r} - v_t] = \frac{1}{r}[T_r v_t - v_t] \to T_t u - u, \text{ as } r \downarrow 0$$

hence $v_t \in D(L)$.

On the other hand

$\frac{v_t}{t} \to u, t \downarrow 0$, yielding an approximation of an arbitrary $u \in B$ by elements $\frac{v_t}{t}$ in $D(L)$. $\qquad \square$

Corollary 1. *If* $T = (T_t)_{t\geq 0}, S = (S_t)_{t\geq 0}$ *are two* C_0-*contraction semigroups on a Banach space with the same generator* L, *then* $T_t = S_t \quad \forall t \geq 0$.

Proof. From Exercise 2 we have easily $\frac{d}{ds}T_{t-s}S_s u = 0, \forall 0 \leq s \leq t, \forall u \in D(L)$ from which $T_t u = S_t u \forall u \in D(L)$ follows, hence $T_t = S_t$, these being bounded and $D(L)$ being dense. $\qquad \square$

The above corollary implies that the usual notation $T_t = e^{tL}, t \geq 0$ for the semigroup with generator L is justified.

The question when a given densely defined linear operator L is the generator of a C_0-contraction semigroup is answered by the theory of Hille-Yosida. For this we recall some basic definitions.

If L is a linear injection (1-1 map), then L^{-1} is defined on $D(L^{-1}) = LD(L)$, by $L^{-1}u = v, u \in D(L^{-1})$, with v s.t. $Lv = u$.

For a linear operator L the resolvent set is defined by:

$\rho(L) \equiv \{\alpha \in \mathbb{K} | \alpha - L : D(L) \to B$ is an injection onto B i.e. $D((\alpha - L)^{-1}) = B$. Moreover $(\alpha - L)^{-1}$ is bounded.$\}$

Exercise 4. Show that if $\rho(L) \neq 0$ then $\rho(L)$ is closed (use that $(\alpha - L)^{-1}$ for $\alpha \in \rho(L)$ is bounded).

The spectrum $\sigma(L)$ of L is by definition the complement in $\mathbb{K}$ of $\rho(L)$. For $\alpha \in \rho(L), G_\alpha \equiv (\alpha - L)^{-1}$ (which exists as a bounded operator on B) is called the resolvent of L at α.

$(G_\alpha)_{\alpha \in \rho(L)}$ is called the <u>resolvent family</u> associated to L.

Exercise 5. Show that $(G_\alpha)_{\alpha \in \rho(L)}$ satisfies the resolvent identity $G_\alpha - G_\beta = (\beta - \alpha)G_\alpha G_\beta = (\beta - \alpha)G_\beta G_\alpha, \forall \alpha, \beta \in \rho(L)$.

Proposition 4. *Let* L *be the generator of a* C_0-*contraction semigroup on a Banach space. Then* $(0, \infty) \subset \rho(L)$ *and for any*

$$Re\alpha > 0 : (\alpha - L)^{-1}u = G_\alpha u = \int_0^{+\infty} e^{-\alpha t}T_t u \, dt$$

(where the integral is in Bochner's sense) and $\|G_\alpha\| \leq \frac{1}{Re\alpha}$.

Proof. Set $R_\alpha \equiv \int\limits_0^{+\infty} e^{-\alpha t} T_t dt$.

It is easily seen that $(\alpha - L)R_\alpha u = u, \forall u \in B, Re\alpha > 0$. Since L is closed for all $u \in D(L) : LR_\alpha u = R_\alpha Lu$, from which one deduces that $\alpha - L$ is injective for $Re\alpha > 0$ (in particular for $\alpha > 0$) and $R_\alpha = G_\alpha$. The bound in Proposition 4 then follows from the definition of R_α. □

Remark 1. G_α is the Laplace transform of T_t (in the sense given by Proposition 4).

Theorem 1. *(Hille-Yosida, for C_0-contraction semigroups):*
Let L be a linear operator in a Banach space B. The following are equivalent:

  i) L is the generator of a C_0-contraction semigroup $T = (T_t)_{t \geq 0}$ on B.
  ii) L is densely defined and
      α) $(0, \infty) \subset \rho(L)$
      β) $\|\alpha(\alpha - L)^{-1}\| \leq 1 \quad \forall \alpha > 0$

Corollary 2. *If ii) is fullfilled then L is closed and uniquely determined.*

Proof. ii) implies i) by Theorem 1 and hence that L is closed by Proposition 2. The rest follows from Corollary 1. □

Proof. (of Theorem 1)
i) $\Rightarrow$ ii): From i) we have L closed, densely defined (Propositions 2,3). That $(0, \infty) \subset \rho(L)$ and ii) holds follows from Proposition 4.
ii) $\Rightarrow$ i): For details we refer to, e.g.[413]. In the proof the following Proposition is useful.

Proposition 5. *Let L satisfy the conditions ii) of Theorem 1. Set $G_\alpha = (\alpha - L)^{-1}, \alpha > 0$. Then*

  i) $\alpha G_\alpha u \to u$ in B, as $\alpha \to +\infty$
  ii) Define $L^{(\alpha)} \equiv -\alpha + \alpha^2 G_\alpha, \alpha > 0$ ("Yosida approximation of L"). Then $L^{(\alpha)}$ is bounded, $D(L^{(\alpha)}) = B, L^{(\alpha)}u \to Lu, \alpha \uparrow +\infty, u \in D(L)$, and $e^{tL^{(\alpha)}}u$ converges as $\alpha \uparrow +\infty$ for all $u \in D(L)$ to $\tilde{T}_t u$, where $\tilde{T}_t$ is a C_0-contraction semigroup, with generator L. Moreover $\tilde{T}_t$ coincides with the semigroup T_t generated by L mentioned in i).

Proof. For $u \in D(L)$ we have

$$\|\alpha G_\alpha u - u\| = \|\alpha(\alpha - L)^{-1}u - (\alpha - L)(\alpha - L)^{-1}u\|$$
$$= \|L(\alpha - L)^{-1}u\|$$
$$= \|(\alpha - L)^{-1}Lu\|$$
$$\leq \frac{1}{\alpha}\|Lu\| \to 0, \alpha \uparrow +\infty$$

(where we used Proposition 4). But αG_α is a contraction by Proposition 4 and $D(L)$ is dense by assumption, hence $\alpha G_\alpha u \to u$ as $\alpha \uparrow +\infty$, for all $u \in B$.

From this it is easy to see that $\alpha G_\alpha L u \to L u, u \in D(L)$, as $\alpha \uparrow +\infty$, and thus $L^{(\alpha)} u = -\alpha u + \alpha^2 G_\alpha u = \alpha G_\alpha L u \to L u$ as $\alpha \uparrow +\infty$.
The rest follows by realizing that

$$e^{tL^{(\alpha)}} u = \sum_{n=0}^{\infty} \frac{t^n}{n!} L^{(\alpha)^n} u = e^{\alpha t} e^{-\alpha^2 G_\alpha} u$$

Remark 2. Another useful "approximation formula" for T_t in terms of the resolvent is the following one:

$$T_t u = \lim_{n \to \infty} \left(\frac{n}{t}\right)^n \left(G_{\frac{n}{t}} u\right)^n, \forall u \in B$$

(see, e.g., [413], p. 33).

Remark 3. In the formulation of Hille-Yosida's theorem i) can be replaced by a statement involving the generator of a C_0-contraction resolvent family according to the following definition.

Definition 4. *A C_0-contraction resolvent family is a family $(G_\alpha)_{\alpha > 0}$ such that*

$$\alpha G_\alpha u \to u, \alpha \uparrow +\infty, \|\alpha G_\alpha\| \leq 1, \alpha > 0$$

and the resolvent identity in Exercise 5 holds.
Hille-Yosida's theorem holds then with i) replaced by:

i') *L is the generator of a C_0-contraction resolvent family $(G_\alpha)_{\alpha > 0}$ in the sense that $G_\alpha = (\alpha - L)^{-1}$ on B. There is a one-to-one correspondence between C_0-contraction semigroups $(T_t)_{t \geq 0}$ and C_0-contraction resolvent families $(G_\alpha)_{\alpha > 0}$ given by the Laplace-transform formula in Proposition 4 (and Remark 1) resp. Proposition 5 or Remark 2 after Proposition 5.*

Hille-Yosida's characterization of generators L involves the resolvent G_α. A pure characterization of L, under some "direct restrictions" on L is given by the Lumer-Phillips theorem, for which we need a definition.

Definition 5. *The duality set $F(u)$ for any element u in a Banach space B is defined by*

$$F(u) \equiv \left\{ u^* \in B^* | \langle u^*, u \rangle = \|u\|^2 = \|u^*\|^2 \right\},$$

where B^* is the dual of B (the space of continuous linear functionals on B) and $\langle,\rangle$ is the dualization between B and B^*.
An operator L is <u>*dissipative*</u> *on B if for any $u \in D(L)$ there exists some $u^* \in F(u)$ such that $\overline{Re\langle u^*, L u\rangle} \leq 0$.*
($-L$ is then said to be <u>*accretive*</u>*).*

Proposition 6. *L is dissipative iff*

$$\|(\alpha - L)u\| \geq \alpha\|u\|, \forall u \in D(L) \forall \alpha > 0$$

Proof. See, e.g. [413] (Theorem 4.2).   □

Proposition 7. *Let L be dissipative. Then L is closed iff Range $(\alpha - L)$ is closed, for all $\alpha > 0$.*

Proof. The proof is left as an exercise (cf,e.g., [413]).   □

We recall that an operator L_0 in a Banach space is said to be closable if there exists at least one closed extension $\tilde{L}_0$ of it, i.e. $\tilde{L}_0$ closed and $\tilde{L}_0 u = L_0 u, \forall u \in D(L_0) \subset D(\tilde{L}_0)$. One calls closure $\overline{L}_0$ of L_0 the minimal closed extension of L_0.

Theorem 2. *(Lumer-Phillips)*
Let L be a linear closable operator in a Banach space. Then the closure $\overline{L}$ of L generates a C_0-contraction semigroup on B iff

a) *$D(L)$ is dense in B*
b) *L is dissipative*
c) *The range of $\alpha_0 - L$ is dense in B, for some $\alpha_0 > 0$.*

Proof. See, e.g., [413] Theorem 4.3   □

Remark 4. If L is the generator of a C_0-contraction semigroup on B then a) holds, c) holds for all $\alpha > 0$ and b) holds, see, e.g. [413], [424].

Remark 5. If L is a linear operator satisfying a),b) then L is closable. This, together with c) gives that $\overline{L}$ generates a C_0-contraction semigroup. See [424],(p.240 and p.345).

1.2 The case of a Hilbert space

We shall consider here the special case where the Banach space B of section 1.1 is a Hilbert space $\mathcal{H}$, with scalar product $(,)$.
We first observe that if R is a contraction then

$$|(Ru, u)| \leq \|Ru\|\|u\| \leq \|u\|^2.$$

Hence $Re(Ru, u)$ and $Im(Ru, u)$ are bounded absolutely by $\|u\|^2$.
If $(T_t)_{t \geq 0}$ is self-adjoint, i.e. $T_t^* = T_t$ (where R^* means the adjoint to R) and T_t is a C_0-contraction semigroup on $\mathcal{H}$ with generator L, then for all $u, v \in D(L)$, using the self-adjointness of T_t :

$$(-Lu, v) = \lim_{t \downarrow 0} \frac{1}{t}(u - T_t u, v)$$
$$= (u, -Lv)$$

i.e. L is symmetric in H (in the sense that L^* is an extension of L or, equivalently, $(u, Lv) = (Lu, v), \forall u, v \in D(L)$).

Remark 6. If A is a symmetric operator in $\mathcal{H}$ we have $(u, Au) = (Au, u), \forall u \in D(A)$. On the other hand $\overline{(u, Au)} = (Au, u)$ (by the properties of the scalar product), hence $(u, Au) = \overline{(u, Au)}$ for symmetric operators, i.e. (u, Au) is real.
For A bounded with $D(A) = B$ we have A symmetric iff A is self-adjoint (but this is not so in general for A unbounded!).
In particular a C_0-contraction semigroup is symmetric iff it is self-adjoint. It is easily seen that the following are equivalent:

i) $(T_t)_{t \geq 0}$ is a symmetric C_0-contraction semigroup
ii) $(G_\alpha)_{\alpha > 0}$ is a symmetric C_0-contraction resolvent family

(use, e.g., the Laplace transformation Proposition 4, resp. Proposition 5).
We also see that if (T_t) is a symmetric C_0-contraction semigroup then

$$|(u, T_t u)| = |(T_t u, u)| \leq \|u\|^2, \quad \text{for all } u \in \mathcal{H}. \tag{1}$$

On the other hand $\lim_{t \downarrow 0} \left(\frac{T_t - 1}{t} u, u\right) = (Lu, u), \forall u \in D(L)$.

But $\left(\frac{(T_t - 1)u}{t}, u\right)$ is real (by the symmetry property) and negative, by (1), hence $(Lu, u) \leq 0$.
One calls a densely defined operator A in a Hilbert space positive if $(u, Au) \geq 0, \forall u \in D(A)$.

Remark 7. A positive implies $-A$ dissipative. The above says that $(-L)$ is positive, or equivalently, that L is negative.
By Lumer-Phillips theorem the range of $\alpha_0 - L$ is dense in H, for some $\alpha_0 > 0$.
Hence we have proven:

Proposition 8. *The generator of a symmetric C_0-contraction semigroup in a Hilbert space is a negative densely defined closed symmetric operator L s.t. the range of $\alpha_0 - L$ is dense, for some $\alpha_0 > 0$.*

Remark 8. One easily shows that the fact that the range of $\alpha_0 - L$ is dense for some $\alpha_0 > 0$ implies that L is self-adjoint (see, e.g. [424]).

Viceversa, if L is linear, symmetric (hence closable) densely defined on $\mathcal{H}$, negative and such that the range of $\alpha_0 - L$ is dense in $\mathcal{H}$ for some $\alpha_0 > 0$ then, by Lumer-Phillips theorem, its closure $\overline{L}$ (which is self-adjoint by the above remark) generates a symmetric C_0-contraction semigroup (symmetry can be seen, e.g., by the symmetry of $G_\alpha = (\alpha - \overline{L})^{-1}$ and the above considerations on the symmetry properties of G_α resp. T_t).

Remark 9. $\overline{L}$ in Remark 8 can be easily replaced by any self-adjoint negative extension $\tilde{L}$ of L. In fact then both $\tilde{L}$ and its adjoint $\tilde{L}^* = \tilde{L}$ are negative hence dissipative and then they generate a C_0-contraction semigroup, see [424],p.248.

Remark 10. Spectral theory also gives a direct relation between self-adjoint properties of generators L and corresponding semigroups, recalling that $L = \int\limits_{\sigma(L)} \lambda dE(\lambda)$, $T_t = \int\limits_{\sigma(L)} e^{t\lambda} dE(\lambda)$, $E(\lambda)$ being the spectral family associated with L. Here $\sigma(L) \subset (-\infty, 0]$.

1.3 Examples

We shall concentrate, in this section, on:

Semigroups in Banach or Hilbert spaces associated with differential operators over finite dimensional spaces.

The typical situation is given by the finite dimensional space $\mathbb{R}^d$ and the ("finite dimensional") differential operator Δ (the Laplacian) acting, e.g. in the Hilbert space $\mathcal{H} = L^2(\mathbb{R}^d)$ resp. on the Banach space $B = C_b(\mathbb{R}^d)$.

Let us first consider the case $\mathcal{H} = L^2(\mathbb{R}^d)$.

We see that $(\Delta, C_0^\infty(\mathbb{R}^d))$ (or, e.g., $(\Delta, \mathcal{S}(\mathbb{R}^d))$ is densely defined and symmetric in $\mathcal{H}$ (as a consequence of an integration by parts).

Let U be the map from $L^2(\mathbb{R}^d)$ into $L^2(\hat{\mathbb{R}}^d)$ defined by L^2-Fourier transform i.e.

$$(Uf)(k) \equiv (2\pi)^{-\frac{d}{2}} \int_{\mathbb{R}^d} e^{ik\cdot x} f(x)dx, k \in \hat{\mathbb{R}}^d$$

($\hat{\mathbb{R}}^d$ a copy of $\mathbb{R}^d$, for the Fourier transform variables). Then U is unitary (by Parseval's theorem), i.e. $U^*U = UU^* = 1$.

Let M be the multiplication operator given by $M\hat{u}(k) \equiv |k|^2\hat{u}(k), k \in \hat{\mathbb{R}}^d, \hat{u} \in L^2(\hat{\mathbb{R}}^d)$, on its natural domain $D(M) \equiv \{\hat{u} \in L^2(\hat{\mathbb{R}}^d) | M\hat{u} \in L^2(\hat{\mathbb{R}}^d)\}$.

M is self-adjoint positive (since $(M + \alpha)$, has dense range for all $\alpha > 0$).

Let us set

$$H_0 = U^*MU$$

with

$$D(H_0) = \{u \in L^2(\mathbb{R}^d) | Uu \in D(M)\}$$
$$= \{U^*D(M)\}$$

(i.e. $u \in D(H_0) \leftrightarrow \hat{u} \in D(M)$).

Remark 11. One easily shows that $D(H_0) = H^{2,2}(\mathbb{R}^d)$ is the Sobolev space obtained by closing $C_0^\infty(\mathbb{R}^d)$ in the norm given by the scalar product

$$(u, v)_2 \equiv \sum_{|\alpha| \leq 2} \int \overline{D^\alpha u} D^\alpha v \, dx.$$

H_0 is self-adjoint positive in $L^2(\mathbb{R}^d)$, being unitary equivalent to the self-adjoint positive operator M (positivity is immediate; self-adjointness follows e.g. by spectral theory, the spectrum of H_0 being the same as the one of M and the spectral family of H_0 being $U^*E_\lambda U$, where E_λ is the spectral family to M).

By Lumer-Phillips theorem (or spectral theory) we have that $e^{-tM}, t \geq 0$, is a symmetric C_0-contraction semigroup on $L^2(\mathbb{R}^d)$, hence

$$e^{-tH_0} = U^*e^{-tM}U, t \geq 0$$

is also a symmetric C_0-contraction semigroup on $L^2(\mathbb{R}^d)$.

Its spectral representation can be obtained by the one of M, in fact since $e^{-tM}\hat{u}(k) = e^{-t|k|^2}\hat{u}(k)$, we have for all $u \in L^2(\mathbb{R}^d)$.

$$e^{-tH_0}u(x) = \int_{\mathbb{R}^d} \pi_t(x,y)u(y)\,dy, \qquad (2)$$

where $\pi_t(x,y) \equiv (4\pi t)^{\frac{-d}{2}}e^{-\frac{|x-y|^2}{4t}}, t > 0$ is the heat kernel density.

(2) holds for $t = 0$ with $\pi_t(x,y)dy$ replaced by the Dirac measure $\delta_x(dy)$ (since $e^{-tH_0}|_{t=0}$ is the unity operator in $L^2(\mathbb{R}^d)$).

Remark 12. Formula (2) easily extends to $t \in \mathbb{C}$ with $Re(t) > 0$.

In particular we have a representation for the unitary group $e^{itH_0}, t \in \mathbb{R}$. This unitary group (uniquely associated to H_0 by Stone's theorem) gives the time evolution in the quantum mechanics of one (non relativistic) particle, see, e.g. [423],[424], [425],[426].

One can ask the question:

do there possibly exist other semigroups $e^{t\tilde{L}}, t \geq 0$ (unitary groups $e^{it\tilde{L}}, t \in \mathbb{R}$) generated by self-adjoint extensions $\tilde{L}$, different from the closure $\bar{L}$ of Δ from $C_0^\infty(\mathbb{R}^d)$ in B?

That the answer is no, for $B = L^2(\mathbb{R}^d)$ (or $C_b(\mathbb{R}^d)$) , can be seen using the following important Theorem, for which we need a definition.

Definition 6. *Let L be a closed linear operator on a Banach space B. A linear subset D in $D(L)$ is called a <u>core</u> for L if $\overline{L \upharpoonright D} = L$ (i.e. the closure of the restriction $L \upharpoonright D$ of L to D is precisely L).*

Theorem 3. *(Nelson)*

Let L be the generator of a C_0-contraction semigroup on a Banach space B. Let $D_0 \subset D_1 \subset D(L), \overline{D_0} = B$, such that e^{tL} maps D_0 into D_1. Then D_1 is a core for L.

Proof. See, e.g., [393], [227] p.17, [424]. For extensions see [501]. $\qquad\qquad\square$

For the application of the theorem to our situation, let us take $e^{tL} = e^{-tH_0}$, with $H_0 = U^*MU$ as above. To see that Nelson's theorem can be applied with $D_0 = D_1 = S(\mathbb{R}^d)$ we observe that $D(L)$ contains $S(\mathbb{R}^d)$ (as seen from the fact that $US(\mathbb{R}^d) = S(\hat{\mathbb{R}}^d)$, and M maps $S(\hat{\mathbb{R}}^d)$ into itself, and $U^*S(\hat{\mathbb{R}}^d) = S(\mathbb{R}^d)$) and by (2) we have $e^{-tH_0}S(\mathbb{R}^d) \subset S(\mathbb{R}^d)$ (the smoothness of the elements of $e^{-tH_0}S(\mathbb{R}^d)$ can be checked directly, using, e.g., dominated convergence). Thus we have shown that $S(\hat{\mathbb{R}}^d)$ is a core for e^{-tH_0}.

To see that also $C_0^\infty(\mathbb{R}^d)$ is a core in $L^2(\mathbb{R}^d)$, let us set $A \equiv -\Delta$ on $C_0^\infty(\mathbb{R}^d)$. Let $v \in D(A^*)$, then

$$(-\Delta u, v) = (Au, v) = (u, A^*v), \forall u \in C_0^\infty(\mathbb{R}^d).$$

Hence, $-\Delta v$,defined by looking at $v \in L^2(\mathbb{R}^d)$ as a distribution, is equal to $A^*v \in L^2(\mathbb{R}^d)$.

Thus $v \in H^{2,2}(\mathbb{R}^d)$ and $A^*v = H_0 v$ (by the fact that $D(H_0) = H^{2,2}(\mathbb{R}^d)$).
This shows that $D(A^*) \subset D(H_0)$ and H_0 is an extension of A^*. Conversely, for $v \in D(H_0)$ we have $H_0 v \in L^2(\mathbb{R}^d)$, hence $(H_0 u, v) = (u, H_0 v) \forall u \in C_0^\infty(\mathbb{R}^d)$, thus $v \in D(A^*), A^*v = H_0 v$, i.e. A^* is an extension of H_0. Thus H_0 must coincide with A^*, and then $A^* = A^{**}$ (since $H_0 = H_0^*$ by self-adjointness), which shows that the closure of A is self-adjoint and coincides with H_0, thus $C_0^\infty(\mathbb{R}^d)$ is a core for H_0, in $L^2(\mathbb{R}^d)$.

Remark 13. From the explicit formula (2) we see that the r.h.s. of (2) also maps the Banach space $B = C_\infty(\mathbb{R}^d)$ (the continuous functions on $\mathbb{R}^d$ vanishing at infinity with supremum norm), into itself, and is a C_0-contraction semigroup $\tilde{P}_t$.
Let us call $\tilde{L}$ the generator of $\tilde{P}_t$.
$D(\tilde{L}) \supset S(\hat{\mathbb{R}}^d)$ as easily verified by the definition of the generator and (2). In fact $\tilde{L} = -\Delta$ on $S(\hat{\mathbb{R}}^d)$ and by Nelson's theorem applied to $D_0 = D_1 = S(\hat{\mathbb{R}}^d), B = C_\infty(\mathbb{R}^d)$ we have that $S(\hat{\mathbb{R}}^d)$ is a core for $\tilde{P}_t$ in $C_\infty(\mathbb{R}^d)$.

Remark 14. P_t and $\tilde{P}_t$ can be identified in the following sense.
P_t and $\tilde{P}_t$ on $C_\infty(\mathbb{R}^d) \cap L^2(\mathbb{R}^d)$, as C_0-contraction semigroups, coincide, hence by the density of $C_\infty(\mathbb{R}^d) \cap L^2(\mathbb{R}^d)$ in $L^2(\mathbb{R}^d)$, $P_t = \tilde{P}_t$ on $L^2(\mathbb{R}^d)$.
Similarly one can show $P_t = \tilde{P}_t$ in $C_\infty(\mathbb{R}^d)$, by exploiting the boundedness of $P_t, \tilde{P}_t$ in $C_\infty(\mathbb{R}^d)$ and their equality on the dense subset $C_\infty(\mathbb{R}^d) \cap L^2(\mathbb{R}^d)$ of $C_\infty(\mathbb{R}^d)$.
In this sense then the heat semigroup e^{-tH_0} can be identified in $C_\infty(\mathbb{R}^d)$ and $L^2(\mathbb{R}^d)$ with the semigroup with generator Δ having $S(\hat{\mathbb{R}}^d)$ (or $C_0^\infty(\mathbb{R}^d)$) as core, both in $C_\infty(\mathbb{R}^d)$ and $L^2(\mathbb{R}^d)$.

2 Closed symmetric coercive forms associated with C_0-contraction semigroups

2.1 Sesquilinear forms and associated operators

Sesquilinear forms Let $\mathcal{H}$ be a Hilbert space over $\mathbb{K} = \mathbb{R}$ or $\mathbb{C}$, with scalar product $(\cdot, \cdot)$ (conjugate linear in the first argument, linear in the second argument), and corresponding norm $\| \cdot \|^2 = (\cdot, \cdot)$.

Let D be a linear subspace of $\mathcal{H}$.

Definition 7. *A map $\mathcal{E} : D \times D \to \mathbb{K}$, conjugate linear in the first argument, linear in the second argument is called a sesquilinear form (on D, in $\mathcal{H}$).*

D is called the domain of $\mathcal{E}$. One writes $(\mathcal{E}, D)$ whenever it is important to specify the domain.

$\mathcal{E}[u] \equiv \mathcal{E}(u, u), u \in D$ is called the associated quadratic form.

Remark 15. For $\mathcal{K} = \mathbb{C}$, $(\mathcal{E}[u], u \in D)$ uniquely determines $(\mathcal{E}, D)$ by the polarization formula

$$\mathcal{E}(u, v) = \tfrac{1}{4}(\mathcal{E}[u + v] - \mathcal{E}[u - v] + i\mathcal{E}[u + iv] - i\mathcal{E}[u - iv]).$$

This is not so, in general, for $\mathbb{K} = \mathbb{R}$ (see, e.g., [495])

Definition 8. *A sesquilinear form $\mathcal{E}$ is said to be underline{symmetric} if $\forall u, v \in D$:*

$$\mathcal{E}(u, v) = \overline{\mathcal{E}(v, u)}$$

(where $-$ stands for complex conjugation).

Remark 16. The quadratic form associated with a symmetric sesquilinear form is real-valued.

Definition 9. *A sesquilinear form $\mathcal{E}$ is said to be lower bounded if there exists $\gamma \in \mathbb{R}$ such that:*

$$\mathcal{E}[u] \geq \gamma \|u\|^2, \quad \forall u \in D(\mathcal{E})$$

One writes then $\mathcal{E} \geq \gamma$. γ is said to be the lower bound for $\mathcal{E}$.

$\mathcal{E}$ is called positive if $\gamma = 0$.

Remark 17. If $\mathcal{E}$ is positive then

$$|\mathcal{E}(u, v)| \leq (\mathcal{E}[u])^{1/2}(\mathcal{E}[v])^{1/2}$$

Proof. This is Cauchy-Schwarz' inequality.

Example 1. Let A be a linear operator with domain $D(A)$ in $\mathcal{H}$. Define for $u, v \in D(A)$:
$$\mathcal{E}(u, v) = (u, Av).$$
Then $\mathcal{E}$ is a sesquilinear form with domain $D(\mathcal{E}) = D(A)$. The following equivalences follow immediately from the definitions.

$\mathcal{E}$ is symmetric iff A is symmetric.

$\mathcal{E} \geq \gamma$ iff $A \geq \gamma$ (in the sense that $(u, Au) \geq \gamma \|u\|^2$ for some $\gamma \in \mathbb{R}$, $\forall u \in D(A)$; in which case one says that A is lower bounded with lower bound γ).

$\mathcal{E} \geq 0$ iff $A \geq 0$ (in which case one says that A is positive).

Closed forms Let $\mathcal{E}$ be a sesquilinear, lower bounded form on $\mathcal{H}$.

Definition 10. *A sequence $(u_n)_{n \in \mathbb{N}}$ is said to be $\mathcal{E}$-convergent to $u \in \mathcal{H}$, for $n \to \infty$, and one writes $u_n \overset{\mathcal{E}}{\to} u$, $n \to \infty$, if $u_n \in D(\mathcal{E})$, $u_n \to u$ (i.e. (u_n) converges to u in $\mathcal{H}$) and $\mathcal{E}[u_n - u_m] \to 0$, $n, m \to \infty$ (i.e. u_n is an "$\mathcal{E}$-Cauchy sequence").*

N.B. u is not required to be in $D(\mathcal{E})$.

Definition 11. *$\mathcal{E}$ is said to be <u>closed</u> if $u_n \overset{\mathcal{E}}{\to} u$, $n \to \infty$, implies $u \in D(\mathcal{E})$ and $\mathcal{E}[u_n - u] \to 0$, as $n \to \infty$.*

Let $\mathcal{E}$ be a symmetric, positive sesquilinear form. Define for any $\alpha > 0$:
$$\mathcal{E}_\alpha(u, v) \equiv \mathcal{E}(u, v) + \alpha(u, v), \quad \forall u, v \in D(\mathcal{E}).$$
Then $D(\mathcal{E})$ taken with the norm given by
$$\|u\|_1 \equiv (\mathcal{E}_1[u])^{\frac{1}{2}}, u \in D(\mathcal{E})$$
is a pre Hilbert space, in the sense that $(D(\mathcal{E}), \| \cdot \|_1)$ has all properties of a Hilbert space, except for completeness. We call $D(\mathcal{E})_1$ this space.

Remark 18. a) $\underline{u} \equiv (u_n)_{n \in \mathbb{N}}$ is $\mathcal{E}$-convergent iff $\underline{u}$ is Cauchy in $D(\mathcal{E})_1$.

b) $u_n \overset{\mathcal{E}}{\to} u$, $n \to \infty$, $u \in D(\mathcal{E})$ iff $\|u_n - u\|_1 \to 0$, $n \to \infty$.

Proposition 9. *A lower bounded form $\mathcal{E}$ is closed iff $D(\mathcal{E})_1$ is complete.*

Proof: This is left as an exercise (cf., e.g., [312], p. 314).

Example 2. Let S be a linear operator with domain $D(S) \subset \mathcal{H}$. Define $\mathcal{E}(u, v) \equiv (Su, Sv)$, $D(\mathcal{E}) = D(S)$. Then $\mathcal{E}$ is a positive, symmetric sesquilinear form. $\mathcal{E}$ is closed iff S is closed (the proof of the latter statement is left as an exercise).

Closed forms

Definition 12. *A sesquilinear lower bounded form $\overset{\circ}{\mathcal{E}}$ is said to be closable if it has a closed extension $\mathcal{E}$, i.e., $\mathcal{E}$ is closed, $D(\mathcal{E}) \supset D(\overset{\circ}{\mathcal{E}})$ and $\mathcal{E} = \overset{\circ}{\mathcal{E}}$ on $D(\overset{\circ}{\mathcal{E}})$.*

Proposition 10. *A sesquilinear lower bounded form $\overset{\circ}{\mathcal{E}}$ is closable iff $u_n \overset{\mathcal{E}}{\rightarrow} 0$, $n \to \infty$ implies $\mathcal{E}[u_n] \to 0$, $n \to \infty$.*

Proof. This is left as an exercise (cf., e.g., [312], p. 315).

Definition 13. *The smallest closed extension of a sesquilinear lower bounded form $\mathcal{E}$ is by definition <u>the closure</u> $\bar{\mathcal{E}}$ of $\mathcal{E}$.*

Example 3. Let $\mathcal{E}$ be as in Example 2, i.e. $\mathcal{E}(u,v) = (Su, Sv)$, $\forall u, v \in D(\mathcal{E}) = D(S)$, S a linear operator on $\mathcal{H}$. Then $\mathcal{E}$ is closable iff S is closable. In the latter case one has $\bar{\mathcal{E}}(u,v) = (\bar{S}u, \bar{S}v)$, where $\bar{S}$ is the closure of the operator S (a linear operator A is said to be closable if it has a closed extension, cf. Definition 3 in Chapter 1 for the concept of closed operators). Moreover one has $\mathcal{E}$ closed iff S is closed.

The proofs are left as execises.

Remark 19. Not every sesquilinear symmetric positive form is closable. Consider, e.g., $\mathcal{H} = L^2(\mathbb{R})$, $\mathcal{E}(u,v) \equiv \bar{u}(0)v(0)$, $u, v \in D(\mathcal{E}) = C_o^\infty(\mathbb{R})$. Then $\mathcal{E}$ is sesquilinear, symmetric, and positive but not closable. In fact take a sequence $u_n \in C_o^\infty(\mathbb{R})$, with $u_n(x) = 0$ for $|x| \geq \frac{c}{n}$, $u_n(0) = 1$, $u_n(x) \leq 1$, $\forall x \in \mathbb{R}$, then we have, (by the mean-value theorem) $\|u_n\| \leq \frac{2c}{n} \to 0$, hence $u_n \to 0$, $n \to \infty$, moreover

$$\mathcal{E}[u_m - u_n] = (\bar{u}_m(0) - \bar{u}_n(0)) \cdot (u_m(0) - u_n(0)) = 0$$

hence $u_n \overset{\mathcal{E}}{\rightarrow} 0$, $n \to \infty$. On the other hand $\mathcal{E}[u_n] = \bar{u}_n(0)u_n(0) = 1$ does not converge to 0 as $n \to \infty$, which shows by Proposition 10 that $\mathcal{E}$ is not closable.

N.B. Concerning closability the situation with forms and densely defined operators is thus very different: every symmetric densely defined operator A is namely closable! (since A symmetric means by definition that the adjoint A^*, which exists uniquely since A is densely defined, is an extension of A, but every adjoint operator is closed, see, e.g. [312], p. 168).

Forms constructed from positive operators

Proposition 11. *Let A be a positive symmetric operator. Then*

$$\overset{\circ}{\mathcal{E}}_A(u,v) \equiv (u, Av), \quad u, v \in D(\overset{\circ}{\mathcal{E}}_A) = D(A)$$

is a sesquilinear, symmetric, positive, closable form.

Proof. $\overset{\circ}{\mathcal{E}}_A$ is clearly sesquilinear, symmetric, positive. To prove the closability, let $u_n \overset{\mathcal{E}}{\to} 0$, $n \to \infty$. We have to show $\overset{\circ}{\mathcal{E}}_A[u_n] \to 0$, $n \to \infty$. But by the triangle inequality resp. Cauchy-Schwarz inequality:

$$\overset{\circ}{\mathcal{E}}_A[u_n] \le |\overset{\circ}{\mathcal{E}}_A(u_n, u_n - u_m)| + |\overset{\circ}{\mathcal{E}}_A(u_n, u_m)|$$
$$\le \overset{\circ}{\mathcal{E}}_A[u_n]^{1/2}[\overset{\circ}{\mathcal{E}}_A[u_n - u_m]]^{1/2} + |(u_n, Au_m)| \tag{3}$$

where for the latter term we have used the definition of $\overset{\circ}{\mathcal{E}}_A$.

But from the assumption $u_n \overset{\mathcal{E}}{\to} 0$, $n \to \infty$, we have for any given $\epsilon > 0$, that there exists $N(\epsilon)$ s.t. for $n, m > N(\epsilon)$:

$$\overset{\circ}{\mathcal{E}}_A[u_n - u_m] \le \epsilon^2. \tag{4}$$

Moreover, by the symmetry of A

$$|(u_n, Au_m)| = |(Au_n, u_m)| \le \|Au_n\|\|u_m\| \overset{m \to \infty}{\to} 0 \tag{5}$$

for any fixed $n \in \mathbb{N}$ since $u_m \overset{\mathcal{E}}{\to} 0$, $m \to \infty$ implies $\|u_m\| \to 0$, $m \to \infty$.

Hence from (3)–(5), for any given $\epsilon > 0$, for some $N(\epsilon)$ large enough,

$$\overset{\circ}{\mathcal{E}}_A[u_n] \le \overset{\circ}{\mathcal{E}}_A[u_n]^{1/2}\epsilon, \quad n > N(\epsilon). \tag{6}$$

For given $n > N(\epsilon)$, either $\overset{\circ}{\mathcal{E}}_A[u_n] = 0$, or $\overset{\circ}{\mathcal{E}}_A[u_n] > 0$, in which case from (6) we deduce $\overset{\circ}{\mathcal{E}}_A[u_n]^{1/2} \le \epsilon$. In both cases $\overset{\circ}{\mathcal{E}}_A[u_n] \le \epsilon$, $n > N(\epsilon)$, which shows that $\overset{\circ}{\mathcal{E}}_A[u_n] \to 0$, $n \to \infty$. $\square$

Positive closed operators from positive symmetric closed forms

Theorem 4 (Friedrichs representation theorem). *Let $\mathcal{E}$ be a densely defined sesquilinear, symmetric, positive, closed form. Then there exists a unique self-adjoint positive operator $A_{\mathcal{E}}$ s.t.*

i) $D(A_{\mathcal{E}}) \subset D(\mathcal{E})$, $\mathcal{E}(u, v) = (u, A_{\mathcal{E}}v)$, $\forall u \in D(\mathcal{E}), v \in D(A_{\mathcal{E}})$.

ii) $D(A_{\mathcal{E}})$ is a core for $\mathcal{E}$ (in the sense that the closure of the restriction of $\mathcal{E}$ to $D(A_{\mathcal{E}})$ coincides with $\mathcal{E}$, i.e. $\overline{\mathcal{E}_{|D(A_{\mathcal{E}})}} = \mathcal{E}$).

iii) $D(\mathcal{E}) = D(A_{\mathcal{E}}^{1/2})$ (where $A_{\mathcal{E}}^{1/2}$ is the unique square root of the positive self-adjoint operator $A_{\mathcal{E}}$, defined, e.g., by the spectral theorem), and:

$$\mathcal{E}(u, v) = (A_{\mathcal{E}}^{1/2}u, A_{\mathcal{E}}^{1/2}v), \quad \forall u, v, \in D(\mathcal{E}).$$

And viceversa: if A is a self-adjoint positive operator, then $\mathcal{E}$ defined by $\mathcal{E}(u,v) \equiv (A^{1/2}u, A^{1/2}v)$ with $D(\mathcal{E}) = D(A^{1/2})$ is a densely defined sesquilinear form. $\mathcal{E}$ is the closure $\overset{\circ}{\mathcal{E}}$ with

$$\overset{\circ}{\mathcal{E}}(u,v) = (u, Av), \quad v \in D(A), \quad u \in D(\overset{\circ}{\mathcal{E}}) = D(A).$$

Remark 20. One says $A_{\mathcal{E}}$ (in the first part of the theorem) is the self-adjoint operator associated with the form $\mathcal{E}$. Viceversa, in the second part of the theorem, $\mathcal{E}$ is the form associated with the operator A.

One often writes $-L_{\mathcal{E}}$ instead of $A_{\mathcal{E}}$

The proof of the first part relies on following

Lemma 1. *Let $\mathcal{H}_1$ be a dense subspace of a Hilbert space $\mathcal{H}$. Let a scalar product $(\cdot, \cdot)_1$ (in general different from the scalar product $(\cdot, \cdot)$ in $\mathcal{H}$) be defined on $\mathcal{H}_1$, so that $(\mathcal{H}_1, (\cdot, \cdot)_1)$ is a Hilbert space. Suppose that there exists a constant $\kappa > 0$ s.t. $\kappa\|u\|^2 \leq \|u\|_1^2$ for all $u \in \mathcal{H}$. Then there exists uniquely a self adjoint operator A in $\mathcal{H}$ s.t. $D(A) \subset \mathcal{H}_1$, $(Au, v) = (u, v)_1, \forall u \in D(A), v \in \mathcal{H}_1$, and, moreover, $A \geq \kappa$.*

A is described by

$$D(A) = \{u \in \mathcal{H}_1 \mid \exists \hat{u} \in \mathcal{H} \mid (u,v)_1 = (\hat{u}, v)\forall v \in \mathcal{H}_1\}, \quad Au = \hat{u}.$$

$D(A)$ is both dense in $\mathcal{H}_1$ with respect to the $\|\cdot\|_1$-norm and in $\mathcal{H}$ with respect to the $\|\cdot\|$-norm.

Proof. (cf. e.g., [495], [427]): We first remark that $\hat{u}$ in the definition of $D(A)$ is uniquely defined, since $\mathcal{H}_1$ is dense in $\mathcal{H}$ by assumption. Moreover, $u \mapsto \hat{u}$ is linear, from the definition, thus A is linear.

Let $J : \mathcal{H} \to \mathcal{H}_1$ with $D(J) = \mathcal{H}_1 \subset \mathcal{H}$, $Jf = f$, $\forall f \in D(J)$. Then J is closed from $D(J) = \mathcal{H}_1 \subset \mathcal{H}$ to $\mathcal{H}_1$ (in the sense that $f_n \in D(J)$, $f_n \to f$, $n \to \infty$, in $\mathcal{H}$, $Jf_n \to h$ in $\mathcal{H}_1$ implies $f \in D(J)$ and $Jf = h$:

in fact $Jf_n = f_n$ and $Jf_n \to h$ in $\mathcal{H}_1$ implies $f_n \to h$ in $\mathcal{H}$ by $\|f_n - h\|^2 \leq \frac{1}{\kappa}\|f_n - h\|_1^2$. But then $Jf_n = f_n \to f$ in $\mathcal{H}$, by assumption, and $f_n \to h$ in $\mathcal{H}_1$, again by assumption, imply $f = h$ in $\mathcal{H}_1 = D(J)$ hence $f \in D(J)$, $Jf = f = h$ by the definition of J and the fact that $f = h$ as elements of $\mathcal{H}_1$.

J is densely defined from $\mathcal{H}$ into $\mathcal{H}_1$, with $D(J) = \mathcal{H}_1$ and closed (a fortiori closable), then J^* is uniquely and densely defined, closed from $\mathcal{H}_1$ into $\mathcal{H}$ (by Th. 5.29 in [312], p. 168).

Set $A_0 = J^*$. Then we have $\forall u \in D(J^*)$, $v \in \mathcal{H}_1$:

$$(A_0 u, v) = (J^* u, v) = (u, Jv)_1 = (u, v)_1.$$

Set $A = A_0$, looked upon as an operator from $\mathcal{H}$ into $\mathcal{H}$. It is then clear that $D(A) \subset \mathcal{H}_1 \subset \mathcal{H}$,

$$(Au, v) = (A_0 u, v) = (u, v)_1 \quad \forall u \in D(A), v \in \mathcal{H}_1. \qquad (*)$$

That $A \geq \kappa$ follows from the fact that $(Au, u) = (u, u)_1 \geq \kappa(u, u)$, $\forall u \in D(A)$, by the definition of $(\cdot, \cdot)_1$. That A is symmetric in $\mathcal{H}$ follows from $(Au, v) = (u, v)_1$, $\forall u \in D(A), v \in \mathcal{H}_1$ and, for $v \in D(A)$:

$$(u, Av) = (u, A_0 v) = (u, J^* v) = (Ju, v)_1 = (u, v)_1.$$

Also the description of $D(A)$ given in the lemma is proven, since $D(A)$ is characterized by the definition of A_0 and J^* as the set of all $u \in \mathcal{H}_1$ s.t.

$$(Au, v) = (A_0 u, v) = (u, v)_1 \quad \forall v \in \mathcal{H}_1.$$

That $D(A)$ is $(\cdot, \cdot)_1$-dense in $\mathcal{H}_1$ is clear from the fact that $D(J^*)$ is $(\cdot, \cdot)_1$-dense in $\mathcal{H}_1$.

. That $D(A)$ is $(\cdot, \cdot)$-dense is also clear from the relation between the $\| \cdot \|_1$ and $\| \cdot \|$-norms.

It remains to show that A is self-adjoint. For this it is enough to prove that the range of A is $\mathcal{H}$ (cf., e.g., [495]). Let us consider $v \in \mathcal{H}, w \in \mathcal{H}_1$:

$$|(v, w)| \leq \|v\| \|w\| \leq \tfrac{1}{\sqrt{\kappa}} \|v\| \|w\|_1$$

where in the latter inequality we used the relation between $\| \cdot \|$ and $\| \cdot \|_1$. This shows that, $\forall v \in \mathcal{H}$, $w \mapsto (v, w)$ is a continuous linear functional on $\mathcal{H}_1$, hence there exists, by Riesz' theorem (see, e.g., [423]) a $\tilde{v} \in \mathcal{H}_1$ s.t.

$$(\tilde{v}, w)_1 = (v, w) \quad \forall w, v \in \mathcal{H}.$$

By the definition $(*)$ of A (used with w replacing v and $\tilde{v}$ replacing u) we have then $(v, w) = (Au, w)$ for any $v \in \mathcal{H}, \forall w \in \mathcal{H}$, which shows that any $v \in \mathcal{H}$ can be written as Au for some $u \in D(A)$, hence the range of A is the whole of $\mathcal{H}$.

The uniqueness of A in the lemma is proven as follows: Let B be self-adjoint in $\mathcal{H}$ s.t.

$$(Bu, v) = (u, v)_1 \quad \forall u \in D(B), v \in \mathcal{H}_1.$$

Then by definition of A, A is an extension of B (i.e. $B \subset A$). But A is self-adjoint so

$$B \subset A = A^* \subset B^*$$

B being itself self-adjoint, this implies $B = A$. This finishes the proof of the lemma and of the theorem. □

2.2 The relation between closed positive symmetric forms and C_0-contraction semigroups and resolvents

The basic relations

Theorem 5. *Le $\mathcal{E}$ be a densely defined positive symmetric sesquilinear form which is closed, in a Hilbert space $\mathcal{H}$. Let $-L_{\mathcal{E}}$ be the associated self-adjoint positive operator given by Theorem 4 (in 2.1) so that*

$$\mathcal{E}(u,v) = ((-L_{\mathcal{E}})^{1/2}u, (-L_{\mathcal{E}})^{1/2}v) \quad \forall u, v \in D(\mathcal{E}).$$

Then $L_{\mathcal{E}}$ generates a C_0-contraction semigroup $T_t = e^{tL_{\mathcal{E}}}$, $t \geq 0$, in $\mathcal{H}$.

And viceversa, if T_t is a symmetric C_0-contraction semigroup, then its generator L is self-adjoint, negative (i.e., $-L$ is positive) and the associated form given by Theorem 4 in 2.1 is positive, symmetric, closed.

One has

$$\lim_{t \searrow 0} \tfrac{1}{t}(u - T_t u, v) = \mathcal{E}(u,v), \quad \forall u, v \in D(\mathcal{E})$$

Proof. The direct way follows from the Theorem 4 given in Chapter 2, 2.1. The viceversa part follows from the fact that L is self-adjoint, negative and the same Theorem 4. $\qquad\square$

Theorem 6. *All statements in Theorem 5 hold with the semigroup $(T_t)_{t \geq 0}$ replaced by the symmetric resolvent family $(G_\alpha)_{\alpha > 0}$, $G_\alpha \equiv (\alpha - L_{\mathcal{E}})^{-1}$, corresponding to $(T_t)_{t \geq 0}$.*

One has for all $u \in \mathcal{H}, v \in D(\mathcal{E})$:

$$\mathcal{E}_\alpha(G_\alpha u, v) = (u,v)$$

(where we recall the definition $\mathcal{E}_\alpha(u,v) \equiv \mathcal{E}(u,v) + \alpha(u,v)$).

Moreover,

$$\mathcal{E}(u,v) = \lim_{\alpha \to +\infty} \alpha(u - \alpha G_\alpha u, v), \quad \forall u, v \in D(\mathcal{E}).$$

Proof. $(G_\alpha)_{\alpha > 0}$ is self-adjoint, by the spectral theorem. The relation for $\mathcal{E}_\alpha$ holds because of

$$\mathcal{E}_\alpha(G_\alpha u, v) = \mathcal{E}(G_\alpha u, v) + \alpha(G_\alpha u, v) \tag{7}$$

(as seen using the definition of $\mathcal{E}_\alpha$, noting the fact that $G_\alpha u \in D(L) \subset D(\mathcal{E})$, for L the operator associated to $\mathcal{E}$ in the sense of Theorem 4 in Chapter 2,2.1). But

$$\mathcal{E}(G_\alpha u, v) = (-LG_\alpha u, v) \tag{8}$$

by the relation between $\mathcal{E}$ and L. The r.h.s. of latter relation can be written as

$$((-L + \alpha - \alpha)G_\alpha u, v) = (u,v) - (\alpha G_\alpha u, v), \tag{9}$$

where we used $G_\alpha = (\alpha - L)^{-1}$. The relation involving $\mathcal{E}_\alpha$ then follows from (7)-(9).

For the limit relation we use (7), the relation just shown for $\mathcal{E}_\alpha$ to get

$$(u, v) = \mathcal{E}(G_\alpha u, v) + \alpha(G_\alpha u, v)$$

hence

$$\alpha(u, v) = \mathcal{E}(\alpha G_\alpha u, v) + \alpha^2(G_\alpha u, v),$$

and the fact that $\alpha G_\alpha \to 1$ as $\alpha \to +\infty$. $\square$

Remark 21. The "relations $\mathcal{E} \leftrightarrow L \leftrightarrow T \leftrightarrow G$" as described in Theorems 4, 5, 6 can be summarized in the following two tables:

Table 1

$B =$ Banach space over $\mathbb{K} = \mathbb{R}, \mathbb{C}$

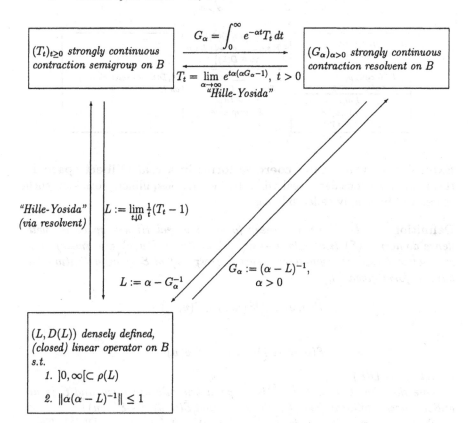

Table 2

$\mathcal{H}$ = Hilbert space over $\mathbb{R}$ with inner product (,) and norm $\| \ \| := (,)^{1/2}$.

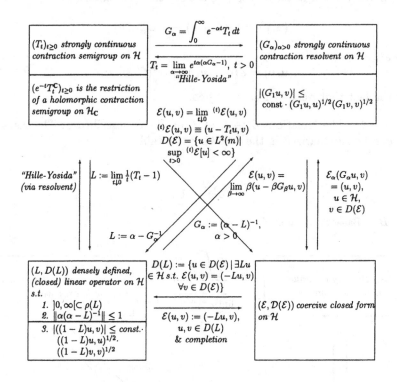

$G_\alpha = \int_0^\infty e^{-\alpha t} T_t \, dt$

$T_t = \lim_{\alpha \to \infty} e^{t\alpha(\alpha G_\alpha - 1)}, \ t > 0$
"Hille-Yosida"

$(T_t)_{t \geq 0}$ *strongly continuous contraction semigroup on* $\mathcal{H}$

$(G_\alpha)_{\alpha > 0}$ *strongly continuous contraction resolvent on* $\mathcal{H}$

$(e^{-t}T_t^c)_{t \geq 0}$ *is the restriction of a holomorphic contraction semigroup on* $\mathcal{H}_\mathbb{C}$

$|(G_1 u, v)| \leq$ const $\cdot (G_1 u, u)^{1/2}(G_1 v, v)^{1/2}$

$\mathcal{E}(u, v) = \lim_{t \downarrow 0} {}^{(t)}\mathcal{E}(u, v)$

${}^{(t)}\mathcal{E}(u, v) \equiv (u - T_t u, v)$
$D(\mathcal{E}) = \{u \in L^2(m)|$
$\sup_{t > 0} {}^{(t)}\mathcal{E}[u] < \infty\}$

"Hille-Yosida" (via resolvent)

$L := \lim_{t \downarrow 0} \frac{1}{t}(T_t - 1)$

$\mathcal{E}(u, v) = \lim_{\beta \to \infty} \beta(u - \beta G_\beta u, v)$

$\mathcal{E}_\alpha(G_\alpha u, v) = (u, v),$
$u \in \mathcal{H},$
$v \in D(\mathcal{E})$

$L := \alpha - G_\alpha^{-1}$

$G_\alpha := (\alpha - L)^{-1},$
$\alpha > 0$

$(L, D(L))$ *densely defined, (closed) linear operator on* $\mathcal{H}$ *s.t.*
1. $]0, \infty[\subset \rho(L)$
2. $\|\alpha(\alpha - L)^{-1}\| \leq 1$
3. $|((1 - L)u, v)| \leq$ const $\cdot$ $((1 - L)u, u)^{1/2} \cdot$ $((1 - L)v, v)^{1/2}$

$D(L) := \{u \in D(\mathcal{E}) \, | \, \exists Lu \in \mathcal{H} \ s.t. \ \mathcal{E}(u, v) = (-Lu, v) \ \forall v \in D(\mathcal{E})\}$

$\mathcal{E}(u, v) := (-Lu, v),$
$u, v \in D(L)$
& completion

$(\mathcal{E}, D(\mathcal{E}))$ *coercive closed form on* $\mathcal{H}$

Extension to the case of coercive forms in a real Hilbert space

In this section we consider a real Hilbert space $\mathcal{H}$. Sesquilinear forms on such spaces will be simply called bilinear.

Definition 14. *Let $\mathcal{E}$ be a bilinear form on a real Hilbert space $\mathcal{H}$, with dense domain $D(\mathcal{E})$ (i.e. both $u \to \mathcal{E}(u, v)$ and $v \to \mathcal{E}(u, v)$ are linear). The symmetric (resp. antisymmetric) part $\tilde{\mathcal{E}}$ (resp. $\check{\mathcal{E}}$) of $\mathcal{E}$ is by definition the bilinear form given by:*

$$\tilde{\mathcal{E}}(u, v) \equiv \tfrac{1}{2}[\mathcal{E}(u, v) + \mathcal{E}(v, u)]$$

resp.

$$\check{\mathcal{E}}(u, v) \equiv \tfrac{1}{2}[\mathcal{E}(u, v) - \mathcal{E}(v, u)],$$

for all $u, v \in D(\mathcal{E})$.

One then has $\mathcal{E} = \tilde{\mathcal{E}} + \check{\mathcal{E}}$. $\tilde{\mathcal{E}}$ is a symmetric bilinear form and $\check{\mathcal{E}}$ is an antisymmetric bilinear form (in the sense that $\check{\mathcal{E}}(u, v) = -\check{\mathcal{E}}(v, u)$).

Suppose $\mathcal{E}$ is positive definite (i.e. $\mathcal{E}(n, n) \geq 0$ for all $u \in D(\mathcal{E})$). Then one says that $\mathcal{E}$ satisfies the weak sector condition with constant $k \geq 0$ if

$$|\mathcal{E}_1(u,v)| \leq k\mathcal{E}_1[u]^{1/2}\mathcal{E}_1[v]^{1/2}.$$

$(\mathcal{E}, D(\mathcal{E}))$ is called a _coercive closed form_ if $\mathcal{E}$ satisfies the weak sector condition and $(\mathcal{E}, D(\mathcal{E}))$ is closed.

The relations $\mathcal{E} \leftrightarrow L \leftrightarrow T \leftrightarrow G$ discussed in Theorems 4, 5, 6 (and in Tables 2.2, 2.2 extend to the case of a real Hilbert space, with the symmetry and positivity in $\mathcal{E}$, $-L$, T, G replaced respectively by:

a) Coerciveness for $\mathcal{E}$.

b) L is closed operator with $\rho(L) \subset (0, \infty)$ s.t. $\|\alpha(\alpha - L)^{-1}\| \leq 1$, for $\alpha > 0$,

$$|((1-L)u,v)| \leq C((1-L)u,u)^{1/2}((1-L)v,v)^{1/2}.$$

c) G is a C_0-contraction resolvent family with

$$|(G_1 u, v)| \leq C(G_1 u, u)^{1/2}(G_1 v, v)^{1/2} \quad u, v \in \mathcal{H}.$$

d) T is a C_0-contraction semigroup s.t. its natural linear extension to the complexification $\mathcal{H}_{\mathbb{C}} = \mathcal{H} + i\mathcal{H}$ of $\mathcal{H}$ satisfies the following condition: the operator $e^{-t}T_t^{\mathbb{C}}$ is the restriction of a holomorphic contraction semigroup on the sector

$$\{z \in \mathbb{C} \mid |Im(z)| \leq \tfrac{1}{k}Re(z)\} \text{ (with } k \text{ as in Definition 14).}$$

Moreover, G, T are accompanied by dual semigroups $\hat{G}$, $\hat{T}$ (that only in the symmetric case coincide with G, T), see, e.g., [312], [427], [367].

Remark 22. The direct relation between $\mathcal{E}$ and T has been discussed in [407] and [136]. E.g. one has the result, relating $\mathcal{E}$ and T:

$$\mathcal{E}(u,v) = \lim_{t \searrow 0} {}^{(t)}\mathcal{E}(u,v), \quad \forall u, v \in D(\mathcal{E}),$$

with ${}^{(t)}\mathcal{E}(u,v) \equiv \frac{1}{t}(u - T_t u, v)$, $D(\mathcal{E}) = \{u \in \mathcal{H} \mid \sup_{t>0} {}^{(t)}\mathcal{E}[u] < \infty\}$.

An example In the whole course, we shall have two basic examples, one in finite dimensions and one in infinite dimensions. Here is the first basic example, the second one will be introduced in Chapter 3, 4.2.

Let μ be a positive Borel measure on $\mathbb{R}^d$ with supp $\mu = \mathbb{R}^d$. Let us consider the bilinear form in $\mathcal{H} = L^2(\mathbb{R}^d, \mu)$:

$$\overset{\circ}{\mathcal{E}}_\mu(u,v) = \int_{\mathbb{R}^d} <\nabla u, \nabla v> d\mu, \quad u, v \in C_0^\infty(\mathbb{R}^d),$$

with $\nabla u(x) = (\frac{\partial u}{\partial x_1}, \dots \frac{\partial u}{\partial x_d})$, $x = (x_1, \dots, x_d) \in \mathbb{R}^d$, $<\nabla u, \nabla v> = \sum_{i=1}^d \frac{\partial u}{\partial x_i} \frac{\partial v}{\partial x_i}$.

We call $\overset{\circ}{\mathcal{E}}_\mu$ with domain $D(\overset{\circ}{\mathcal{E}}_\mu) = C_0^\infty(\mathbb{R}^d)$ the classical pre-Dirichlet form given by μ. $\overset{\circ}{\mathcal{E}}_\mu$ is symmetric and positive. The basic question is: For which μ is $\overset{\circ}{\mathcal{E}}_\mu$ closable?

In Proposition 11 in 2.1, we have already indicated a condition for a positive symmetric sesquilinear form to be closable, namely that it can be expressed by a symmetric operator B s.t.

$$\overset{\circ}{\mathcal{E}}_\mu(u,v) = (u, Bv)_\mu, \quad u,v \in D(B) = D(\overset{\circ}{\mathcal{E}}_\mu)$$

(where $(u,v)_\mu$ stands for the $L^2(\mathbb{R}^d, \mu)$-scalar product). When can we find such a B? The problem is solved when we can derive an "integration by parts formula"("IP"), writing

$$\overset{\circ}{\mathcal{E}}_\mu(u,v) = -\int_{\mathbb{R}^d} u\Delta v d\mu - \int_{\mathbb{R}^d} u < \beta_\mu, \nabla v > d\mu, \tag{10}$$

with $< \beta_\mu(x), y > d\mu(x) = $ "$< \nabla_x, y > d\mu(x)$", $y \in \mathbb{R}^d$, whenever μ is differentiable in a suitable sense. For this it suffices, e.g., that μ is absolutely continuous with respect to Lebesgue measure, with density ρ s.t.

$$\frac{1}{\rho}\frac{\partial\rho}{\partial x_i} \in L^2(\mathbb{R}^d, \mu), \quad i = 1, \ldots, d$$

because then

$$\beta_\mu = \frac{\nabla\rho}{\rho} = \nabla \ln\rho. \tag{11}$$

In this case

$$\overset{\circ}{\mathcal{E}}_\mu(u,v) = (u, A_\mu v)_\mu \tag{12}$$

with

$$A_\mu = -\Delta - < \beta_\mu(x), \nabla_x >$$

on $C_0^\infty(\mathbb{R}^d)$ or shortly $A_\mu = -\Delta - \beta_\mu \cdot \nabla$ (thus the operator B we were seeking is this A_μ). A_μ is called the classical pre-Dirichlet operator given by μ. The quantity β_μ in (11) is the logarithmic derivative of ρ. More generally: whenever there is a measurable vector field $\beta_\mu = (\beta_\mu^1, \ldots, \beta_\mu^d)$ s.t.

$$\int \frac{\partial u}{\partial x_i} d\mu = -\int u \beta_\mu^i d\mu, \quad \forall u \in C_0^\infty(\mathbb{R}^d) \tag{13}$$

one says that β_μ is the logarithmic derivative of μ (in $L^2(\mathbb{R}^d, \mu)$) whenever $\beta_\mu^i \in L^2(\mathbb{R}^d, \mu), \forall i = 1, \ldots, d$.

In this case one easily sees that (10) holds, hence $\overset{\circ}{\mathcal{E}}_\mu$ is closable. Detailed conditions for closability to hold have been worked out, based on the above idea. E.g. for $d = 1$, $\overset{\circ}{\mathcal{E}}_\mu$ is closable iff μ has a density $\rho \in L^1_{loc}(\mathbb{R})$ with respect to the Lebesgue measure s.t. $\rho = 0$ a. e. on $S(\rho) \equiv \mathbb{R} - R(\rho)$, where

$$R(\rho) \equiv \{x \in \mathbb{R} \mid \int_{x-\epsilon}^{x+\epsilon} \frac{dy}{\rho(y)} < +\infty, \exists \epsilon > 0\}$$

is the regular set for ρ ("Hamza's condition") (cf. [244],[119]).

Remark 23. A condition of this type is also necessary and sufficient for the "partial classical pre-Dirichlet form" $\int_{\mathbb{R}^d} \frac{\partial u}{\partial x_k} \frac{\partial v}{\partial x_k} d\mu$, $u, v \in C_0^\infty(\mathbb{R}^d)$ to be closable, for all $k = 1, \ldots, d$, see [119]. For an example where the above condition is not satisfied, yet the total classical pre-Dirichlet form is closable, see [258].

Definition 15. *If the classical pre-Dirichlet form $\overset{\circ}{\mathcal{E}}_\mu$ given by μ is closable, the closure $\mathcal{E}_\mu$ is called* <u>*classical Dirichlet form associated with μ.*</u> *The corresponding self-adjoint negative operator L_μ s.t.*

$$((-L_\mu)^{1/2}u, (-L_\mu)^{1/2}v)_\mu = \mathcal{E}_\mu(u, v)$$

is called the <u>*classical Dirichlet operator associated with μ.*</u>

The corresponding classical Dirichlet form shares with other forms an essential "contraction property", which shall be discussed in Chapter 3 to which we refer also for other comments on classical Dirichlet forms.

Let us however discuss already at this stage briefly why classical Dirichlet forms are important, relating them to (generalized) Schrödinger operators.

Let μ be absolutely continuous with respect to the Lebesgue measure on $\mathbb{R}^d$ with density ρ. To start with we assume $\rho(x) > 0$ for all $x \in \mathbb{R}^d$ and ρ smooth.

Let us consider the map $W : L^2(\mathbb{R}^d) \to L^2(\mathbb{R}^d, \mu)$ given by

$$Wu \equiv \frac{u}{\sqrt{\rho}}, \quad u \in L^2(\mathbb{R}^d)$$

(where $L^2(\mathbb{R}^d)$ denotes the space of square summable functions with respect to the Lebesgue measure). W is unitary (and $W^*v = \sqrt{\rho}v$, $v \in L^2(\mathbb{R}^d, \mu)$, as seen from the construction). Let A_μ be given by (12) and consider on the domain $W^*D(A_\mu)$, the operator A:

$$A \equiv W^*A_\mu W.$$

If $D(A_\mu)$ is dense in $L^2(\mathbb{R}^d)$ (which is the case discussed above, where $D(A_\mu) \supset C_0^\infty(\mathbb{R}^d)$) then, by unitarity, $W^*D(A_\mu)$ is dense in $L^2(\mathbb{R}^d, \mu)$, hence A is densely defined in $L^2(\mathbb{R}^d, \mu)$. We have, for $u, v \in W^*D(A_\mu)$,

$$(Wu, A_\mu Wv) = (u, W^*A_\mu Wv) = (u, Av)$$

where $(\cdot, \cdot)$ is the scalar product on $L^2(\mathbb{R}^d)$.

Set $\sqrt{\rho} = \varphi$, assuming $\varphi(x) > 0$, $\forall x \in \mathbb{R}^d$, $\varphi \in C^\infty(\mathbb{R}^d)$. We compute the l.h.s of the above equality for $u, v \in C_0^\infty(\mathbb{R}^d)$ (observing that then also

$\frac{u}{\varphi}, \frac{v}{\varphi} \in C_0^\infty(\mathbb{R}^d)$ by our assumption on φ) using the relations between A_μ and $\overset{\circ}{\mathcal{E}}_\mu$ given by (12) and the definition of $\overset{\circ}{\mathcal{E}}_\mu$:

$$(Wu, A_\mu Wv)_{L^2(\mu)} = \overset{\circ}{\mathcal{E}}_\mu(Wu, Wv) = \int_{\mathbb{R}^d} \nabla\left(\frac{u}{\varphi}\right)\nabla\left(\frac{v}{\varphi}\right)\varphi^2 dx. \quad (14)$$

But $\nabla\left(\frac{u}{\varphi}\right) = \frac{1}{\varphi}\nabla u - \frac{1}{\varphi^2}(\nabla\varphi)u$ and correspondingly with u replaced by v. Inserting these equalities into (14) we get the following four terms:

a)

$$\int_{\mathbb{R}^d} \frac{1}{\varphi}(\nabla u)\frac{1}{\varphi}(\nabla v)\varphi^2 dx = -\int_{\mathbb{R}^d} u(\triangle v)dx$$

(where we first observed that the φ-terms cancel, and then we integrated by parts, using $u, v \in C_0^\infty(\mathbb{R}^d)$),

b)

$$\int_{\mathbb{R}^d} \frac{1}{\varphi^2}(\nabla\varphi)u\frac{1}{\varphi^2}(\nabla\varphi)v\varphi^2 dx = \int_{\mathbb{R}^d} u\left(\tfrac{1}{4}\beta_\mu^2\right)v dx$$

(where we used that by (11): $\beta_\mu = \nabla\ln\varphi^2 = 2\frac{\nabla\varphi}{\varphi}$),

c) Two mixed terms:

$$-\int_{\mathbb{R}^d} \frac{\nabla\varphi}{\varphi}u(\nabla v)dx - \int_{\mathbb{R}^d} \frac{\nabla\varphi}{\varphi}v(\nabla u)dx$$

$$= -\int_{\mathbb{R}^d} \frac{\nabla\varphi}{\varphi}\nabla(uv)dx = \int_{\mathbb{R}^d} \nabla\left(\frac{\nabla\varphi}{\varphi}\right)u\,v\,dx = \frac{1}{2}\int_{\mathbb{R}^d}(div\,\beta_\mu)\,u\,v\,dx$$

(where we have used again integration by parts and $\beta_\mu = 2\frac{\nabla\varphi}{\varphi}$).

In total we then get:

$$(Wu, A_\mu Wv)_\mu = (u, Av), \quad (15)$$

with $A = -\triangle + V(x)$, with

$$V(x) \equiv \frac{1}{4}\beta_\mu(x)^2 + \frac{1}{2}div\beta_\mu(x).$$

Remark 24. The operator A is a Schrödinger operator with potential term V, acting in $L^2(\mathbb{R}^d)$ (in this interpretation it is appropriate to take $L^2(\mathbb{R}^d)$ as the complex Hilbert space of square integrable functions). A is densely defined (on $C_0^\infty(\mathbb{R}^d)$) in $L^2(\mathbb{R}^d)$, positive and symmetric as seen from (15) and the positivity and symmetry of A_μ (coming from corresponding properties of $\overset{\circ}{\mathcal{E}}_\mu$, see (14)). It is well known that Schrödinger operators describe the dynamics of quantum systems. More precisely: any self-adjoint extension H of A can be taken to define the dynamics of a quantum mechanical particle moving in the force field given by the potential V (having made the inessential convention $\frac{\hbar^2}{2m} = 1$, where $\hbar$ is the reduced Planck's constant and m the mass of the particle; obviously a "scaling" of variables will remove this restriction). The corresponding Schrödinger equation is then

$$i\frac{\partial}{\partial t}\psi = H\psi, \quad t \geq 0,$$

with $\psi_{|t_0} = \psi_0 \in D(H) \subset L^2(\mathbb{R}^d)$; it is solved by $e^{-itH}\psi_0$ (the unitary group, when we let $t \in \mathbb{R}$, generated by H via Stone's theorem).

Assumptions on V are known s.t. H is uniquely determined by its restriction A to $C_0^\infty(\mathbb{R}^d)$ (i.e. A is essentially self-adjoint on $C_0^\infty(\mathbb{R}^d)$; see, e.g. [423] for this concept), e.g. V bounded is enough (but in fact H is uniquely determined in much more general situations, see, e.g. [426]).

Remark 25. We can write, using $\beta_\mu = 2\frac{\nabla\varphi}{\varphi}$:

$$V = \left(\frac{\nabla\varphi}{\varphi}\right)^2 + div\left(\frac{\nabla\varphi}{\varphi}\right) = \frac{(\nabla\varphi)^2}{\varphi^2} + \frac{\varphi\Delta\varphi - (\nabla\varphi)^2}{\varphi^2} = \frac{\Delta\varphi}{\varphi}.$$

This shows

$$(-\Delta + V)\varphi = 0$$

i.e. φ is a "$(-\Delta + V)$-harmonic function" ".

The passage $A_\mu \to A$ can thus be seen as a particular case of "Doob's transform" technique, going from an operator A_μ in $L^2(\mathbb{R}^d, \mu)$, with 1 as harmonic function (1 is in the domain, if μ is finite) to an operator A in $L^2(\mathbb{R}^d)$, here $-\Delta + V$, with φ as harmonic function (see, e.g. [217] for the discussion of Doob's transform). Doob's transform is also called, in the context of the above operator, "ground state transformation" (see, e.g. [426], [467]).

Viceversa: Let us consider a "stationary Schrödinger equation" of the form

$$(-\Delta + \tilde{V})\varphi = E\varphi$$

for some $\tilde{V} : \mathbb{R}^d \to \mathbb{R}$, $E \in \mathbb{R}$. If there is a solution $\varphi \in L^2(\mathbb{R}^d)$ s.t. $\varphi(x) > 0$ $\forall x \in \mathbb{R}^d$, then setting $V \equiv \tilde{V} - E$ we get $V = \frac{\Delta\varphi}{\varphi}$. If, e.g., $D(V) \supset C_0^\infty(\mathbb{R}^d)$, we can define $A = -\Delta + V$ on $C_0^\infty(\mathbb{R}^d)$, as a linear densely defined symmetric operator. The associated sesquilinear form

$$\overset{\circ}{\mathcal{E}}(u, v) = (u, Av)$$

defined for $u, v \in C_0^\infty(\mathbb{R}^d)$ is then symmetric and closable in $L^2(\mathbb{R}^d)$. Moreover, if

$$\|\nabla\varphi\|_2^2 + (\varphi, V\varphi) \geq 0$$

(which is, e.g. the case whenever $V \geq 0$), then $\overset{\circ}{\tilde{\mathcal{E}}}$ is positive. We also observe that, defining $A_\mu = -\Delta - \beta_\mu \cdot \nabla$ with $\beta_\mu = \frac{\nabla \rho}{\rho} = 2\nabla ln(\varphi)$ with $\rho = \varphi^2$, $d\mu = \rho \, dx$, we have

$$\overset{\circ}{\tilde{\mathcal{E}}}(u,v) = (Wu, A_\mu W v)_\mu,$$

with $A_\mu \equiv WAW^*$, $u,v \in C_0^\infty(\mathbb{R}^d)$.

Defining on the other hand

$$\overset{\circ}{\tilde{\mathcal{E}}}_\mu(u,v) = \int <\nabla u, \nabla v> d\mu$$

we see, as in the beginning of this section, that

$$\overset{\circ}{\tilde{\mathcal{E}}}_\mu(u,v) = (u, A_\mu v)_\mu = (u, Av) = \overset{\circ}{\tilde{\mathcal{E}}}(u,v).$$

Note that $\overset{\circ}{\tilde{\mathcal{E}}}_\mu$ is a form in $L^2(\mathbb{R}^d, \mu)$, whereas $\overset{\circ}{\tilde{\mathcal{E}}}$ is a form in $L^2(\mathbb{R}^d)$.

Thus to a generalized Schrödinger operator of the form $-\Delta + \tilde{V}$ in $L^2(\mathbb{R}^d)$ we have associated a classical pre-Dirichlet form $\overset{\circ}{\tilde{\mathcal{E}}}_\mu$ and its closure $\tilde{\mathcal{E}}_\mu$ and corresponding associated densely defined operators L_μ° resp. L_μ. Even though $\overset{\circ}{\tilde{\mathcal{E}}}$ and A in $L^2(\mathbb{R}^d)$ have a more direct physical interpretation, rather than their corresponding objects $\mathcal{E}_\mu$ resp. L_μ in $L^2(\mathbb{R}^d, \mu)$, the latter are more appropriate whenever discussing "singular interactions" (see, e.g. [44], [104], [20], [106], [40], [152], [157], [164], [176], [193], [183], [266], [503] or the case where $\mathbb{R}^d$ is replaced by an infinite dimensional space E (like in quantum field theory), see, below and, e.g., [278], [15], [41] (in fact in the latter case there are interesting probability measures μ on E, whereas no good analogue of Lebesgue measure on E exists). In this sense operating with $\mathcal{E}_\mu$, L_μ is more natural and general than operating with Schrödinger operators.

Remark 26. In the above discussion of closability and Doob's transformation we have assumed φ, V, $\tilde{V}$ to be smooth and $\varphi > 0$. These assumptions can be strongly relaxed. Moreover, the considerations extend to cover the cases where, instead of A resp. A_μ, general elliptic symmetric operators with L_{loc}^p-coefficients are handled, see e.g. [367], [499], [104], [106], [20], [22], [174], [364], [375] where various questions (including, e.g. closability) are discussed (we shall give more references in Chapter 4).

3 Contraction properties of forms, positivity preserving and submarkovian semigroups

3.1 Positivity preserving semigroups and contraction properties of forms- Beurling-Deny formula.

In this whole section the Banach respectively Hilbert space of Chapter 2 will be the functional space $L^2(m) \equiv L^2(E, \mathcal{B}, m; \mathbb{K})$, where $(E, \mathcal{B}, m)$ is a measure space and $L^2(E, \mathcal{B}, m; \mathbb{K})$ stands for the $\mathbb{K}$-valued (equivalence class of) functions over E which are square integrable with respect to m. Let $(,)$ be the scalar product in $L^2(m)$, $\| - \|$ the corresponding norm.

Definition 16. *A linear operator A in $L^2(m)$ is said to be positivity preserving (p.p.) iff $u \geq 0$ m-a.e. implies $Au \geq 0$ m-a.e., for any $u \in L^2(m)$. A semigroup $T = (T_t)_{t \geq 0}$ is said to be positivity preserving if T_t is p.p. for all $t \geq 0$.*

Proposition 12. *Let L be a self-adjoint operator in $L^2(m)$ s.t. $L \leq 0$ (i.e. $-L \geq 0$). The corresponding symmetric C_0-contraction semigroups $(e^{tL})_{t>0}$ is positivity preserving iff the corresponding resolvent family $(\alpha - L)^{-1} = G_\alpha$, $\alpha > 0$, is positivity preserving.*

This is an immediate consequence of the Laplace-transform formula, see chapter 1:
$G_\alpha u = \int_0^\infty e^{-\alpha t} e^{tL} u \, dt$, $u \in L^2(m)$ resp. the Yosida approximation formula

$$e^{tL} = \lim_{\alpha \to +\infty} e^{tL^{(\alpha)}}$$

with $L^\alpha \equiv -\alpha + \alpha^2 G_\alpha$.

Theorem 7. *(Beurling-Deny representation for positivity preserving semigroups)*
Let $\mathbb{K} = \mathbb{R}$ and let L be a self-adjoint operator in $L^2(m)$, with $L \leq 0$. Then the following statements are equivalent:

  i) $(e^{tL})_{t \geq 0}$ is positivity preserving
  ii) $\mathcal{E}([|u|]) \leq \mathcal{E}[u]$ for all $u \in D(\mathcal{E})$
      (where $\mathcal{E}$ is the positive, symmetric, closed form associated with L, according to Friedrichs representation theorem of Chapter 2)

Remark 27. This is an important theorem since it expresses the positivity preserving property of semigroups (and in probability theory, as we shall see, one is primarily interested in such semigroups, e.g., transition semigroups) through corresponding properties of associated forms, which are often easier to verify directly.

Proof. i) → ii)

One first remarks that $|e^{tL}u| \leq e^{tL}|u|$ (which follows from $e^{tL}(|u| \pm u) \geq 0$, by the assumption in i)).

But for all $u \in L^2(m)$, using the semigroup property: $(u, e^{tL}u) = \|e^{\frac{t}{2}L}u\|^2$, which is then bounded from above by $\|e^{\frac{t}{2}L}|u|\|^2 = (|u|, e^{tL}|u|)$, where in the latter equality we again used the semigroup property.

From this we deduce

$$(u, (e^{tL} - 1) u) \leq (|u|, (e^{tL} - 1) |u|). \qquad \forall t \geq 0$$

For $u \in D(L)$ the left hand side divided by $t > 0$ by Theorem 5 in chapter 2 section 2.2 converges for $t \downarrow 0$ to $(u, Lu) = -\mathcal{E}[u]$.

The right hand side is non positive, since e^{tL} is positivity preserving, thus:

$$-\mathcal{E}[u] \leq \frac{1}{t} \left(|u|, (e^{tL} - 1) |u|\right) \leq 0 \qquad \forall t \geq 0.$$

Hence the limit for $t \downarrow 0$ of $\frac{1}{t} \left(|u|, (e^{tL} - 1) |u|\right)$ exists by subsequences, and by the definition of $D(L)$ we must have then that this is equal to $(|u|, L|u|)$.

Thus $-\mathcal{E}[u] \leq (|u|, L|u|) = -\mathcal{E}[|u|] \leq 0$ for all $u \in D(L)$. Since $D(L)$ is a core for $D(\mathcal{E})$, and L is closed, this extends by continuity to all $u \in D(\mathcal{E}) = D((-L)^{1/2})$, showing that i) → ii).

ii) → i):

We have, for $u \in D(L), v \in D(L)$:

$$\mathcal{E}_\alpha[u + v] = \mathcal{E}_\alpha[u] + \mathcal{E}_\alpha[v] + 2((\alpha - L)u, v).$$

Replace now u by $w = (\alpha - L)^{-1}u$ ($\in D(L)$!), and take $v \geq 0$, $v \in D(\mathcal{E})$, then $\mathcal{E}_\alpha[w + v] = \mathcal{E}_\alpha[w] + \mathcal{E}_\alpha[v] + 2(u, v)$. Taking $u \geq 0$, v as before, we get:

$$\mathcal{E}_\alpha[w + v] \geq \mathcal{E}_\alpha[w] + \mathcal{E}_\alpha[v]. \qquad (16)$$

But

$$\mathcal{E}_\alpha[|w| + v] = \mathcal{E}_\alpha[w + (|w| - w) + v] = \mathcal{E}_\alpha[w + v'], \qquad (17)$$

with $v' \equiv |w| - w + v$. Applying (16) on the right hand side, with v replaced by v' we get:

$$\mathcal{E}_\alpha[w + v'] \geq \mathcal{E}_\alpha[w] + \mathcal{E}_\alpha[v']. \qquad (18)$$

Hence from (17), (18):

$$\mathcal{E}_\alpha[|w| + v] \geq \mathcal{E}_\alpha[w] + \mathcal{E}_\alpha[v']. \qquad (19)$$

By assumption ii) the r.h.s. is bounded from below by

$$\mathcal{E}_\alpha[|w|] + \mathcal{E}_\alpha[v']. \tag{20}$$

Taking now $v = 0$, we get from (19):

$$\mathcal{E}_\alpha[|w|] \geq \mathcal{E}_\alpha[|w|] + \mathcal{E}_\alpha[|w| - w] \tag{21}$$

(where we inserted the definition of v'). On the other hand, by $\mathcal{E}_\alpha \geq 0$:

$$\mathcal{E}_\alpha[|w| - w] \geq 0. \tag{22}$$

Hence, from (21), (22):

$$\mathcal{E}_\alpha[|w| - w] = 0$$

thus $|w| - w = 0$ i.e. $|w| = w$, in particular $w \geq 0$. From the definition of w, then:

$$(\alpha - L)^{-1} u \geq 0 \ \ for \ u \geq 0,$$

which proves i), using Proposition 12.

Remark 28. 1) There are several other statements which are equivalent to the ones of Theorem 7, see e.g., [367],[368], [136].

2) In a similar way one proves the Theorem for the complex Hilbert space $L_{\mathbb{C}}^2(m)$. It is also easy to show that i),ii) are equivalent with iii), e^{tL} is reality preserving and $\mathcal{E}[u^+] \leq \mathcal{E}[u]$ for all real $u \in D(\mathcal{E})$, with $u^+ \equiv \sup(u, 0)$. This is left as an exercise (cf. [423]).

3.2 Beurling-Deny criterium for submarkovian contraction semigroups

Theorem 8. *Let $\mathbb{K} = \mathbb{R}$ and let L be a self-adjoint operator in $L^2(m)$, with $L \leq 0$, s.t. $T_t u \geq 0$ for $u \geq 0$, with $T_t = e^{tL}$, $t \geq 0$. Then the following are equivalent:*

a) e^{tL} is a contraction on $L^\infty(m) \cap L^2(m)$, $\forall t \geq 0$.
b) For $u \geq 0$, $u \in D(\mathcal{E})$, then $\mathcal{E}[u \wedge 1] \leq \mathcal{E}[u]$ (with $u \wedge 1 \equiv \inf(u, 1)$).

Proof. The proof of b) $\rightarrow$ a) is similar to the one of i) $\rightarrow$ ii) in Theorem 7. For a) $\Rightarrow$ b) the idea is to show

$$(u \wedge 1, (1 - T_t)(u \wedge 1)) \leq (u, (1 - T_t)u) \ \forall t \geq 0. \tag{23}$$

Once this is shown a) follows easily by dividing the inequality by $t > 0$ and going to the limit $t \downarrow 0$, which yields, for $u \in D(L)$

$$\mathcal{E}[u \wedge 1] \leq (u, (1 - L)u) = \mathcal{E}[u].$$

b) then follows by the fact that $D(L)$ is a core for $\mathcal{E}$. The proof of (23) is obtained by proving it first for special step functions and then going to the limit, exploiting the fact that $F(u) \equiv u \wedge 1$ satisfies $|F(u)| \leq |u|$, $F(u) - F(v) \leq |u - v|$, see, e.g., [258], [367], [423] for details.

3.3 Dirichlet forms

Let (E, B, M) be a σ-finite measure space, and let $\mathbb{K} = \mathbb{R}$.

Definition 17. *Let S be a linear operator in $L^2(m)$ with $D(S) = L^2(m)$, S bounded. S is called* <u>*submarkovian*</u> *if $0 \leq Su \leq 1$ m-a.e., whenever $0 \leq u \leq 1$ m-a.e., for all $u \in L^2(m)$*

Remark 29. If S is positivity preserving in $L^2(m)$ and $1 \in L^2(m)$, then $Su \leq 1$ for $u \leq 1$ m-a.e., $u \in L^2(m)$ follows from $S(u - 1) \leq 0$ and linearity, since $u - 1 \in L^2(m)$.

Definition 18. *Let $\mathcal{E}$ be a positive, symmetric, bilinear closely defined form in $L^2(m)$ (not necessarily closed). Given $\Phi : \mathbb{R} \to \mathbb{R}$, we say that $\mathcal{E}$ contracts under Φ if $u \in D(\mathcal{E})$ implies $\Phi(u) \in D(\mathcal{E})$ and $\mathcal{E}[\Phi(u)] \leq \mathcal{E}[u]$. If $\Phi(t) = (0 \vee t) \wedge 1$, $t \in \mathbb{R}$, then we call Φ the "unit contraction". Let, for all $\epsilon > 0$, Φ_ϵ be such that $\Phi_\epsilon(t) = t$, $t \in [0, 1]$*

$$-\epsilon \leq \Phi_\epsilon(t) \leq 1 + \epsilon \ \forall t \in \mathbb{R},$$

$$0 \leq \Phi_\epsilon(t_2) - \Phi_\epsilon(t_1) \leq t_2 - t_1, \ t_1 \leq t_2,$$

$$\Phi_\epsilon(u) \in D(\mathcal{E}),$$

for $v \in D(\mathcal{E})$ and $\liminf_{\epsilon \downarrow 0} \mathcal{E}[\Phi_\epsilon(u)] \leq \mathcal{E}[u]$, then we call Φ_ϵ "ϵ-approximation" of the unit contraction and we say that $\mathcal{E}$ is <u>*submarkovian*</u>.

Theorem 9. *Let $\mathcal{E}$ be a positive, symmetric, bilinear, closed densely defined form and let T_t, G_α be the associated C_0-contraction semigroup resp. contraction resolvent. Then the following assertions are equivalent:*

a) T_t is submarkovian for all $t \geq 0$
b) αG_α is submarkovian for all $\alpha > 0$
c) $\mathcal{E}$ is submarkovian
d) $\mathcal{E}$ contracts under the unit contraction
e) The infinitesimal generator L of T_t has the "Dirichlet contraction property" $(u, L(u - 1)^+) \leq 0 \ \forall u \in D(L)$.

Proof. a) $\to$ b): Proposition 12 in Chapter 3.3.1
d) $\to$ c): take $\Phi_\epsilon = unit$
b) $\to$ d): this is similar to the proof of i) $\to$ ii) Theorem 7 in Chapter 3.3.1
c) $\to$ b): this is similar to the proof of b) $\to$ a) in Theorem 8 in Chapter 3.3.2.
For the remaining parts see [244], [258], [367], [172]. $\square$

For the applications the following "Addendum" is useful.

Proposition 13. *Let $\overset{\circ}{\mathcal{E}}$ be a symmetric, positive, closable, densely defined bilinear form. Assume moreover that $\overset{\circ}{\mathcal{E}}$ is submarkovian on $D(\overset{\circ}{\mathcal{E}})$, then the closure $\mathcal{E}$ of $\overset{\circ}{\mathcal{E}}$ satisfies a)-e) of Theorem 9.*

Proof. (Sketch): Let $u \in D(\mathcal{E})$, since $\overset{\circ}{\mathcal{E}}$ is closable by assumption, there exists a sequence $u_n \in D(\overset{\circ}{\mathcal{E}})$, s.t. $u_n \to u$ in $L^2(m)$, for $n \to \infty$, and $\overset{\circ}{\mathcal{E}}[u_n] \to \mathcal{E}[u]$. Choose $\varPhi_{\frac{1}{n}}$ as $\epsilon = \frac{1}{n}$- approximation of the unit contraction, for $n \in \mathbb{N}$. Then $\sup \overset{\circ}{\mathcal{E}}_1[\varPhi_{\frac{1}{n}}(u_n)] \leq \sup \overset{\circ}{\mathcal{E}}_1[u_n]$ and this is finite, by the convergence of $\overset{\circ}{\mathcal{E}}_1[u_n]$. $(D(\mathcal{E}), \mathcal{E}_1)$ is a Hilbert space and by the Banach-Saks theorem (cf. e.g. Th 2.2. in App.2 in [367]), $g_n \equiv \frac{1}{n} \sum_1^n \varPhi_{\frac{1}{j}}(u_j)$ converges strongly for $n \to \infty$ to $g \in D(\mathcal{E})$ and $\overset{\circ}{\mathcal{E}}[g_n] \to \mathcal{E}[g]$. On the other hand: $g_n \to \varPhi_0[u]$ as $n \to \infty$ $\varPhi_\epsilon \to \varPhi_0$ m-a.e. as $\epsilon \downarrow 0$. So by these convergences we get $\varPhi_0[u] = g$ m-a.e., $\varPhi_0 \in D(\mathcal{E})$ and $\mathcal{E}(\varPhi_0[u])^{\frac{1}{2}} = \lim_{n\to\infty} \overset{\circ}{\mathcal{E}}[g_n]^{\frac{1}{2}}$. By the triangle inequality and the definition of g_n the r.h.s. is bounded from above by $\limsup \frac{1}{n} \sum_{j=1}^n \overset{\circ}{\mathcal{E}}[\varphi_{\frac{1}{j}}(u_j)]^{\frac{1}{2}}$. Since the Cesaro-limit and the ordinary limits coincide whenever they exist, we get that this is $\mathcal{E}[u]^{\frac{1}{2}}$ and thus $\mathcal{E}$ contracts under the unit contraction. $\qquad\square$

Definition 19. *A form $\mathcal{E}$ with the properties as in Theorem 9 (i.e. bilinear, positive, symmetric, closed, densely defined and submarkovian) is called a (symmetric) Dirichlet form.*

Remark 30. Proposition 13 says that the closure $\overline{\overset{\circ}{\mathcal{E}}} = \mathcal{E}$ of the form $\overset{\circ}{\mathcal{E}}$ in Proposition 13 is a Dirichlet form. In applications the forms $\overset{\circ}{\mathcal{E}}$ are usually given on a nice subset D of $L^2(m)$ where they are closable. Proposition 13 permits then to pass from the contraction properties of $\overset{\circ}{\mathcal{E}}$ on D to corresponding ones for its closure $\mathcal{E}$.

Remark 31. Dirichlet forms are studied extensively in [150], [244], [258], [468], [469], [367], [309] (and references therein).

3.4 Examples of Dirichlet forms

Standard finite dimensional example Let $E = \mathbb{R}^d$, $B = \mathcal{B}(\mathbb{R}^d)$, $m = \mu$, $\mu(dx) = \rho(x)\,dx$, $r \geq 0$, (where dx is Lebesgue measure on $\mathbb{R}^d$).

Consider $\overset{\circ}{\mathcal{E}}_\mu(u,v) \equiv \int \nabla u \cdot \nabla v\,d\mu$ in $L^2_{\mathbb{R}}(\mathbb{R}^d, \mu) \equiv L^2(\mathbb{R}^d, \mu; \mathbb{R}) \equiv L^2(\mu)$, $u, v \in C_0^\infty(\mathbb{R}^d)$ (with $\nabla u(x) \equiv (\frac{\partial}{\partial x_1}u, \cdots, \frac{\partial}{\partial x_d}u)(x)$, $x \in \mathbb{R}^d$).

The basic question to be answered is whether $\overset{\circ}{\mathcal{E}}_\mu$ is closable. In Chapter 1 we saw that for this to be the case it is enough to find a (symmetric, positive) operator $\overset{\circ}{A}_\mu$ in $L^2(\mu)$ s.t. $\overset{\circ}{\mathcal{E}}_\mu(u,v) = (u, \overset{\circ}{A}_\mu v)$.

By "integration by parts" we see that $\overset{\circ}{A}_\mu = -\Delta - \beta_\mu \cdot \nabla$ on $C_0^\infty(\mathbb{R}^d)$. This is well defined whenever

$$\beta_{\mu,i}(x) \equiv \frac{1}{\rho}\partial_i\rho(x) \in L^2(\mu), \qquad\qquad \forall i \in \{1, ..., d\}$$

with $\partial_i \equiv \frac{\partial}{\partial x_i}$ (taken in the distributional sense).

"Optimal" conditions for closability of $\overset{\circ}{\mathcal{E}}$ have been given in [119], see also [223], [474].

The second basic question to be answered is whether the closure $\mathcal{E}_\mu \equiv \overset{\widetilde{\circ}}{\mathcal{E}}_\mu$ of $\overset{\circ}{\mathcal{E}}_\mu$ is a Dirichlet form.

The answer is yes, since by using Proposition 13 it is enough to verify that $\overset{\circ}{\mathcal{E}}_\mu$ is submarkovian. Let us take $\epsilon > 0$, $\Phi_\epsilon \in C^\infty$, Φ_ϵ-contraction, an ϵ-approximation of the unit contraction $|\Phi'_\epsilon(\cdot)| \leq 1$. Then:

$$\overset{\circ}{\mathcal{E}}[\Phi_\epsilon(u)] = \int \nabla\Phi_\epsilon(u)\nabla\Phi_\epsilon(u)d\mu = \int \Phi'_\epsilon(u)\nabla u \cdot \Phi'_\epsilon(u)\nabla u d\mu = \int |\Phi'_\epsilon(u)|^2 |\nabla u|^2 d\mu \leq$$

$\int |\nabla u|^2 d\mu = \overset{\circ}{\mathcal{E}}[u]$, $\forall u \in C_0^\infty(\mathbb{R}^d)$, where we used the definition of $\overset{\circ}{\mathcal{E}}$, Leibniz differentiation formula in the second equality, and $|\Phi'_\epsilon| \leq 1$ for the latter inequality.

This shows that $\overset{\circ}{\mathcal{E}}$ contracts under Φ_ϵ and by Proposition 13 we then have that $\mathcal{E}_\mu$ is a Dirichlet form.   □

The standard infinite dimensional example Let E be a separable real Banach space, E' its topological dual. Let μ be the probability measure on the Borel subsets B in E generated by all open subsets of E, and suppose $supp\mu = E$ (i.e. $\mu(U) > 0$ for all $U \subset E$, U open, $U \neq \emptyset$).

Let $FC_b^\infty \equiv FC_b^\infty(E) = \{u : E \to \mathbb{R} \mid \exists m, \exists f \in C_b^\infty(\mathbb{R}^m), \exists l_1, \cdots, l_m \in E' : u(z) = f(< z, l_1 >, \cdots, < z, l_m >)\}$. Then FC_b^∞ is dense in $L^2(\mu) \equiv L_\mathbb{R}^2(E, \mu)$ (this is essentially a form of the Stone-Weierstrass theorem, see, e.g. [396], [277], [290]).

For $u \in FC_b^\infty(E)$, $z \in E$, $k \in K \subset E$ (K a linear subspace of E) we define the (Gâteaux-)derivative in the direction k by:

$$\frac{\partial u}{\partial k}(z) \equiv \frac{d}{ds}u(z + sk)|_{s=0} \left(= \sum_{i=1}^{m} \partial_i f(< z, l_1 >, \cdots, < z, l_m >) < k, l_i > \right).$$

We define

$$\overset{\circ}{\mathcal{E}}_{\mu,k}(u, v) \equiv \int \frac{\partial u}{\partial k} \cdot \frac{\partial v}{\partial k} d\mu,$$

for $u, v \in FC_b^\infty$.

We easily see that $\overset{\circ}{\mathcal{E}}_{\mu,k}$ is a symmetric, positive, bilinear form on $L^2(\mu)$, with domain $D(\tilde{\mathcal{E}}_{\mu,k}) = FC_b^\infty$ (dense in $L^2(\mu)$!). As before the first question to be answered is: for which μ is $\overset{\circ}{\mathcal{E}}_{\mu,k}$ defined on FC_b^∞ closable? A sufficient condition for this is (analogously as in the finite dimensional case) the existence of $\beta_{\mu,k} \in L^2(\mu)$ s.t.

$$\int \frac{\partial u}{\partial k} v d\mu = -\int u \frac{\partial v}{\partial k} d\mu - \int uv\beta_{\mu,k}d\mu, \quad \forall u, v \in FC_b^\infty. \tag{24}$$

Remark 32. This is an "integration by parts" formula, in the spirit of those in the theory of smooth measures [205], [59], [61], [62], [65], [119] and in "Malliavin calculus" (originally on Wiener space, see [372], [396]).
Assuming (24) we have then that

$$\overset{\circ}{L}_{\mu,k} \equiv \frac{\partial^2}{\partial k^2} + \beta_{\mu,k} \frac{\partial}{\partial k}$$

is a well defined linear operator with $D(\overset{\circ}{L}_{\mu,k}) = FC_b^\infty$, symmetric and such that

$$\overset{\circ}{\mathcal{E}}_{\mu,k}(u,v) = (u, (-\overset{\circ}{L}_{\mu,k})v)$$

(with $(\ ,\ )$ the $L^2(\mu)$-scalar product). From Proposition 11 in Chapter 2 we see that $\overset{\circ}{\mathcal{E}}_{\mu,k}$ is closable in $L^2(\mu)$.

The next question is whether the closure $\mathcal{E}_{\mu,k} = \overset{\bar{\circ}}{\mathcal{E}}_{\mu,k}$ of $\overset{\circ}{\mathcal{E}}_{\mu,k}$ in $L^2(\mu)$ is a Dirichlet form. This is proven similarly as in the finite dimensional case. In fact let Φ_ε be an ε-approximation of the unit contraction which is C^∞ on $\mathbb{R}$. Then, by Leibniz formula, $u \to \Phi_\varepsilon(u) \in FC_b^\infty$ if $u \in FC_b^\infty$ and

$$\overset{\bar{\circ}}{\mathcal{E}}_{\mu,k}[\Phi_\varepsilon(u)] \leq \int |\Phi_\varepsilon'(u)|^2 |\nabla u|^2 d\mu \leq \int |\nabla u|^2 d\mu = \overset{\circ}{\mathcal{E}}_{\mu,k}[u].$$

By using Proposition 13 this yields the proof that $\overset{\circ}{\mathcal{E}}_{\mu,k}$ is a Dirichlet form. We assume that E and its dual E' are s.t.

$$E' \subset H' \cong H \subset E,$$

where H is a Hilbert space, densely contained in E, H' (isomorphic to H) is the dual of H (and E' is densely contained in H'), moreover the embedding of H in E (and of E' in H') is continuous. (We remark that this assumption is not strong, in fact it can be realized in great generality, see e.g. [301], [462], [428]).
For $u \in FC_b^\infty$, $k \in E'$ we have $\frac{\partial u}{\partial k}(z) = \underset{E'}{<} k, \nabla u(z) \underset{E}{>}$ (with $\underset{E'}{<}$, $\underset{E}{>}$ the dualisation between E and E').

Definition 20. *For $u, v \in FC_b^\infty$ we define*

$$\overset{\circ}{\mathcal{E}}_\mu(u,v) \equiv \int_E <\nabla u, \nabla v >_H d\mu,$$

where $<$, $>_H$ is the scalar product in H.
Let K be an orthonormal basis in H consisting of elements in E'. We have:

$$\overset{\circ}{\mathcal{E}}_\mu(u,v) = \sum_{k \in K} \overset{\circ}{\mathcal{E}}_{\mu,k}(u,v).$$

We remark that $\overset{\circ}{\mathcal{E}}_\mu$ is a bilinear positive symmetric form. $\overset{\circ}{\mathcal{E}}_\mu$ is closable if all $\overset{\circ}{\mathcal{E}}_{\mu,k}$ are closable (we leave the verification of this as an exercise: one can use, e.g., Proposition 11 in Chapter 2). In this case the closure $\mathcal{E}_\mu$ of $\overset{\circ}{\mathcal{E}}_\mu$ exists and is a positive symmetric closed bilinear form. It is also easy to verify that $\mathcal{E}_\mu$ is a Dirichlet form. $\mathcal{E}_\mu$ is called the "classical Dirichlet form given by μ".

Remark 33. For $E = H = \mathbb{R}^d$ and FC_b^∞ replaced by $C_0^\infty(\mathbb{R}^d)$, $\mathcal{E}_\mu$ coincides with the classical Dirichlet form given by μ (over $\mathbb{R}^d$) as defined in Definition 15.

By Friedrichs' representation theorem to $\mathcal{E}_\mu$ there is a uniquely associated self-adjoint, negative operator L_μ s.t. $\mathcal{E}_\mu(u,v) = ((-L_\mu)^{\frac{1}{2}}u, (-L_\mu)^{\frac{1}{2}}v)$. L_μ is called the Dirichlet operator associated with μ.

Remark 34. $\mathcal{E}_\mu$ should not be confused with the "maximal Dirichlet form given by μ", $\mathcal{E}_\mu^m$, obtained from the closed extension of $\tilde{\mathcal{E}}_\mu$ with domain $D(\mathcal{E}_\mu^m) = \bigcap_{k \in K}\{u \in L^2(\mu)| \int |\frac{\partial u}{\partial k}|^2 d\mu < \infty\}$. In general $\mathcal{E}_\mu^m$ is a strict extension of $\mathcal{E}_\mu$, they coincide exactly when $\tilde{L}_\mu$ is essentially self-adjoint on FC_b^∞, see, e.g., [119], [88], [359] for results of this type.

Exercise 6. Show that
$L_\mu = \overset{\circ}{L}_\mu$ on FC_b^∞, with $\overset{\circ}{L}_\mu = \sum_k \overset{\circ}{L}_{\mu,k}$ on FC_b^∞, where

$$\overset{\circ}{L}_{\mu,k} \equiv \frac{\partial^2}{\partial k^2} + \beta_{\mu,k}(\cdot)\frac{\partial}{\partial k},$$

$$\Delta_H \equiv \sum_{k \in K}\frac{\partial^2}{\partial k^2} \quad, < \beta_\mu(z), \nabla > u(z) = \sum_{k \in K} < \beta_\mu(z), k > \frac{\partial}{\partial k}u(z),$$

$u \in FC_b^\infty$, $z \in E$.
Moreover $\overset{\circ}{\mathcal{E}}_\mu(u,v) = \sum_{k \in K}(u, \overset{\circ}{L}_{\mu,k}v) \ \forall u, v \in FC_b^\infty$.

Exercise 7. When $E = C_{(0)}([0,t];\mathbb{R})$, (Wiener space of continuous functions from $[0,t]$ to $\mathbb{R}$ vanishing at time zero) we have

$$H = H^{1,2}([0,t];\mathbb{R}) = \{w \in E| \int_0^t |\dot{w}(s)|^2 ds < \infty\},$$

where μ is the Wiener measure: show that β_μ is a linear function (cf. [172],[365], [396]).
We know by the general theory of Chapter 3.1 that L_μ generates a symmetric submarkovian C_0-contraction semigroup $L^2(\mu)$. We shall show that L_μ generates a diffusion process. For this we first describe shortly the general structure theory of Dirichlet operators and forms.

*)

3.5 Beurling-Deny structure theorem for Dirichlet forms

We first consider the case of a locally compact separable metric space E (not necessarily a linear space). Let m be a Radon measure on E (see, e.g., [462] for Radon measures).

Definition 21. *A bilinear symmetric positive densely defined form $\mathcal{E}$ in $L^2_{\mathbb{R}}(E, m) \equiv L^2(m)$ is called regular if $D(\mathcal{E}) \cap C_0(E)$ is dense in $D(\mathcal{E})$ with respect to the $\mathcal{E}_1$-norm and dense in $C_0(E)$ with respect to the supremum-norm.*

Definition 22. *A bilinear form $\mathcal{E}$ in $L^2(m)$ is called local if $\mathcal{E}(u,v) = 0$, for all $u, v \in D(\mathcal{E})$ s.t. $supp[u] \cap supp[v] = 0$ (for some representatives $[u]$, $[v]$ of u resp. v as element of $L^2(m)$, with $supp[u]$, $supp[v]$ compact).*

Examples are:

a) $\mathcal{E}(u,v) = \int u\, v\, dm, \quad u, v \in C_0^\infty(\mathbb{R}^d)$,
b) $\mathcal{E}(u,v) = \int \nabla u \cdot \nabla v\, dm, \quad u, v \in C_0^\infty(\mathbb{R}^d)$.

Definition 23. *A bilinear form $\mathcal{E}$ in $L^2(m)$ is called strong local if $\mathcal{E}(u,v) = 0 \;\forall u, v \in D(\mathcal{E})$ such that v is constant in a neigborhood of $supp[u]$.*

Remark 35. Strong local implies local, but not viceversa, e.g. in the examples above a) is local but not strong local.

Theorem 10. *(Beurling-Deny structure theorem)*
Let $\mathcal{E}$ be a regular Dirichlet form. Then $\mathcal{E}$ can be written as

$$\mathcal{E} = \mathcal{E}^c + \mathcal{E}^j + \mathcal{E}^k$$

on $D(\mathcal{E}) \cap C_0(E)$, with $\mathcal{E}^c$ strong local,

$$\mathcal{E}^j(u,v) = \int_{E \times E}' [u(x) - u(y)][v(x] - v(y)]J(dxdy),$$

where $(E \times E)' \equiv \{(x,y) \in E \times E \mid x \neq y\}$ and J is a symmetric Radon measure on $(E \times E)'$,

$$\mathcal{E}^k(u,v) = \int uv\, dk,$$

dk being a Radon measure on E. The parts $\mathcal{E}^c$ (diffusion part), $\mathcal{E}^j$ (jump part), $\mathcal{E}^k$ (killing part) are uniquely determined by $\mathcal{E}$.

Proof. For the proof see [244], [258]. □

If E is a manifold or $E = U$ with U an open subset of $\mathbb{R}^d$ then $\mathcal{E}^c$ can be further specified.

Theorem 11. *(Beurling-Deny structure theorem for an open subset of $\mathbb{R}^d$)*
Let U be an open subset of $\mathbb{R}^d$, m a Radon measure on U. Then any submarkovian symmetric positive bilinear form $\overset{\circ}{\mathcal{E}}$ on $L^2(U,m)$ with $D(\overset{\circ}{\mathcal{E}}) = C_0^\infty(U)$ s.t. $\overset{\circ}{\mathcal{E}}$ is closable and one has that $\mathcal{E} = \overset{\circ}{\mathcal{E}}$ is a regular Dirichlet form and for $\mathcal{E}$ (and hence $\overset{\circ}{\mathcal{E}}$) on $C_0^\infty(\mathbb{R}^d)$ we have

$$\overset{\circ}{\mathcal{E}} = \mathcal{E}^c + \mathcal{E}^j + \mathcal{E}^k,$$

with $\mathcal{E}^c(u,v) = \sum_{i,j=1}^d \int \frac{\partial u}{\partial x_i}(x) \frac{\partial v}{\partial x_j}(x) d\nu_{ij}(x)$, where $\nu_{ij}(\cdot)$ is a random measure (for all $i,j = 1, \cdots, d$), with $(\nu_{ij}(K))_{i,j=1}^d$ is a positive definite, symmetric matrix for all compact $K \subset U$.

Proof. Since $\overset{\circ}{\mathcal{E}}$ is submarkovian and closable, the closure is a Dirichlet form. That $\mathcal{E}$ is regular follows easily, see [244]. Then we can apply Theorem 1 in chapter 1 to get $\mathcal{E}^c, \mathcal{E}^j, \mathcal{E}^k$. For the formula for $\mathcal{E}^c$ see [244].

Remark 36. There is an extension of Theorem 10 to infinite dimensional spaces, with regularity replaced by "quasi-regularity" (a concept we shall discuss in Chapter 4) see [111], [112].

3.6 A remark on the theory of non symmetric Dirichlet forms

There exists an extension of the entire theory of Dirichlet forms to the case of bilinear forms which are closable and contract under suitable "contraction operations", but are not necessarily symmetric.

Definition 24. *A bilinear form $\mathcal{E}$ in a real Hilbert space H is said to satisfy the <u>weak sector condition</u> if $\exists k > 0$ s.t.*

$$|\mathcal{E}_1(u,v)| \le k\mathcal{E}_1[u]^{\frac{1}{2}}\mathcal{E}_1[v]^{\frac{1}{2}}$$

$\mathcal{E}$ is a <u>coercitive closed form</u> if $D(\mathcal{E})$ is dense in H, $\mathcal{E}$ satisfies the weak sector condition and the symmetric part $\overset{\circ}{\mathcal{E}}$ of $\mathcal{E}$ (defined as $\overset{\circ}{\mathcal{E}}(u,v) \equiv \frac{1}{2}[\mathcal{E}(u,v) + \mathcal{E}(v,u)]\forall u,v \in D(\mathcal{E}), with D(\overset{\circ}{\mathcal{E}}) = D(\mathcal{E}))$ is closed.

Analogs of the relation between symmetric forms, contraction semigroups and associated operators, resolvents, discussed in Chapter 2.3, exist also for coercive closed forms. The main difference consists in the fact that instead of a single semigroup $(T_t)_{t\ge0}$ (resp. resolvent family $(G_\alpha)_{\alpha>0}$) there are two semigroups $(T_t, \hat{T}_t)_{t\ge0}$ (resp. two resolvents $((G_\alpha, \hat{G}_\alpha)_{\alpha>0})$): in duality (i.e. adjoint to each other in the case of a space $L^2(m)$). See the table below. For a Beurling-Deny structure theorem in this case see [378].

Table 3

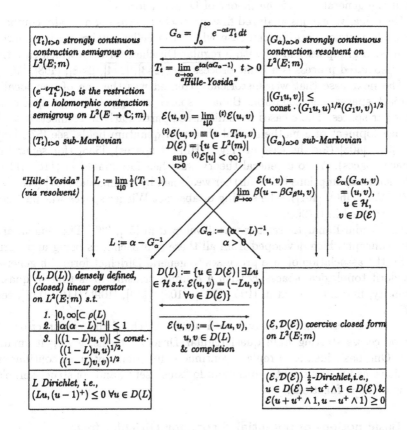

	$G_\alpha = \int_0^\infty e^{-\alpha t} T_t \, dt$	
$(T_t)_{t>0}$ strongly continuous contraction semigroup on $L^2(E;m)$	$T_t = \lim_{\alpha \to \infty} e^{t\alpha(\alpha G_\alpha - 1)}, \; t > 0$ "Hille-Yosida"	$(G_\alpha)_{\alpha>0}$ strongly continuous contraction resolvent on $L^2(E;m)$
$(e^{-t}T_t^C)_{t>0}$ is the restriction of a holomorphic contraction semigroup on $L^2(E \to \mathbb{C}; m)$	$\mathcal{E}(u,v) = \lim_{t \downarrow 0} {}^{(t)}\mathcal{E}(u,v)$	$|(G_1 u, v)| \leq$ const $\cdot (G_1 u, u)^{1/2}(G_1 v, v)^{1/2}$
$(T_t)_{t>0}$ sub-Markovian	${}^{(t)}\mathcal{E}(u,v) \equiv (u - T_t u, v)$ $D(\mathcal{E}) = \{u \in L^2(m) \mid \sup_{t>0} {}^{(t)}\mathcal{E}[u] < \infty\}$	$(G_\alpha)_{\alpha>0}$ sub-Markovian

"Hille-Yosida" (via resolvent) $\quad L := \lim_{t \downarrow 0} \frac{1}{t}(T_t - 1)$

$\mathcal{E}(u,v) = \lim_{\beta \to \infty} \beta(u - \beta G_\beta u, v)$

$\mathcal{E}_\alpha(G_\alpha u, v) = (u, v), \; u \in \mathcal{H}, \; v \in D(\mathcal{E})$

$L := \alpha - G_\alpha^{-1}$

$G_\alpha := (\alpha - L)^{-1}, \; \alpha > 0$

| $(L, D(L))$ densely defined, (closed) linear operator on $L^2(E;m)$ s.t. 1. $]0, \infty[\subset \rho(L)$ 2. $\|\alpha(\alpha - L)^{-1}\| \leq 1$ 3. $|((1-L)u,v)| \leq$ const $\cdot ((1-L)u,u)^{1/2} \cdot ((1-L)v,v)^{1/2}$ | $D(L) := \{u \in D(\mathcal{E}) \mid \exists Lu \in \mathcal{H} \text{ s.t. } \mathcal{E}(u,v) = (-Lu, v) \; \forall v \in D(\mathcal{E})\}$ $\mathcal{E}(u,v) := (-Lu, v), \; u, v \in D(L)$ & completion | $(\mathcal{E}, D(\mathcal{E}))$ coercive closed form on $L^2(E;m)$ |
| L Dirichlet, i.e., $(Lu, (u-1)^+) \leq 0 \; \forall u \in D(L)$ | | $(\mathcal{E}, D(\mathcal{E})) \; \frac{1}{2}$-Dirichlet, i.e., $u \in D(\mathcal{E}) \Rightarrow u^+ \wedge 1 \in D(\mathcal{E})$ & $\mathcal{E}(u + u^+ \wedge 1, u - u^+ \wedge 1) \geq 0$ |

4 Potential Theory and Markov Processes associated with Dirichlet Forms

4.1 Motivations

Let E be a topological space, m be a σ-finite measure on E. Let $T \equiv (T_t)_{t \geq 0}$ be a sub-Markov semigroup in $L^2(m) \equiv L^2_{\mathbb{R}}(E, m)$. We shall discuss the following questions:

1) Is it possible to associate a "nice process" to T such that T is its transition semigroup (analogously as the Brownian motion process on $\mathbb{R}^d$ is associated with the heat semigroup as transition semigroup)?

2) Which kind of Dirichlet forms correspond to such "nice processes"?

As is well known, in the case where E is a locally compact separable metric space, an association of this type is surely possible if the transition semigroup has the additional property of being a Feller semigroup, but what about the general case of the theory of Dirichlet forms?

Historically, one has analyzed first precisely the case of a locally compact separable metric space for the case where the Feller semigroup is replaced by the semigroup associated with a regular Dirichlet form, in which case the nice associated process is a Hunt process, see [243], [244], [258], [468], [469].

The next case that was historically treated was the one of non locally compact spaces, e.g. separable Banach spaces or rigged Hilbert spaces or conuclear spaces. These cases were discussed particularly in connection with the development of a mathematical theory of quantum fields, see [57], [58], [59], [61], [408] (for the case of rigged Hilbert spaces), [335] (for separable Banach spaces). A more general theory was then formulated in [118], [119], [120], some assumptions being further weakened by Schmuland [457] (see also [458], [459], [460], [461]). The setting of abstract Wiener spaces was particularly discussed in [172].

A non standard analytic setting was developed in [39], [36]. The central analytic concept which developed from all these approaches as being appropriate for the association of nice processes to general Dirichlet forms on general Hausdorff topological spaces E with a σ-finite measure m is that of quasi-regularity, first introduced in [105], [107], [108], [114], [109], [110], [113], see [367].

We shall here limit ourselves to a short sketch of the construction of nice processes starting from quasi regular Dirichlet forms, giving its main ideas. One basic idea is to replace continuous functions by quasi-continuous functions, as functions continuous modulo "small sets", and construct kernels acting on such functions.

4.2 Basic notions of potential theory for Dirichlet forms

$\mathcal{E}$-exceptional sets Let E be a Hausdorff topological space, $\mathcal{B}$ the σ-algebra of its Borel subsets, m a σ-finite measure.

Let $(\mathcal{E}, D(\mathcal{E}))$ be a Dirichlet form on $L^2(m) = L_\mathbb{R}^2(E, m)$.

Definition 25. *An increasing sequence* $(F_k)_{k \in \mathbb{N}}$ *of closed subsets of E is called an* $\underline{\mathcal{E}\text{-nest}}$ *if* $\bigcup_k D(\mathcal{E})_{F_k}$ *is* $\tilde{\mathcal{E}}_1^{1/2}$*-dense in* $D(\mathcal{E})$ *(where* $D(\mathcal{E})_F \equiv \{u \in D(\mathcal{E}) \mid u = 0 \text{ m-a.e. on the complement } F^c \text{of} F\}$; $\tilde{\mathcal{E}}_1^{1/2}$ *is the norm given by the scalar product in* $L^2(m)$ *defined by* $\tilde{\mathcal{E}}_1$).

Definition 26. $N \subset E$ *is called "$\mathcal{E}$-exceptional" if* $N \subset \bigcap_k F_k^c$ *for some $\mathcal{E}$-nest* $\{F_k\}_{k \in \mathbb{N}}$. *We say that a certain property of points in E holds "$\mathcal{E}$-quasi everywhere" (q.e.) if it holds outside some $\mathcal{E}$-exceptional subset of E.*

There is an important relation between $\mathcal{E}$-exceptional sets and sets of small capacity; for this we first have to introduce the concept of capacity (an extension to our setting of the concept of capacity in classical potential theory, see, e.g. [162], [276], [160]).

Definition 27. *Let U be an open subset of E. Define the capacity cap U of U by*

$$cap \, U \equiv \inf_{u \in \mathcal{L}_U} \mathcal{E}_1[u]$$

where

$$\mathcal{L}_U \equiv \{u \in D(\mathcal{E}) \mid u \geq 1 \; m\text{-}a.e. \text{ on } U\}$$

(with $\inf\{.\} = +\infty$ if $\mathcal{L}_U = \emptyset$). For any $A \subset E$ define

$$A \equiv \inf_{A \subset U, U \text{ open}} cap \, U.$$

Remark 37. We have that cap $(U) = 0$ implies $m(U) = 0$.

Proof. We have

$$\mathcal{E}_1[u] \geq \|u\|^2 \geq m(U) \quad \forall u \in \mathcal{L}_U,$$

hence cap $U = \inf_{u \in \mathcal{L}_U} \mathcal{E}_1[u] \geq m(U)$.   □

Proposition 14. *Let E, m, $\mathcal{E}$ as above. If $(A_n)_{n \in \mathbb{N}}$ is an increasing sequence of sets then 1) cap $(\bigcap_n A_n) = \sup_n cap \, A_n$*

Moreover,

2) $(F_k)_{k \in \mathbb{N}}$ is an $\mathcal{E}$-nest $\iff \lim_{k \to \infty} cap \, F_k^c = 0$

3) N is $\mathcal{E}$-exceptional $\iff cap \, N = 0$.

Proof. (Sketch):
1) is easy (see, e.g., [244]). 2) $\Rightarrow$ 3) is obvious. To prove 2) one uses the following

Lemma 2. *Let $U \subset E$ be open, s.t. $\mathcal{L}_U \neq \emptyset$, then*
a) there exists uniquely an element $e_U \in \mathcal{L}_U$ (called "equilibrium potential") s.t. $\mathcal{E}_1[e_U] = Cap \, U$.
b) $0 \leq e_U \leq 1$ m-a.e. on E and $U \mapsto e_U$ is monotone increasing.

Proof. For a) one uses that $\mathcal{L}_U$ is convex (which we leave as an exercise).

b) Take $u = (0 \vee e_U) \wedge 1$. $\mathcal{E}$ being a Dirichlet form, it contracts under the unit contraction, i.e. for $u \in D(\mathcal{E})$

$$\mathcal{E}_1[u] \leq \mathcal{E}_1[e_U] = Cap \, U$$

which implies, by the definition of capacity, that $u = e_U$ q.e., hence $0 \leq e_U \leq 1$ m-a.e.. That $u \mapsto e_U$ is monotone increasing is easily seen using $\mathcal{E}_\infty(\sqcap^+, \sqcap^-) \leq \prime$ for any $u \in D(\mathcal{E})$, see [244].   □

Definition 28. *Let $(F_k)_{k \in \mathbb{N}}$ be an $\mathcal{E}$-nest. Set*

$$C(\{F_k\}) \equiv \Big\{u : A \to \mathbb{R}, \text{ for some } A \subset E \text{ s.t. } \bigcup_k F_k \subset A \subset E, u_{|F_k}$$

$$\text{continuous } \forall k \in \mathbb{N}\Big\}.$$

An $\mathcal{E}$-q.e. defined function u on E is called $\mathcal{E}$-quasi-continuous (q.c.) if there exists an $\mathcal{E}$-nest $\{F_k\}_{k \in \mathbb{N}}$ s.t. $u \in C(\{F_k\})$. $C(\{F_k\})$ is then called the set of quasi-continuous functions associated with the nest $(F_k)_{k \in \mathbb{N}}$.

Remark 38. One shows that u is $\mathcal{E}$-q.c. if for any $\epsilon > 0$ $\exists U \subset E$ open s.t. Cap $U < \epsilon$, $u_{|E-U}$ continuous, see, e.g. [367].

4.3 Quasi-regular Dirichlet forms

Definition 29. *Let E be a Hausdorff topological space, m a σ-finite measure on E, and let $\mathcal{B}$ the smallest σ-algebra of subsets of E with respect to which all continuous functions on E are measurable.*

A Dirichlet form $\mathcal{E}$ is called quasi-regular if

   i) there exists an $\mathcal{E}$-nest $(F_k)_{k\in\mathbb{N}}$ of compact subsets of E;

  ii) there exists an $\tilde{\mathcal{E}}_1^{1/2}$-dense subset of $D(\mathcal{E})$ whose elements have $\mathcal{E}$-q.c. m-versions.

 iii) there exists $u_n \in D(\mathcal{E})$, $n \in \mathbb{N}$, with $\mathcal{E}$-q.c. m-versions $\tilde{u}_n$ and there exists an $\mathcal{E}$-exceptional subset N of E s.t. $\{\tilde{u}_n\}_{n\in\mathbb{N}}$ separates the points of $E - N$.

Remark 39. Thinking of Stone-Weierstrass type results, "point separation" means richness of elements...

Remark 40. We leave as an exercise to prove that if E is a locally compact separable metric space then $\mathcal{E}$ regular implies $\mathcal{E}$ quasi-regular but not viceversa (in general).

4.4 Association of "nice processes" with quasi-regular Dirichlet forms

$(E, \mathcal{B}, m)$ be as in the preceding section. Let $\mathcal{E}$ be a symmetric Dirichlet form in $L^2(m)$ and $(T_t = e^{tL})_{t\geq 0}$ the associated symmetric submarkovian C_0-contraction semigroup on $L^2(m)$.

Definition 30. *Let $(p_t)_{t\geq 0}$ be a submarkovian semigroup acting in $C_b(E)$ s.t. $(p_t u)(x) = \int p_t(x, dy)u(y), u \in C_b(E), p_t$ being a submarkovian semigroup of kernels i.e. $p_t(x, A) \in [0, 1], x \to p_t(x, A)$ is measurable $\forall x \in E, A \in \mathcal{B}$, $A \to p_t(x, A)$ is a measure on $(E, \mathcal{B})$ with $p_t(x, E) \leq 1$(see e.g., [142]). p_t is said to be associated with $\mathcal{E}$ (or $(T_t)_{t\geq 0}$ or with the infinitesimal generator L) if $p_t u$ is an m-version of $T_t u, \forall t > 0$, for all $u \in \mathcal{B}_b(E) \cap L^2(m)$. Let $X \equiv (X_t)_{t\geq 0}$ be a (sub-) Markov process with state space E and transition semigroup $(p_t)_{t\geq 0}$, s.t.*

$$(p_t u)(z) = E^z[u(X_t)]$$

$\forall z \in E, t \geq 0$ *(with E^z the expectation for the process with start measure $\delta_z(.)$) is said to be associated with $\mathcal{E}$ (or $(T_t)_{t\geq 0}$ or the infinitesimal generator*

L) if p_t is associated with $\mathcal{E}$.
X is said to be __properly associated__ with $\mathcal{E}$ if in addition $p_t u$ is $\mathcal{E}$-quasi
continuous (in the sense of the definition in IV.3) for all $t > 0$.

Remark 41. In a sense the $\mathcal{E}$-quasi-continuity of $p_t u$ is replaces the Feller-
property (for the latter see, e.g., [162], [201], [220]).
The following theorem gives the basic relation between certain properties of
symmetric Dirichlet forms and corresponding properties of associated (sub-)
Markov processes; the meaning of the latter properties will be shortly
discussed afterwards.

Theorem 12. *Let E be a topological Hausdorff space. Then:*

a) *$\mathcal{E}$ is a quasi-regular Dirichlet form iff X is an m-tight special standard*
   process.
b) *$\mathcal{E}$ is a local quasi-regular Dirichlet form iff X is an m-tight special standard*
   process and it is a diffusion i.e.
   $P^z \{t \to X_t$ continuous on $[0, \zeta)\} = 1$ for all $z \in E$
   (for some random variable ζ, with values $[0, +\infty]$, the life time of the
   process).

The basic concept on the r.h.s. is defined as follows:

Definition 31. *(right process)*
Let Δ be one pointset, disjoint from E, and define $E_\Delta \equiv E \cup \Delta$.

i) *Let $X \equiv (X_t)_{t \geq 0}$ be a (family of) stochastic process(es) on a probability*
   space $(\Omega, \mathcal{M}, \bar{P}^z)_{z \in E_\Delta}$, with state space E, life time ζ, a measurable map
   $\Omega \to [0, +\infty]$ (if $\zeta = +\infty$ one can forget about Δ), s.t.

$$X_t(\omega) \in E \text{ for } t < \zeta(\omega),$$

$$X_t(\omega) \in \Delta \text{ for } t \geq \zeta(\omega).$$

   $\forall \omega \in \Omega$. Assume $(t, \omega) \to X_t(\omega)$ is measurable.
   Let $\mathcal{M}_t$ be a filtration in $\mathcal{M}$ s.t. X_t is $\mathcal{M}_t$- adapted.
ii) *Assume that $z \to P^z(B)$ is measurable for all $B \in \mathcal{M}$ and the Markov*
    property holds:

$$P^z(X_{s+t} \in B | \mathcal{M}_s) = P^{X_s}(X_t \in B),$$

   $\forall s, t \geq 0, P^z$-a.s. , $z \in E_\Delta, B \in \mathcal{B}(E_\Delta)$.
   X is then called a Markov process.
iii) *X is called normal if $P^z(X_0 = z) = 1, \forall z \in E_\Delta$.*
iv) *X is said to be right continuous if $t \to X_t(\omega)$ is right continuous,*
    $\forall t \geq 0, \forall \omega \in \Omega$.

v) X is said to be strong Markov if $\mathcal{M}_t$ is right continuous and

$$P^\mu \left(X_{\sigma+t} \in \mathcal{B} | \mathcal{M}_\sigma \right) = P^{X_\sigma} (X_t \in \mathcal{B}),$$

P^μ-*a.s.*, $\forall \mathcal{M}_t$-*stopping times σ, for any probability measure μ on E, and with $P^\mu(.) \equiv \int P^z(.)\mu(dz)$.*

X is said to be a right process if i)-v) hold.
An "m-tight special standard process" is shortly described as a right process which is "concentrated on compacts" (m-tightness), has almost surely left limits and is almost-surely left quasi-continuous, see [367] for details.

Remark 42. 1) An analogon of the above theorem holds also for non symmetric Dirichlet forms $\mathcal{E}$ in the sense discussed in [367]. In the same way as to $\mathcal{E}$ there are associated two submarkov semigroups $(T_t, \hat{T}_t)_{t \geq 0}$ in duality, there are two corresponding (properly)- associated "nice processes" $X, \hat{X}$.

2) There also exists, in the same spirit as in the above theorem, an analytic characterization (in terms of Dirichlet forms) of Hunt processes (see [104], [367]).

3) The main consequence of the above theorem is that concepts of analytic potential theory, like capacity and equilibrium potential become related to concepts of probabilistic potential theory, like hitting distributions and entrance times, e.g. one has for an open set U:

$$\text{cap}(U) = \mathcal{E}_1[e_U]$$

with e_U the "1-equilibrium potential".
Moreover $\text{cap}(U) < \infty$ iff $e_U(z)$ is a quasi-continuous version of $E^z(e^{-\sigma_U})$, with σ_U the entrance time of the associated process in U.
In this way capacity gets related with hitting distribution and, e.g., $\text{cap}(N) = 0$ iff N is an $\mathcal{E}$-exceptional set and this is so iff $P^m(\sigma_{\tilde{N}} < \infty)$, for some Borel subset $\tilde{N} \supset N$ of E; see below and [244], [258], [367] for details.

A hint to the proof of Theorem 12:
From the experience with the construction of Markov processes in the simpler case of a locally compact separable space ("finite dimensional case"), we know it is easy to construct a (sub-) Markov process if one disregards detailed path properties. In contrast the proof of the existence of a version of it with "nice properties" is quite hard.
In fact the case of a general Hausdorff space E in the above theorem is indeed reduced to the case of E locally compact, separable, metric, which had been treated before by Fukushima [246] and Silverstein [468]. This reduction has to be done in such a way as to preserve, e.g., the quasi-regularity property (in fact it gets transformed into the regularity property under compactification). In the case where E is locally compact separable metric one realizes that the

Feller property of the transition semigroup leads to strong Markov processes and eventually to Hunt processes (cf. [162]).
On the other hand one shows:

Lemma 3. *Each $u \in D(\mathcal{E})$ (for $\mathcal{E}$ a regular Dirichlet form, in the sense of [246], [258]) has a quasi-continuous modification s.t. $u \upharpoonright (E_\Delta - G)$ is quasi-continuous, for any open subset G of E (cf. Theorem 3.1.3 in [246])*

Proof. A) One has Cap $G_\lambda^u \leq \frac{1}{\lambda^2} \mathcal{E}_1[u], \forall \lambda > 0, u \in D(\mathcal{E}) \cap C(E))$ with
$G_\lambda^u \equiv \{x \in E | |u(x)| > \lambda|\}$
(seen by realizing that G_λ^u is open, $|u|/\lambda \in \mathcal{L}_{G_\lambda^u}$, for $u \in D(\mathcal{E}) \cap C(\mathcal{E})$, with $\mathcal{L}_{l,\lambda}$ as in the definition of capacity, and using the "Dirichlet property" $\mathcal{E}_1[|u|] \leq \mathcal{E}_1[u].$)

B) By the regularity of $\mathcal{E}$ and A) one can find $u_n \in D(\mathcal{E}) \cap C_0(E)$ s.t. $u_n \to u$. By passing if necessary to a subsequence, denoted again by u_n, one then has $\mathcal{E}_1[u_{n+1} - u_n] \leq \frac{1}{2^{3n}}$
Then by A): $\mathrm{Cap} G_{2^n}^{u_{n+1}-u_n} \leq \frac{1}{2^n}$.
Hence $F_k \equiv \bigcap_{l=k}^{\infty} \left(G_{2^n}^{u_{n+1}-u_n} \right)^c$ is an $\mathcal{E}$-nest.

But $|u_n(x) - u_m(x)| \leq \sum_{\nu=N+1}^{\infty} |u_{\nu-1}(x) - u_\nu(x)|$, for any $x \in F_k$, $n, m > N \geq k$.
Setting $\tilde{u}(x) \equiv \lim_{n \to \infty} u_n(x), x \in \bigcup_k F_k$ we have that $\tilde{u} \upharpoonright F_k$ is continuous and $\tilde{u}$ is quasi-continuous. $\qquad \square$

The next observation consists in showing that one can construct a countable set $B_0 \subset D(\mathcal{E}) \cap C_0(E)$ dense in sup-norm in $C_0(E)$ s.t. B_0 is linear and closed under $|.|$ (as seen by approximation, using the regularity of $\mathcal{E}$, see [Fu], proof of Lemma 6.1.2).
Set $H_0 \equiv \bigcup_{t \in \mathbb{Q}_+} T_t(B_0) \cup G_1(B_0)$ (where $(T_t)_{t \geq 0}$ is the Markov semigroup associated with $\mathcal{E}$ and G_1 is the corresponding resolvent G_α evaluated at $\alpha = 1$).
Let $u \in H_0$ and let $\tilde{u}$ be its quasi-continuous modification given by the above Lemma.
Set $\tilde{H}_0 = \{\tilde{u} | u \in H_0\}$. One shows (using that $\tilde{H}_0$ is countable!) that there exists a regular nest $\{F_k^0\}$ on E s.t. $\tilde{H}_0 \subset C(\{F_k^0\})$ (where $C(\{F_k^0\})$)
denotes the functions which are continuous in $(F_k^0)^c$). Let $\bigcup F_k^0 \equiv Y_0$. One sets for $u \in L^2(m), x \in Y_0, t \in \mathbb{Q}_+ : \tilde{p}_t u(x) \equiv T_t u(x)$.
$\tilde{p}_t$ is not yet a semigroup, but has a submarkovian kernel $\tilde{p}_t(x, B), x \in Y_0$, $B \in \mathcal{B}$ (cfg. [246], proof of Lemma 6.1.2).
One extends $\tilde{p}_t$ to E by setting $\tilde{p}_t(x, B) = 0, \forall x \in E - Y_0$. One shows $\tilde{p}_t C_\infty(E) \subset C(\{F_k^0\})$, and that $\tilde{p}_t$ is a quasi-continuous version of $T_t u$, for any $u \geq 0$, Borel, $u \in L^2(m), t \in \mathbb{Q}_+$ (cf. Lemma 6.1.2 in [246]), and that $\tilde{p}_t$ is a semigroup of Markovian kernels on $(E, \mathcal{B})$.
One then uses $\tilde{p}_t$ to get by Ionescu-Tulcea-Kolmogorov's construction a

Markov process on X_Δ. The crucial step then consists in showing that this Markov process has a càdlàg version on some Borel subset $\tilde{Y}, \tilde{Y} \subset Y_0$, s.t. $\text{Cap}(E - \tilde{Y}) = 0$. This relies on an interplay of analytic and probabilistic methods, where regularity again plays an important role (see [246], Lemma 6.2.3). The same ingredients then enable us to show that the process is a Hunt process.

By the reduction of the infinite dimensional case to the locally compact case, one then completes the construction in the general case, see [114], [367] for details.

Remark 43. 1) The m-tight special standard process X properly associated with a quasi-regular symmetric Dirichlet form is m-symmetric (in the sense that its transition semigroup is symmetric in $L^2(m)$), and has m as an invariant measure (by construction).

   In general it has a finite life time ζ, but if $T_t 1 = 1$ (T_t being the Markov semigroup associated with $\mathcal{E}$) and $1 \in L^2(m)$, then the life time is infinite. The process X can always be taken as a canonical process (cfg. [367]).

2) Let X be a right process properly associated with a quasi-regular Dirichlet form $\mathcal{E}$ (in the general case where E is a topological Hausdorff space and m is a σ-finite measure on E).

   One shows that there exists an $\mathcal{E}$-nest (F_k), with F_k compact measurable subsets in E and a locally compact separable metric space $E^\#$ containing densely $Y \equiv \bigcup_k F_k$ with $\mathcal{B}(Y) \equiv \{A \in \mathcal{B}(E^\#) | A \subset Y\}$ and, moreover, there exists a Hunt process $\overline{X}$ on $E^\#$, the "natural extension" of $X \upharpoonright (E - N), N \subset E$,

   N invariant, $\mathcal{E}$-exceptional, s.t. $\overline{X}$ is properly associated with the regular Dirichlet form $\mathcal{E}^\#$, the image of $\mathcal{E}$ in $L^2(E^\#, m^\#)$. This observation is exploited in the "regularization method", see [244], [114], [465].

4.5 Stochastic analysis related to Dirichlet forms

Let $(E, \mathcal{B}, m)$ be as in chapter 4.1

Definition 32. *(Additive functional)*
Let $X = (X_t)_{t \geq 0}$ be a right process associated with a quasi regular Dirichlet form $\mathcal{E}$ (according to Theorem 12 in chapter 4.4).
$A \equiv (A_t)_{t \geq 0}$ is called an additive functional associated with X if A_t is $\mathcal{M}_t$-measurable, càdlàg and such that $A_{t+s}(\omega) = A_t(\omega) + A_s(A_t(\omega))$ for all $s, t \geq 0, \omega$ in the underlying probability space Ω.
A is called a continuous additive functional if $t \to A_t(\omega)$ is continuous $\forall \omega \in \Omega$.

Definition 33. *Let A be an additive functional. The energy of A is by definition*

$$e(A) \equiv \lim_{t\downarrow 0} \frac{1}{2t} E^m(A_t^2),$$

if the limit exists, m-a-s.

Definition 34. *Let* $\mathfrak{M} = \{M \mid M$ *additive functional,* $E^z(M_t^2) < \infty,$
$E^z(M_t) = 0, \mathcal{E} - q.e., z \in E, \forall t \geq 0\}.$
One shows that any $M \in \mathfrak{M}$ *is a martingale under* P^x, *for any* $x \in E - N$,
*N being a properly exceptional set (depending on M, in general). Thus the
elements of* $\mathfrak{M}$ *are called martingale additive functionals.*
*One shows ([246] p.135, by a method of P.A. Meyer) that for any martingale
additive functional M, there exists uniquely a positive continuous additive
functional* $\langle M \rangle$ *s.t. for any* $t > 0$:
$E^z(\langle M \rangle_t) = E^z(M_t^2)$, *q.e.* $z \in E, \langle M \rangle_t$ *is then by definition the quadratic
variation of M.*
Let $\mathfrak{N}^c = \{N$ *is a continuous additive functional of zero energy i.e.*
$e(N) = 0$ *and s.t.* $E^z(|N_t|) < \infty$ *for q.e.* $z \in E\}.$

Theorem 13. *(Fukushima's decomposition)*
*Let X be the right process associated with a quasi regular symmetric Dirichlet
form on a Hausdorff topological space. If* $u \in D(\mathcal{E})$ *then there exists uniquely
a martingale additive functional of finite energy* $M^{[u]}$ *and an element of zero
energy* $N^{[u]} \in \mathfrak{N}^c$ *s.t. for any quasi-continuous version* $\tilde{u}$ *of u:*

$$\tilde{u}(X_t) = \tilde{u}(X_0) + M_t^{[u]} + N_t^{[u]}.$$

Remark 44. $N^{[u]}$ is not necessarily of bounded variation (but it has zero
energy).
For the proof of this theorem see [244], [258], [367], [114].

5 Diffusions and stochastic differential equations associated with classical Dirichlet forms

5.1 Diffusions associated with classical Dirichlet forms

We consider the example discussed in chapter 4, 4.2 of the classical Dirichlet
form $\mathcal{E}_\mu$, associated with a probability measure μ on a separable Banach space
E s.t.

$$E' \subset \mathcal{H}' \cong \mathcal{H} \subset E$$

We suppose as in chapter 4,4.2 that there exists $\beta_\mu \in L^2(\mu)$ s.t. the integration by parts formula holds, i.e.

$$\int \frac{\partial u}{\partial k} d\mu = - \int u \beta_{\mu,k} d\mu \qquad \forall k \in K. \subset E', u \in FC_b^\infty,$$

K consisting of elements which form an orthonormal basis in $\mathcal{H}$.
Let $\beta_\mu : E \to E$ s.t., for all $z \in E$:

$$\langle\, k, \beta_\mu(z)\,\rangle_{E'\ \ E} = \beta_{\mu,k}(z),$$

where $\langle\,,\,\rangle_{E\ E'}$ is the dualization between E and E'.
Let L_μ be the self-adjoint operator with $-L_\mu \geq 0$ associated with $\mathcal{E}_\mu$ (Dirichlet operator associated with μ). Then, on FC_b^∞:

$$L_\mu = \triangle_\mathcal{H} + \beta_\mu(z) \cdot \nabla_\mathcal{H},$$

where $\triangle_\mathcal{H} = \sum_{k\in K} \partial_k^2$ is the Gross-Laplacian associated with $\mathcal{H}$, $\nabla_\mathcal{H}$ the natural gradient associated with $\mathcal{H}$, so that

$$\beta_\mu(z) \cdot \nabla_\mathcal{H} = \sum_{k\in K} \beta_{\mu,k}(z)\partial_k.$$

Proposition 15. *The classical Dirichlet form $\mathcal{E}_\mu$ given by μ is quasi-regular.*

Proof. One has to verify the properties i),ii),iii) in Definition 29 of quasi-regularity for Dirichlet forms. For i) it is enough to show that there exist compacts $F_k \uparrow E$ with Cap $(E - F_k) \downarrow 0$ ("tightness of the capacity"), we leave this as an exercise (cf. [367]).

ii) The subset FC_b^∞ is $\widetilde{\mathcal{E}}_1$-dense in $D(\mathcal{E}_\mu)$ by the construction of $\mathcal{E}_\mu$ as the closure of $\overset{\circ}{\mathcal{E}}_\mu$. Its elements are continuous.
iii) FC_b^∞ separates the points of E (and hence also of $E - N$) since E is a separable Banach space (use, e.g., the theorem of Hahn-Banach, cf. [367], p.119)

$\square$

Proposition 16. *$\mathcal{E}_\mu$ is local.*

Proof. We have to show that $\mathcal{E}_\mu(u,v) = 0$ if $\operatorname{supp} u, \operatorname{supp} v$ are compact, $\operatorname{supp} u \cap \operatorname{supp} v = \emptyset$.
This is obvious for $u, v \in FC_b^\infty$. Now for arbitrary $u, v \in D(\mathcal{E}_\mu)$ we can find (by the $\widetilde{\mathcal{E}}_1$-density of FC_b^∞) $u_n, v_n \in FC_b^\infty$ s.t. $\mathcal{E}_1([u_n - u]) \to 0$, $\mathcal{E}_1([v_n - v]) \to 0$, as $n \to \infty$, and the Proposition is proven. $\square$

By the general theory of association, cf. Chapter 4, to $\mathcal{E}_\mu$ there is properly associated an m-tight special standard process $X_t, 0 \leq t \leq \zeta$, which by the locality of $\mathcal{E}_\mu$ is a diffusion process.
Since the C_0-contraction Markov semigroup $T_t = e^{tL}, t \geq 0$, associated with $\mathcal{E}_\mu$ moreover satisfies $T_t 1 = 1$, it follows that $X = (X_t)_{t\geq 0}$ is a μ-symmetric conservative process s.t. $\zeta = +\infty$.
An application of Fukushima's decomposition theorem (Chapter 4, Theorem 12) to the present case yields:

Theorem 14. *For any* $u \in D(\mathcal{E}_\mu), t \geq 0$:

a) if $u \in D(L_\mu)$ *then*

$$N_t^{[u]} = \int_0^t (L_\mu u)(X_s)ds.$$

b) $\langle M_t^{[u]} \rangle = 2 \int_0^t \langle \nabla u(X_s), \nabla u(X_s) \rangle_{\mathcal{H}} ds$, *where* $\langle \cdot \rangle_t$ *denotes the quadratic variation process (so that* $E^z \left(\langle M \rangle_t \right) = E^z(M_t^2)$ *,q.e.* $z \in E, t > 0, M_t$ *a martingale additive functional.)*

Proof. a) The proof is based on an extension to quasi-regular Dirichlet forms on general Hausdorff topological spaces (cf. [367]) of the following Lemma:

Lemma 4. *Let* $\mathcal{E}$ *be a regular Dirichlet form on* $L^2(m)$ *on a locally compact separable space* E, *and let* X *be a properly associated right process. Then for any* $g \in L^2(m)$:

$\int_0^t g(X_s)ds$ *is a continuous additive funtional of zero energy.*

The proof of the Lemma is left as an exercise (hint: use the Markov property of X).
Now take $u \in D(L_\mu)$, so that $u = G_1 f, f \in L^2(m)$. Set $g \equiv u - f$
Then:

$$\int_0^t g(X_s)ds = \int_0^t (u(X_s) - f(X_s)) \, ds$$

$$= \int_0^t L_\mu u(X_s)ds,$$

where for the latter equality we have used

$$L_\mu u = [L_\mu - 1 + 1]u = -f + G_1 f.$$

Applying the above Lemma we get that
$\int_0^t L_\mu u(X_s)ds$ is a continuous additive functional of 0 energy and by the
uniqueness in Fukushima's decomposition theorem this is then $N_t^{[u]}$.
b) To give an idea of the proof of this point, let us look at the finite dimensional case $E = U, U$ an open subset of $\mathbb{R}^d$, with a classical pre-Dirichlet form $\overset{\circ}{\mathcal{E}}_\mu$.

From the finite dimensional theory we know that if X is the Markov process properly associated with the closure $\mathcal{E}_\mu$ of $\overset{\circ}{\mathcal{E}}_\mu$, then by the Beurling-Deny decomposition

$$\langle M^{[u]} \rangle = 2 \int_t^t \langle \nabla u(X_s), \nabla u(X_s) \rangle ds$$

For the infinite dimensional case see [367].

$\square$

Remark 45. a) holds for any quasi-regular Dirichlet form.
b) can be generalized to

$$M_t^{[u]} = \int_0^t \rho(X_s)\, ds, \text{ with}$$

$\rho(x) = L_\mu u^2(x) - 2u(x)L_\mu u(x).$
In this form the proof of b) can found in [208] for the case of locally compact spaces and [126], [451], [452], [453] for quasi regular Dirichlet forms.

Remark 46. For $E = \mathbb{R}^d, \mu$ the Lebesgue measure on U, so that $\mathcal{E}_\mu$ is the classical Dirichlet form uniquely associated with $-\Delta$ we have, taking $u(x) \equiv u_i(x) \equiv x_i (\in L^2(\mu_K))$ (with μ_K the restriction of μ to the interior $\overset{\circ}{K}$ of a compact subset K in $\mathbb{R}^d$). But for any $u \in D(\mathcal{E}_{\mu_K}) \exists w \in D(\mathcal{E}_\mu)$ s.t. $w = u$ m-a.e. on $\overset{\circ}{K}$ and $M_t^{[w_i]} = M_t^{[u_i]}$ for $t < \sigma_{\mathbb{R}^d - \overset{\circ}{K}}$ (the hitting time of $\mathbb{R}^d - \overset{\circ}{K}$). On the other hand

$$\langle M^{[u_i]} \rangle = 2 \int_t^t ds = 2t.$$

By a local version of Levy's characterization theorem we then have $M^{[u_i]} = W^i$, with W^i the i-th component of a Brownian motion in $\mathbb{R}^d$.

Remark 47. In particular we see from the preceding remark that the finite energy additive functional $M^{[u_i]}$ is just the i-th component of Brownian motion.
In general, finite energy additive functionals of a quasi-regular Dirichlet form can be represented by stochastic integrals, see [367].

5.2 Stochastic differential equations satisfied by diffusions associated with classical Dirichlet forms

Proposition 17. *Let μ be a probability measure on $S'(\mathbb{R}^d)$, as in Chapter 5.5.1, s.t.*

$$u_k(\cdot) \equiv \underset{S}{\langle} k, \cdot \underset{S'}{\rangle} \in L^2(\mu).$$

Then $u_k \in D(L_\mu) \subset D(\mathcal{E}_\mu)$ (where $\mathcal{E}_\mu$ is the classical Dirichlet form given by μ). Moreover $L_\mu u_k = \beta_{\mu,k} (\mu\text{-a.s.})$

Proof. We have, for any $v \in FC_b^\infty$:

$$\mathcal{E}_\mu(u_k, v) = ((-L_\mu)u_k, v) = \int_{\mathcal{H}} \langle k, \nabla v \rangle \, d\mu,$$

where we used the relation between $\mathcal{E}_\mu$ and L_μ, the definition of $\beta_{\mu,k}$, and the integration by parts formula. $\qquad\square$

Theorem 15. *Let $X \equiv (X_t)_{t\geq 0}$ be the diffusion process associated with the classical Dirichlet form given by μ as in Proposition 17. Then X satisfies "componentwise", in the weak probabilistic sense, the stochastic differential equation:*

$$\langle k, X_t \rangle = \langle k, X_0 \rangle + \int_0^t \beta_{\mu,k}(X_s) ds + w_t^k, t \geq 0, P^z\text{-a.s., q.e } z \in E.$$

Hereby $(w_t^k, \mathcal{F}_t, P^z)_{t\geq 0}$ is a 1-dimensional Brownian motion starting at 0 (for $\|k\|_{\mathcal{H}}$ suitably normalized).

Proof. By the above Fukushima decomposition formula we have

$$\langle M^{[u_k]} \rangle_t = 2 \int_0^t \langle \nabla_{u_k}(X_s), \nabla_{u_k}(X_s) \rangle_{\mathcal{H}} \, ds = 2t\|k\|_{\mathcal{H}}^2$$

(because of $\nabla_{u_k} = \langle k, . \rangle$).
Hence by Levy's characterization of Brownian motion:

$$\left(M^{[u_k]} \right)_t = w_t^k.$$

Moreover:

$$N_t^{[u_k]} = \int_0^t L_\mu u_k(X_s) ds$$

$$= \beta_{\mu,k}$$

because $L_\mu u_k(X_s) = (\Delta_{\mathcal{H}} + \beta_\mu \cdot \nabla_{\mathcal{H}})u_k = \beta_{\mu,k}$,
where in the latter step we have used that $u_k(.) = \langle k, . \rangle$ is linear. $\qquad\square$

Let us now vary k along an orthonormal basis K in $E' \subset \mathcal{H}$. Then $w_t^k, w_t^{k'}$ are independent for $k \neq k'$.

Assume that there exists a probability measure μ_t on E s.t., for all $k \in \mathcal{H}$:

$$\hat{\mu}_t(k) = \exp\left(-\frac{1}{2}t\|k\|_{\mathcal{H}}^2\right),$$

(in which case there exists a Brownian motion on E with unit covariance given by the scalar product in $\mathcal{H}$: see, e.g., [334], [428], [118]).
Then we have the following

Theorem 16. *Under above assumption about the existence of μ_t there exist maps $W, N : \Omega \to C([0, \infty], E)$ s.t. $t \to W_t(\omega), t \to N_t(\omega)$ are $\mathcal{F}_t$-B-measurable, for all non negative t.*
W_t is such that for q.e. $z \in E$, under P^z, it is an $\mathcal{F}_t$- Brownian motion starting at 0, with covariance given by the scalar product in $\mathcal{H}$.
Moreover,

$$\langle k, N_t \rangle = N_t^{[u_k]} = \int_0^t \beta_{\mu,k}(X_s)ds.$$

One has:
$X_t = z + W_t + N_t, t \geq 0, P^z$-a.s.,q.e. $z \in E$.

Remark 48. X also solves the martingale problem for $D \subset D(L_\mu) \subset L^2(E_\mu)$ in the sense that X is a μ-symmetric, right process s.t.

$$\tilde{u}(X_t) - \tilde{u}(X_0) - \int_0^t L_\mu u(X_s)ds, t \geq 0$$

is $\mathcal{F}_t$-measurable under P^z, for some quasi-continuous, right continuous modification $\tilde{u}$, independent of N, and independent of the μ-version of the class $L_\mu u \in L^2(\mu)$.

Remark 49. A particular case of the above results concerns

$$E = C_{(0)}([0, t]; \mathbb{R})$$

(Wiener space),

$$\mathcal{H} = H^{1,2}([0, t]; \mathbb{R})$$

(Cameron-Martin-space),
μ is the standard Wiener measure on E.
In this case

$$\overset{\circ}{\mathcal{E}}_\mu(u, v) = \int \langle \nabla u, \nabla v \rangle_{\mathcal{H}} d\mu,$$

$$\mathcal{E}_\mu(u,v) = \int \langle \overline{\nabla} u, \overline{\nabla} v \rangle_{\mathcal{H}} d\mu,$$

with $\overline{\nabla}$ Malliavin's closed gradient (the verification of the latter is left as an exercise; see also, e.g. [396],[365], [329],[57], and references therein).
In this case we have:

$$\beta_{\mu,k}(.) = \langle k, . \rangle$$

(since $\int \frac{\partial u}{\partial k} d\mu = - \int u \beta_{\mu,k} d\mu, u \in FC_b^\infty$, as seen from the following computation

$$\int \frac{\partial}{\partial s} u(.+sk)|_{s=0} d\mu = \int u(.) \frac{d\mu(.-sk)}{d\mu(.)} d\mu(.)$$

$$= \int v(.) e^{-\langle k,.\rangle} e^{-\frac{1}{2}\|k\|_{\mathcal{H}}^2} d\mu(.))$$

Incidentally: the computation in parenthesis of the Cameron-Martin density under translations of Wiener measure is the basis of a corresponding computation for the quasi-invariance of a natural measure on loop-groups, see [57], [373], [68] and is used in an essential way in the representation of related infinite dimensional Lie groups (cf. [68]).

Remark 50. A similar computation can be done for other Gaussian measures, of the form $\mu = N(0; A^{-1}), A \geq c1, c > 0, A$ a Hilbert Schmidt operator in the Hilbert space $\mathcal{H}$.
In this case we have

$$\beta_{\mu,k} = \langle Ak, . \rangle,$$

see [119], [269]. See also [165], [134], [318] for other results on infinite dimensional Ornstein-Uhlenbeck processes.

5.3 The general problem of stochastic dynamics

Given a (probability) measure on some space one can ask the ("inverse problem") question whether there exists a Markov process X with corresponding transition semigroups P_t, μ-symmetric (in the sense that $P_t^* = P_t$ in $L^2(\mu)$), having μ as P_t-invariant measure, in the sense that

$$\int T_t u d\mu = \int u d\mu$$

for all $u \in L^2(\mu)$.
One then says that X is the "stochastic dynamics" (or "Glauber dynamics") associated with μ.

Remark 51. 1) If μ is a probability measure then we have:
μ is P_t-invariant iff $P_t 1 = 1$
    (with 1 the function identically 1 in $L^2(\mu)$)
    (we leave the proof as an exercise).

2) There is a notion of measure μ infinitesimal invariant with respect to $(T_t)_{t\geq 0}$, e.g. this has been discussed in connection with hydrodynamics (cf. [35]):
μ is namely called infinitesimal invariant under a C_0-semigroup $T_t = e^{tL}, t \geq 0$ if $\int Lu\, d\mu = 0, \forall u \in D_0 \subset D(L), D_0$ dense in $L^2(\mu)$.
In general μ infinitesimal invariant is strictly weaker than μ T_t- invariant , unless T_t is μ-symmetric and $1 \in D(L), L1 = 0$ (because then from $\int Lu\, d\mu = 0$ one deduces $\int L^n u\, d\mu = 0$, for all n, hence $\int T_t u\, d\mu = 0$)
For recent work on invariant and infinitesimally invariant measures see, e.g., [21], [37], [38], [95], [128], [166], [167], [168], [201].

The converse problem to the above "inverse problem" is the following "direct problem": given a Markov process X find a probability measure μ s.t. μ is an invariant measure for X.
In this case one says that μ is the invariant measure to the stochastic dynamics described by X.
Connected with this direct problem are the following ones:

1) Existence of the classical Dirichlet form $\mathcal{E}_\mu$ associated with μ (closability problem for the pre-Dirichlet form $\overset{\circ}{\mathcal{E}}_\mu$ in $L^2(\mu)$). If this is solved then one can construct a diffusion Y having μ as invariant measure (Y in general can be different from X).
2) Does the logarithmic derivative $\beta_\mu = (\beta_{\mu,k})_{k\in K}$ of μ exist e.g. as an element in $L^2(\mu)$?
3) Does X satisfy a stochastic differential equation?

Further associated questions are, e.g.:

4) What is the asymptotic behavior for $t \to \infty$ of X_t, and of the semigroup $T_t = e^{tL_\mu}, t \geq 0$, associated with $\mathcal{E}_\mu$?
5) Is a solution of the martingale problem for L_μ on a closed domain D strictly contained in $D(L_\mu)$ already uniquely determined by the knowledge of L_μ on D? This is the "Markov uniqueness problem", cf. [147], [223], [88], [119] and, for the related "strong uniqueness problem", i.e. the essential-self-adjointness resp. maximal dissipativity of (L_μ, D) see these references and, e.g., [85], [86], [359].

Other problems are, e.g.:

6) When does T_t have the Feller property and thus permit a more direct construction of an associated "nice process"? (see, e.g., [82], [214], [262], [444])
7) Is the invariant measure μ to X unique?

Problems of this type are often encountered, e.g., in the study of processes associated with "Gibbs measures", e.g., in quantum field theory, statistical mechanics (on lattices and in the continuum, in problems connected with the geometry and analysis of configuration spaces), quantum statistical mechanics (and connected problems of geometry and analysis on loop spaces), in the study of self-intersection functionals of diffusion processes and polymer models, in models of population dynamics, see below, chapter 6, for further discussion and for the construction of stochastic dynamics in some examples. First we shall briefly discuss in general the large time asymptotics of processes associated with Dirichlet forms.

5.4 Large time asymptotics of processes associated with Dirichlet forms

Let E be a topological Hausdorff space, m a σ-finite measure on E, as in chapter 3-4 on Dirichlet forms.

Definition 35. *A general Dirichlet form $\mathcal{E}$ in $L^2(m)$ is said to be _irreducible_ if $u \in D(\mathcal{E})$ and $\mathcal{E}[u] = 0$ imply that m is constant m-a.e.*

Definition 36. *Let $(T_t), t \geq 0$ be a submarkovian C_0-contraction semigroup in $L^2(m)$. T_t is called _irreducible_ if $T_t(uf) = uT_t f, \forall t > 0, \forall f \in L^\infty(m)$ implies $u = $ const m-a.e.*

Definition 37. *Let $(T_t)_{t \geq 0}$ be a (submarkovian) C_0-contraction semigroup in $L^2(m)$. $(T_t)_{t \geq 0}$ is said to be $L^2(m)$-ergodic if $T_t u \to \int u dm$ as $t \to \infty$ in $L^2(m), \forall u \in L^2(m)$.*

In [90] (see also [91], [64]) the following Theorem is proven:

Theorem 17. *For symmetric Dirichlet forms $\mathcal{E}$ and associated symmetric submarkovian C_0-contraction semigroups $T_t = e^{tL}, t \geq 0$, in $L^2(m)$, the following are equivalent:*

a) $\mathcal{E}$ is irreducible
b) $(T_t)_{t \geq 0}$ is irreducible
c) $T_t u = u \forall t > 0, u \in L^2(m)$ implies $u = $ const m-a.e.
d) $(T_t)_{t \geq 0}$ is $L^2(m)$-ergodic
e) $u \in D(L)$ and $Lu = 0$ imply $u = $ const m-a.e.

Proof. $b) \to a)$ is immediate, using the contraction property of $\mathcal{E}$. The rest is left as an exercise (cf. [90]). □

It is also interesting to connect asymptotic properties of semigroups with corresponding properties of associated processes.

Definition 38. *Let X be a right process on a topological Hausdorff space, properly associated with a quasi-regular Dirichlet form $\mathcal{E}$. Let for any $\mu \in \mathcal{P}(E)$ (the linear space of probability measures on E):*

$$P^\mu \equiv \int\limits_E P^z \mu(dz),$$

where P^z is the probability measure on the paths of X corresponding to a starting point $z \in E$.

(X, P^μ) is said to be _time-ergodic_ if for any $G : \Omega \to \mathbb{R}$, s.t. G is Θ_t invariant $\forall t \geq 0$ one has $G = \text{const}, P^\mu$-a.s. (that G is Θ_t-invariant means that $G(\Theta_t\omega) = G(\omega)\forall t \geq 0, \Theta_t$ being the natural shift in path space given by $\Theta_t\omega(s) = \omega(t+s), \forall 0 \leq s, t \leq \zeta(\omega))$.

One has then the following

Theorem 18. _("Fukushima's ergodic theorem")_
The classical Dirichlet form $\mathcal{E}_\mu$ _on a topological Hausdorff space_ (X, μ) _(with_ μ σ-finite) is irreducible in $L^2(\mu)$ iff (X, P^μ) is time-ergodic.
Moreover, the transition semigroup P_t to X (so that X_t is properly associated to P_t, $t \geq 0$) is such that $P_t u \to \int u d\mu$ as $t \to +\infty, \mathcal{E}_\mu$-q.e., $\forall u \in B_b(E)$._

Proof. The proof is given in [87] (for previous work see, e.g., [57], [390], [244]). □

Corollary 3. _Let_ $\mathcal{E}_\mu$ _be as in above theorem. Then_ $\mathcal{E}_\mu$ _is irreducible if_ μ _is the only_ P_t-_invariant probability measure on_ $B(E)$ _which does not charge_ $\mathcal{E}_\mu$-_exceptional subsets of_ E.

Proof. See [244]. □

For measures μ which are quasi-invariant with respect to suitable subspaces K of E, which we shall call "space quasi-invariant", one has an interesting relation between above time ergodicity of E and "space ergodicity", i.e. ergodicity with respect to K. This is the context of next section.

5.5 Relations of large time asymptotics with space quasi-invariance and ergodicity of measures

Let E be a Hausdorff topological space, which is a locally convex topological vector space with the topology of a Souslin-space. (cf. [462] for this concept, E can be, e.g., a Banach space, or a space like $\mathcal{S}'(\mathbb{R}^d)$).
Assume there is a Hilbert space $\mathcal{H}$ s.t. $E' \subset \mathcal{H} \subset E$ where the embeddings are dense and continuous.

Definition 39. _A probability measure_ μ _on_ E _is said to be_ K-_quasi invariant if_ $\mu(.) \ll \mu(. + tk), \forall k \in K, \forall t \in \mathbb{R}$ _(where_ $\ll$ _means absolutely continuous)._

Remark 52. An example of a quasi invariant measure is given by $E = C_0, ([0,t]; \mathbb{R}^d), \mu$ the Wiener measure on E, with $K = H^{1,2}([0,t]; \mathbb{R}^d)$ the Cameron-Martin space, cf. Chapter 5.5.2.
As clearly mentioned in Chapter 5.5.2, the non commutative analogon of this setting, with $\mathbb{R}^d$ replaced by a compact Lie group, is the basis for the representation theory of loop groups and algebras, see, e.g., [57], [68].

Definition 40. *A probability measure μ on E is said to be K-ergodic if μ is K-quasi invariant and for any $u \in L^2(\mu)$ one has that $u(z + tk) = u(z)$ μ-a.e. ($\forall z \in E, t \in \mathbb{R}$) implies $u = const$, μ- a.e.*

Given a probability measure μ on E, one might ask whether there is a relation between the irreducibility of the corresponding classical Dirichlet form $\mathcal{E}_\mu$ and the K-ergodicity of μ. This has already been discussed in [57]. The surprise is that a close relation of this type involves another Dirichlet form $\mathcal{E}_\mu^{\max}$ (with larger domain than $\mathcal{E}_\mu$), rather than $\mathcal{E}_\mu$.
In order to define $\mathcal{E}_\mu^{\max}$ let us first define its domain:

$$D(\mathcal{E}_\mu^{\max}) \equiv \left\{ u \in \bigcap_{k \in K} D(\mathcal{E}_{\mu,k}), \sum_k \mathcal{E}_{\mu,k}[u] < \infty \right\}$$

One sees then easily that

$$D(\mathcal{E}_\mu^{\max}) \supset D(\mathcal{E}_\mu).$$

In general however one can have $D(\mathcal{E}_\mu^{\max}) \neq D(\mathcal{E}_\mu)$ (see e.g. [246] for finite dimensional examples, with E replaced by a bounded subset of $\mathbb{R}^d$).

Remark 53. $D(\mathcal{E}_\mu^{\max})$ is an infinite dimensional weighted space analogue of the Sobolev space $H^{1,2}(\mu)$ whereas $D(\mathcal{E}_\mu)$ is an infinite dimensional weighted space analogue of $H_0^{1,2}(\mu)$.
One defines $\mathcal{E}_\mu^{\max}$ to be equal to $\mathcal{E}_\mu$ on $D(\mathcal{E}_\mu)$. One can then show that $\mathcal{E}_\mu^{\max}$ so defined has a unique closed extension to a Dirichlet form with domain exactly equal to $D(\mathcal{E}_\mu^{\max})$ as defined above, see [99], [101], [223].

Remark 54. It is an important open problem to establish whether $\mathcal{E}_\mu^{\max}$ is quasi-regular in general, hence whether to it there can be properly associated a right process.
The advantage of $\mathcal{E}_\mu^{\max}$ over $\mathcal{E}_\mu$ is that irreducibility for it implies K-ergodicily of μ i.e. the following theorem holds

Theorem 19. *If $\mathcal{E}_\mu^{max}$ is irreducible then μ is K-ergodic.*

Proof. Let $u : E \to \mathbb{R}$ be $B(E)$-measurable and in $L^2(\mu)$. Suppose u is k-invariant, $k \in K$. Then one can show that $\frac{\partial}{\partial k} u = 0$, see [119], hence $u \in D(\mathcal{E}_{\mu,k})$.
By the definition of $D(\mathcal{E}_\mu^{\max})$ this implies $u \in D(\mathcal{E}_\mu^{\max})$ and $\mathcal{E}_\mu^{\max}[u] = 0$.
By the irreducibility of $\mathcal{E}_\mu^{\max}$ this implies $u = const$, μ-a.e., which by the definition of K-ergodicity of μ yields that μ is K-ergodic.   □

Remark 55. One has $\mathcal{E}_\mu^{\max}$ irreducible $\Rightarrow \mathcal{E}_\mu$ irreducible (but the converse is not true in general, see, e.g. [119]).

Remark 56. The question whether $\mathcal{E}_\mu^{\max} = \mathcal{E}_\mu$ for a given setting $(E, \mu, \mathcal{H}, K)$ is called the "Markov uniqueness question". One can namely show in general:

$\mathcal{E}_\mu^{\max} = \mathcal{E}_\mu \Leftrightarrow$ the only Dirichlet form extending $(\mathcal{E}_\mu, D(\mathcal{E}_\mu))$ is $\mathcal{E}_\mu$
$\Leftrightarrow$ if $\mathcal{E}$ is a Dirichlet form and $\mathcal{E} = \mathcal{E}_\mu$ on FC_b^∞ then $\mathcal{E} = \mathcal{E}_\mu$
$\Leftrightarrow$ Let $(T_t)_{t\geq 0}$ be the submarkov semigroup with generator coinciding on FC_b^∞ with the classical Dirichlet operator L_μ given by μ (i.e. L_μ is the operator associated to $\mathcal{E}_\mu$ in the sense of the representation theorem) Then $T_t = e^{tL_\mu}$.

In general it is known that Markov uniqueness is weaker than "strong uniqueness" or "L^2-uniqueness", which is the property that L_μ is essentially self-adjoint on FC_b^∞ on $L^2(\mu)$.

The Markov and strong uniqueness problems are thoroughly discussed in [223]. We mention here some further basic work by [85], [86], [89], [499], [471], [185], [37], [38], [440], [441] (connected with applications in various domains).

To give an idea of these connections let us mention shortly what happens in the finite dimensional case $E = \mathbb{R}^d$: for $\mu(dx) = \rho(x)dx$, $\sqrt{\rho} \in H_{\text{loc}}^{1,2}$ one has Markov uniqueness in general (see [359]).

For U a bounded region and $\rho = 1$, $\mathcal{E}_\mu^{\max}$ is the Dirichlet form describing reflected Brownian motion, $\mathcal{E}_\mu$ describes absorbing Brownian motion and there are infinitely many other forms between $\mathcal{E}_\mu$, $\mathcal{E}_\mu^{\max}$ describing Brownian motion with other types of boundary behaviour.

It is also known that there are other closed symmetric positive bilinear forms with associated generators of symmetric C_0-contraction semigroups in $L^2(\mu)$ with generators coinciding with L_μ (here Δ) on $C_0^\infty(U)$ but which are not submarkovian, e.g. the Krein extension of $\Delta \upharpoonright C_0^\infty(U)$, see [244].

A "concrete" (probabilistic and analytic) classification of all extensions in the infinite dimensional case is a very interesting open problem.

Remark 57. There exists a partial converse to the previous theorem.

Theorem 20. *If μ is K-quasi invariant and a "strictly positive" measure on E (in the sense that its Radon-Nikodym derivatives in the directions of K are strictly positive) and moreover μ is K-ergodic, then $\mathcal{E}_\mu^{\max}$ is irreducible*

Proof. See [87]. □

We shall now see that for special μ called "Gibbs measures", one has a close relation between irreducibility and K-ergodicity.
Let

$$\mathcal{G}^b \equiv \{\mu \in P(E) | \mu \text{ satisfies } (IP)^b\},$$

where $(IP)^b$ is the following "integration by parts formula with resp. to μ and the direction b":

$$\int \frac{\partial u}{\partial k}d\mu = -\int ub_k d\mu$$

$\forall u \in FC_b^\infty, \forall k \in K$.

Any element μ of $\mathcal{G}^b$ is called a "b-Gibbs state" " and $b \equiv (b_k)_k \in K$ is the logarithmic derivative of μ.

Remark 58. We shall see below how to relate this definition of b-Gibbs state with other definitions of Gibbs states.

Remark 59. Let us look at a probability measure on $E = \mathbb{R}^n$ of the form

$$d\mu(z) = Z^{-1} e^{-S(z)} dz,$$

where dz is Lebesgue measure on E, S is a lower bounded measurable function on $\mathbb{R}^n$, Z a normalization constant s.t. μ is a probability measure on $\mathbb{R}^n$.

The $(IP)^b$-formula holds with $K = \mathbb{R}^n$ and $b_k = -d_k S(z)$, d_k being the derivative in the direction k, i.e. b_k is the logarithmic derivative of the measure μ.

In this sense it is often inspiring to think of μ also in the case of an infinite dimensional E as a measure of the above form (of course there is no good analogue of Lebesgue measure on infinite dimensional spaces, so this way of thinking has to be understood "cum grano salis", e.g. as limit of finite dimensional measures, see, e.g., [16]).

Remark 60. a) $\mathcal{G}^b$ is a convex set, in the sense that any $\mu \in \mathcal{G}^b$ can be written as an integral with respect to $\nu \in (\mathcal{G}^b)_{ex}$, with $(\mathcal{G}^b)_{ex}$ the set of extreme elements in $\mathcal{G}^b$, see [90].

We have the following

Theorem 21. *Let μ be in $\mathcal{G}^b$.*
Consider the following statements:

   i) $\mu \in \mathcal{G}_{ex}^b$
   ii) $\mathcal{E}_\mu^{max}$ *irreducible*
   iii) $\mathcal{E}_\mu$ *irreducible*
   iv) (X, P^μ)-*time ergodic (with X a right process properly associated with $\mathcal{E}_\mu$)*

   Then: i) $\leftrightarrow$ *ii)* $\rightarrow$ *iii)* $\leftrightarrow$ *iv)*
If $\mathcal{E}_\mu^{max} = \mathcal{E}_\mu$ (i.e. one has Markov uniqueness) then i),ii),iii),iv) are all equivalent.

Proof. ii) $\rightarrow$ *iii)* is clear
iii) $\leftrightarrow$ *iv* was discussed above. For the rest of the proof see [90], [91].   □

Remark 61. $\mathcal{E}_\mu$ acts as a rate function for the large deviation of occupation densities of X from the ergodic behaviour, as shown in [390].

6 Applications

The applications of the theory of Dirichlet forms are so numerous and belong to so many different areas that it would be impossible to give here even a sketchy but balanced overview.

We shall restrict ourselves to some examples, mainly taken from physics, which illustrate some of the basic advantages of the approach and where the analysis has been pushed forward most intensively in recent years, in particular concerning the stochastic processes involved, which are difficult to obtain (if at all) by other methods, and where in any case the theory of Dirichlet forms has played a pioneering role.

In most of the cases we shall consider the processes which have invariant measures of the form of "Gibbs measures", i.e. measures heuristically given by the formula

$$d\mu(z) = Z^{-1}e^{-S(z)}dz, \quad z \in E \tag{1}$$

(E being the state space, cf. Remark 59 in Chapter 5). For somewhat complementary references where problems connected with the ones discussed here see also, e.g., [2], [3], [9], [116].

6.1 The stochastic quantization equation and the quantum fields

Let us consider a classical relativistic scalar field (as a simpler analogue of the classical electromagnetic vector field) over the d-dimensional Minkowski space-time ($d = s + 1, s =$ space dimension, the physical case being for $s = 3$). φ is the (real-valued) solution of the non-linear Klein-Gordon (or massive wave) equation:

$$\Box\varphi + m^2\varphi + v'(\varphi) = 0 \tag{2}$$

(with $\Box = \frac{\partial^2}{\partial t^2} - \Delta_{\vec{x}}$ the d'Alembert wave operator, $m \underset{(-)}{>} 0$ being the mass parameter, v a real valued differentionable function on $\mathbb{R}$ called "(self-)interaction", $t \geq 0$ is the time variable, $\vec{x} \in \mathbb{R}^s$ is the space variable).

Inspired by Feynman's heuristic "path integrals" quantization procedures (we refer to [60], [23], [24], [25], [26], [73], [16], [53], [278], [279], [280], [49] for work implementing this in related situations), Symanzik formulated a program of constructing a quantization of the solution of (2), in terms of a measure of the heuristic form (1) with $S(z)$ an "action functional" of the form

$$S(z) = S_0(z) + \int_{\mathbb{R}^d} v(z(x))dx,$$

with

$$S_0(z) = \frac{1}{2} \int_{\mathbb{R}^d} (-z(x)) \triangle z(x) dx + m^2 \int_{\mathbb{R}^d} |z(x)|^2 dx,$$

with $z = z(x), x = (t, \vec{x}) \in \mathbb{R} \times \mathbb{R}^s = \mathbb{R}^d$. In this case then E should be a space of maps from $\mathbb{R}^d$ into $\mathbb{R}$. The reason for this is that the moment functions $\int z(x_1)..z(x_n)\mu(dz)$ of μ heuristically defined by (1) with such an S give, after analytic continuation $x = (t, \vec{x}) \rightarrow (it, \vec{x})$, the "correlation functions in the vacuum"

$$\langle \varphi_Q(t_1, \vec{x_1})...\varphi_Q(t_n, \vec{x_n}) \rangle$$

of the relativistic quantum field $\varphi_Q(t, \vec{x})$ corresponding to the classical Klein-Gordon field $\varphi(t, \vec{x})$.

From the perspective of Chapter 5.3, the construction of μ is related to the construction of a process $(X_\tau)_{\tau \geq 0}$ on E s.t.

$$dX_\tau = \beta_\mu(X_\tau)d\tau + dw_\tau$$

with w_τ a Brownian motion on E with covariance given by a suitable Hilbert space $\mathcal{H}$, with $\beta_\mu(z) = -\nabla_{\mathcal{H}} S(z)$. For $\mathcal{H} = L^2(\mathbb{R}^d)$ we get heuristically

$$-\nabla_{\mathcal{H}} S(z(x)) = -(-\triangle_x + m^2)z(x) - v'(z(x)),$$

so X_τ satisfies heuristically

$$dX_\tau(x) = (\triangle_x - m^2)X_\tau(x)d\tau - v'(X_\tau(x))d\tau + \eta(\tau, x)d\tau \qquad (3)$$

with $\eta(\tau, x)$ a Gaussian white noise in all variables $\tau \in \mathbb{R}, x \in \mathbb{R}^d$, s.t. heuristically,

$$\frac{d}{d\tau} w_\tau(x) = \eta(\tau, x)$$

with $(w_\tau(\cdot))$ a (cylindrical) Brownian motion on $\mathcal{S}'(\mathbb{R}^d)$ with covariance given by the scalar product in $L^2(\mathbb{R}^d)$.

(3) is called the "stochastic quantization equation". It has been discussed by Parisi-Wu as a computational, Monte-Carlo type method for the construction of μ (τ being a "computer time"). This equation has since received a lot of attention, both in physics and mathematics, after the pioneering work of Jona-Lasinio and coworkers [302].

As for the definition of the measure $\mu \equiv \mu^v$, heuristically given by (1) with S as above, one starts from the case $v = 0$. In this case, as realized by E. Nelson [393] (see also, e.g. [10], [15], [428],[429] for other connections) $\mu^0 \equiv \mu^{v=0}$ is realized rigorously as the normal distribution with mean zero and covariance $(-\triangle_x + m^2)^{-1}$ (which, by Minlos theorem, is a well defined measure, e.g., in $\mathcal{S}'(\mathbb{R}^d)$, with support e.g. in $\mathcal{H}^{-1,2}(\mathbb{R}^d)$) (this is called Nelson's free field measure).

For $d = 1, \mu^v$ has been constructed for large classes of v as weak limit as

$\Lambda \uparrow \mathbb{R}$ of measures of the form $\mu_\Lambda^v(dz) = Z_\Lambda^{-1} e^{-\int_\Lambda v(z(x))dx} \mu^0(dz)$ (see [293], [450], [197]).

A direct analogous procedure for $d = 2$ fails, since μ_Λ^v is ill defined, since for $z \in \text{supp}\mu^0$, $\int_{\mathbb{R}^d} v(z(x))dx$ is infinite μ^0-a.s. (this is due to the singularity of the covariance of μ^0 on the diagonal $(x = y)$ of the type $|x - y|^{-(d-2)}$ for $d \geq 3$ and $-2\pi\ln|x - y|$ for $d = 2$), with :: being the Wick ordering.

But, for $d = 2$, replacement of v by $: v :$ (so that e.g. for $v(y) = y^n$, $\langle k, : v : (z) \rangle$, $k \in \mathcal{S}(\mathbb{R}^2)$ is an element in the n-th chaos subspace in $L^2(\mu_0)$) yields (by a fundamental estimate of Nelson, see, e.g. [466]) a well-defined probability measure μ_Λ^v (heuristically given by $Z^{-1} e^{-\int_\Lambda :v:(z(x))dx} \mu^0(dz)$), and one shows then that μ_Λ^v converges weakly, under some assumptions on v, for $\Lambda \uparrow \mathbb{R}^2$, to a well defined probability measure μ^v on $\mathcal{S}'(\mathbb{R}^2)$ (see [466], [265], [39]). μ^v is then by definition the "$v(\varphi)_2$-model" of (Euclidean) quantum field theory (for v a polynomial P one has the "$P(\varphi)_2$-model").

Remark 62. The problem whether the coordinate process X with distribution μ^v is a global Markov field was open for a long time and was solved in works by Albeverio, Gielerak, Høegh-Krohn, Zegarlinski, see, e.g., [132] and references therein.

Looked upon as an $\mathcal{S}'(\mathbb{R}^{d-1})$-valued symmetric Markov process $t \to X_t(f)$, $f \in \mathcal{S}(\mathbb{R}^{d-1})$, $t \geq 0$, it has a generator which coincides on a dense set, e.g. FC_b^∞, with the $\mathcal{S}'(\mathbb{R}^{d-1})$-valued diffusion process $X_t(f)$, $t \geq 0$ associated with the classical Dirichlet form given by μ_0^v, where μ_0^v is the restriction of μ^v to the σ-algebra $\sigma(X_0(f), f \in \mathcal{S}(\mathbb{R}^{d-1})$. Wether $X_t(f)$ and $\widetilde{X_t}(f)$ have generators coinciding on their full domain is an open question for $v \neq 0$ ("Markov uniqueness" for the process associated with μ_0^v) (for $v = 0, X_t(f) = \widetilde{X_t}(f) = $ Nelson's free field at time t and with test function f. Its generator is the Hamiltonian of the relativistic free field). The corresponding problem for the diffusion generated by the analogue $\mu_{(0),\Lambda_0}^v$, of $\mu_{(0)}^v$ in a bounded region Λ_0 of $\mathbb{R}^d$ has been solved in [359].

Let us now come back to the stochastic quantization equation (SQE): it has been verified in [119], [255], [421] that μ^v is (for $d = 2$ and a large class of v's) such that the classical Dirichlet form μ^v given by it exists and that the properly associated diffusion $X = (X_\tau)_{\tau \geq 0}$ indeed solves the SQE (3), componentwise (in fact $\beta_{\mu,k} \in L^2(\mu)$) and on E itself.

Recent work on pathwise solutions of the SQE is in [52] and [199], see also [384] for a discussion of the impossibility to use a Girsanov transformation to produce solutions, even on a bounded domain Λ in $\mathbb{R}^2$.

Remark 63. a) The problem of the necessity of the renormalization $v \to: v :$ in order to avoid "triviality" is discussed in [42], [52], [51], [47].

b) For a discussion of the Markov resp. strong uniqueness problem for μ_Λ^v see
[359], [499] (the case where $\Lambda = \mathbb{R}^2$, i.e for μ^v is still open , see [121] for
a partial result and [37], [38] for a related problem in hydrodynamics).

c) Despite the unproven Markov uniqueness, the space of Gibbs states for
μ^v (in the sense discussed, e.g., in [72]) can be identified with $\mathcal{G}^b$, and b
given on FC_b^∞ by the expression β_{μ^v}, i.e.

$$b_k(z) = \underset{\mathcal{S}}{\langle} (\Delta_x - m^2)k, z \underset{\mathcal{S}'}{\rangle} - \underset{\mathcal{S}}{\langle} k, : v'(z) : \underset{\mathcal{S}'}{\rangle}$$

(with $\underset{\mathcal{S}\ \mathcal{S}'}{\langle , \rangle}$, the dualization between $\mathcal{S}(\mathbb{R}^2)$ and $\mathcal{S}'(\mathbb{R}^2)$). It is important
to realize that b is independent of the Gibbs state, it only depends on μ^0
and v in the support of k.

Using the ergodic theory briefly exposed in Chapter 5.4, it has been shown
in [90], [91] that for $\mu \in \mathcal{G}_{ex}^b$ one has that $\mathcal{E}_\mu$ is irreducible and the solution
X of the SQE (3) is time-ergodic. This is a result which has been hard to
obtain, and holds, e.g., for $v(y)$ an even degree polynomial, with leading
term of the form $\lambda^2 y^n, \lambda > 0$ sufficiently small ($y \in \mathbb{R}$).

d) More work has been done on a stochastic quantization equation with regularization $\varepsilon > 0$, denoted by $(SQE)_\varepsilon$, obtained from (3) by replacing on
the r.h.s. X_τ by $A^{1-\varepsilon}X_\tau, w_\tau$ by $A^{-\frac{\varepsilon}{2}}w_\tau$ and $: v : (X_\tau)$by$A^{-\varepsilon} : v' : (X_\tau)$,
with $A \equiv -\Delta + m^2$. μ^v is heuristically still invariant for $(SQE)_\varepsilon$, for any
$\varepsilon > 0$, this has been shown rigorously in [302], [169], [120], [199] (see
also [384] and references therein). Markov uniqueness for $\mu_\Lambda^{v_\varepsilon}, \Lambda \subset \mathbb{R}^d$ has
been also shown in [121], [123]; L^p-uniqueness of $\mu_\Lambda^v, \Lambda \in \mathbb{R}^d$ in the sense
of [223] (the case $p = 2$ being strong uniqueness) has been shown in [359],
[200], see also [223]. The problem of corresponding uniqueness results for
$\mathbb{R}^d$ instead of Λ is still open.

e) Log-Sobolev inequalities for $\mu^{v_\varepsilon}, (\varepsilon \geq 0)$ would yield exponential ergodicity of X, but this is still an open problem (even for $d = 1$) (the analogous
problem in lattice statistical mechanics is solved in [89], see Section 2
below).

Remark 64. For $d = 3$ only a construction of the analogue of μ^v works for
$v \neq 0$ in the special case $v(y) = y^4$. It is not known whether one can associate
a Markov process to any of the $\mu^v, \mu_\Lambda^v, \mu_0^v, \mu_{0,\Lambda_0}^v, \Lambda \subset \mathbb{R}^3, \Lambda_0 \subset \mathbb{R}^2$, for a
negative result see [133].

For further discussions of these topics and related ones see also [11], [12], [13],
[15], [54], [55], [56], [196], [198], [201], [219], [221], [228], [260], [241], [272],
[291], [303], [310], [318], [328], [496], [497], [498], [499], [500], [360], [394], [395],
[430], [431], [432], [433], [435], [437], [438], [439], [440], [441], [442], [478], [479],
[494].

6.2 Diffusions on configuration spaces and classical statistical mechanics

We shall present here shortly a probabilistic construction of diffusion processes on configuration spaces, see [92], [93], [94] for details (and, e.g., [325], [202], [436], [369], [401], [434] for continuation of the latter work; for previous related work on stochastic analysis related to Poisson processes see, e.g., [179], [305], [420]).

Let M be a connected oriented C^∞ Riemanian manifold such that $\mathrm{Vol}(M) = +\infty$ (where Vol is the volume measure). Let

$\Gamma \equiv$ configuration space (of locally finite configurations) over M

$= \{\gamma \subset M | |\gamma \cap K| < \infty \text{ for each compact } K \subset M\}.$

$\gamma \in \Gamma$ can be identified with the $\mathbb{Z}_+$-valued Radon measure $\sum_{x \in \gamma} \varepsilon_x$, we shall not distinguish in the following γ and the corresponding Radon measure $\sum_{x \in \gamma} \varepsilon_x$.

Any $f \in C_0^\infty(M)$ can be lifted to the map from Γ to $\mathbb{R}$ given by

$$\langle f, \gamma \rangle = \sum_{x \in \gamma} f(x) = \int_M f d\gamma.$$

One can "lift the geometry from M to Γ", e.g., given $v \in V_0(M) \equiv \{\text{smooth vector fields on } M\}$, one gets a flow ϕ_t^v on M, and this flow is lifted to the flow $\tilde{\phi}_t^v$ on Γ, defined by

$$\tilde{\phi}_t^v(\gamma) = \{\phi_t^v(x) | x \in \gamma\}.$$

Let $T_x M$ be the tangent space to M at $x \in M$ and let TM be the tangent bundle $(T_x M)_{x \in M}$. Let ∇_v^M the derivation on M given by

$$\nabla_v^M f(x) = \langle \nabla^M f(x), v(x) \rangle_{T_x M},$$

∇^M being the gradient operator associated with M.

One can also lift the operations $\nabla_v^M, \nabla^M, \tilde{\nabla}_v^M, \tilde{\nabla}^M$ from quantities associated with M to quantities associated with Γ by defining first the space FC_b^∞ of smooth bounded cylinder functions on Γ by

$$FC_b^\infty \equiv \{u : \Gamma \to \mathbb{R} | u(\gamma) = g(\langle f_1, \gamma \rangle, ..., \langle f_n, \gamma \rangle)) | \exists n \in \mathbb{N}, f_i \in C_0^\infty(M), g \in C_b^\infty(\mathbb{R}^n)\}$$

and setting for u as in the definition of FC_b^∞, $\tilde{\nabla}_u^M(\gamma) = \int \sum_i \partial_i g(x) \nabla^M f_i(x) \gamma(dx)$

Moreover:

$$\tilde{\nabla}_v^\Gamma u(\gamma) = \langle \tilde{\nabla}^M u(\gamma, x), v(x) \rangle_{L^2(M \to TM; \gamma)}.$$

In this way we let also correspond to the tangent bundle TM the tangent bundle $T\Gamma \equiv (T_\gamma \Gamma)_{\gamma \in \Gamma}$ with metric given by the inner product in $L^2(M \to$

$TM; \gamma$). This gives then a lift of ∇^M as acting on smooth functions on M to $\tilde{\nabla}^\Gamma$ as acting on smooth cylinder functions on Γ.

Similarly one can define a lift div $^\Gamma$ of div^M to an operation on vector fields on Γ. Defining then $\Delta^\Gamma \equiv \text{div}^\Gamma \nabla^\Gamma$ on FC_b^∞, we have a lift of the Laplace-Beltrami operator Δ^M on functions on M to a Laplace-Beltrami operator on functions over Γ.

A natural question here is for which measures μ on Γ does one have div$^\Gamma = (-\nabla^\Gamma)^*$, the adjoint being taken in $L^2(\mu)$. The following theorem was proven in [94].

Theorem 22. *For μ a probability measure μ on Γ with finite first absolute moments, i.e., s.t.*

$$\int |\langle f, \cdot \rangle| d\mu(\cdot) < \infty \qquad \forall f \in C_b^\infty(M)$$

the following are equivalent:

i) *div$^\Gamma = (-\nabla^\Gamma)^*$,*

ii) *μ is a mixed Poisson measure, i.e. there exists a σ-finite measure λ on $\mathbb{R}_+$ s.t. $\mu = \int\limits_0^\infty \pi_{z\sigma(\cdot)}\lambda(z)$, where $\sigma \equiv \text{Vol}(\cdot)$, and π_σ is the Poisson measure on Γ with intensity measure σ so that*

$$\hat{\pi}_\sigma(f) = \int_\Gamma e^{i\langle f, \gamma \rangle} \pi_\sigma(d\gamma) = e^{\int_M (e^{if(\cdot)}-1)d\sigma(\cdot)}. \qquad \forall f \in C_0^\infty(M)$$

Remark 65. It follows for μ as in the above theorem:

1) μ is the volume measure on Γ (in the natural sense of being a "flat measure" on Γ)

2) μ is quasi-invariant with respect to $\gamma \to \phi(\gamma), \phi \in \text{Diff}_0(M)$ (the diffeomorphisms which are identically the unit outside some compact subset of M).

A stochastic dynamics can be associated with the classical (quasi regular local) Dirichlet form $\mathcal{E}_\mu$, in the form of a diffusion process, satisfying a differential equation of the form given in Theorem 14, in Chapter 5, with a drift coefficient in $L^2(\mu)$. This process is generated by Δ^Γ, moreover, ergodicity and strong uniqueness hold, see [87].

There is an extension of this work to the case where μ is replaced by a "Gibbs measure" (see, e.g., [263]), called again μ, describing a system of particles, in the sense of classical statistical mechanics, for a general class of interactions including "physically realistic" ones, see [93], [333], [325], [403], [505], [48]. Correspondingly in this case one has div$^\Gamma = (-\nabla^\Gamma)^*$ but with $*$ taken with respect to the "non flat" measure μ, i.e. $(\nabla^\Gamma)^* = \nabla^\Gamma - \beta_\mu$, with β_μ a drift term (the logarithmic derivative of μ).

Also the case where $\mathbb{R}^d$ is replaced by a (non compact) manifold M has been handled and a corresponding Hodge type L^2-cohomology theory has been developed, see [29], [30], [34], [32], [33]. For relations with representation theory of infinite dimensional groups see, e.g., [292], [349].

6.3 Other applications

In this section we briefly mention some other applications of the method of classical Dirichlet forms and associated diffusions for defining and studying stochastic processes.

Classical spin systems In this case one studies random variables associated with points on a lattice $\mathbb{Z}^d$ (or other discrete structures), with values in $\mathbb{R}$ (or, e.g., a compact Lie group M), with distributions of the "Gibbs type", i.e. of the form

$$\mu(dz) = \text{``}Z^{-1}e^{-S(z)}dz\text{''}$$

with $z = (z_k)_{k \in \mathbb{Z}^d}, z_k \in \mathbb{R}$ (or M).

$S(z)$ describes the interaction between the spins in the "spin configuration" z. Also in this case the diffusion properly associated with the classical Dirichlet form $\mathcal{E}_\mu$ satisfies a stochastic differential (SDE) equation with drift in $L^2(\mu)$, and it is ergodic, if μ is an extreme state. As opposite to the cases discussed before, in Sect. 6.1, 6.2, log-Sobolev inequalities for classical spin systems have been proven, so that exponential ergodicity holds.
The solution process to the corresponding SDE has a drift in $L^2(\mu)$. A dynamical theory of phase transitions can be developed. See [98], [82], [83], [84], [89], [189], [213], [271], [356], [446], [447] for references and also for current work.

Natural measures and diffusion processes associated with individual and lattice loop spaces Let (M, g) be a (compact) Riemannian manifold. Let $E = LM = C(S^1, M)$ be the corresponding free loop space, and let $L_x M = \{\gamma \in LM | \gamma(0) = x \in M\}$ be the corresponding x-based loop space. Let μ resp. μ_x be the pinned Wiener measure on LM resp. $L_x M$, associated with a Brownian loop in M, with initial distribution the Høegh-Krohn-Bismut measure $\text{Vol}(\cdot)P_t(x, x)$ resp. the Dirac measure in $x \in M$. On $L^2(\nu)$ ($\nu = \mu$ resp. μ_x) we consider the classical Dirichlet form given by ν:

$$\mathcal{E}_\nu = \overline{\overset{\circ}{\mathcal{E}}_\nu}, \text{ with } \overset{\circ}{\mathcal{E}}_\nu(u, v) = \int_{\mathcal{H}} \langle \nabla u, \nabla v \rangle \, d\nu,$$

$u, v \in FC_b^\infty$, (FC_b^∞ being defined as an analogon of the smooth bounded cylinder functions on M) and $\mathcal{H}$ the Cameron-Martin space associated with E, consisting of loops with finite kinetic energy. This diffusion on E has been constructed and discussed in [103] (full loops) and [218] (based loops).

For a long time the problem of log-Sobolev inequalities has been open, see, e.g. [270], it was discussed recently in the negative by Eberle [222], [224] (for positive results for the case where a "potential" is added see [267], [268], [270] and references therein).

For the problem of uniqueness see [4], [8], [5], [6], [7], [412].

For other problems related to loop spaces and strings see [350], [352], [351], [353], [473].

For applications to quantum statistical mechanics see [78], [79], [80], [81], [97], [321], [322], [326], [357].

6.4 Other problems, applications and topics connected with Dirichlet forms

We mention here some topics that - although of great interest - have unfortunately not been covered in these lectures. They illustrate other aspects of the usefulness and power of the method of Dirichlet forms.

Polymers The construction and study of diffusions with polymer measures as invariant measures has been made possible using methods of Dirichlet forms in [124], resp. [125] (for the case of polymer measures of the Edwards-Westwater-type in 2 resp. 3 dimensions). An open problem here is the ergodicity of the process constructed in the 3-dimensional case. One notes that in two dimensions the drift is in $L^2(\mu)$ but this is not so in three dimensions. The stochastic differential equation satisfied by the diffusion is studied by other methods in [27].

Non-symmetric Dirichlet forms and generalized Dirichlet forms Although the theory of non symmetric resp. generalized Dirichlet forms could only be mentioned shortly in this course, it has lead to important new developments in the theory of singular (finite and infinite dimensional) processes. The main attention has been given to the local forms associated with diffusion processes, see, e.g. [110], [112], [314], [346], [348], [347], [383], [367].

Whereas non symmetric Dirichlet forms have first order terms essentially dominated by the symmetric part, generalized Dirichlet forms allow the inclusion of general first order terms in the generators [474], [475], [476], [477]. The latter lead to non proper associated processes which have found striking applications in the study of stochastic PDE's (see [?] for the Gaussian noise and [355] for Lévy noise. See also [492] for further developments). It should also be mentioned that the theory of generalized Dirichlet forms include also time dependent Dirichlet forms, cf. 6.4, below.

Complex-valued Dirichlet forms A theory of such Dirichlet forms has been developed in [117], [131], with applications to quantum theory. It has also lead to a new approach to some aspects of non symmetric Dirichlet forms [386]. See also [317] for further developments in connections with "open system".

Invariant measures for singular processes A theory of such invariant measures has been developed especially in the case of diffusions, e.g. [21],[330]. For jump processes see [130],[128], [129].

Subordination of diffusions given by Dirichlet forms A theory of subordination of diffusions given by Dirichlet forms has been developed in [127] (see also [129], based on Lévy processes on Banach spaces [126]). For previous work on subordination and Dirichlet forms see [294], [295], [296], [170], [171], [230], [231], [232], [299], [229], [298], [284], [285], [287], [288], [289], [382], [400]; for relations with relativistic Schrödinger operators see [181], [464]. For other relation to jump processes see [311], [376], [397].

Time dependent Dirichlet forms A theory of time-dependent Dirichlet forms leads in particular to processes which satisfy S(P)DE's with time dependent coefficients see [406], [404], [405], [474], [475]. The case of Nelson's diffusions is covered in [474], [475], [173]. For related work see also [300].

Differential operators and processes with boundary conditions Examples of processes described by Dirichlet forms in finite dimensions, including complicated boundary behavior, are given in [258], see also, e.g., [159], [177], [204] (and [188a] for systems of elliptic equations). In infinite dimensions not so many examples have been developed until now, see however [507].

Convergence of Dirichlet forms The problem of when a sequence of Dirichlet forms converges in such a way that the limit is again a Dirichlet form has been discussed originally, in the symmetric case, in [70], [64], [102], [143], [343], [417], [409], [443], [442], [486], [493], [398], [397]. The study of such questions in the non symmetric case has been initiated in [377], [378], [379], [380].

Dirichlet forms and geometry In the sections 6.2 and 6.3 in Chapter 6 we already mentioned some work involving Dirichlet forms and geometry (loop spaces, configuration spaces). For work in other directions, in particular in connection with differential geometry in finite dimensions resp. on special infinite dimensional manifolds see [225], [139], [154], [156], [187], [180], [29], [144], [146], [247], [250], [251], [254], [161], [148], [281], [282], [286], [319], [226], [283], [186], [261], [374].

Further problems involving classical Dirichlet forms For relations with hyperbolic problems see [320] and for scattering problems [324], [323], [175], [182], [184], [327]. For control problems see Nagai [392], [391]. For problems of filter theory see [385]. For Dirichlet forms associated with Lévy Laplacian see [1], [17]. For problems of homogenization theory see [18], [257]. For an inverse problem in stochastic differential equations see [19]. For a small time asymptotics for Dirichlet forms see [422]. For a Girsanov transformation for Dirichlet forms on infinite dimensional spaces see [122]. Structural questions about Dirichlet forms and associated spaces are discussed in [472], [149], [163], [178], [215], [216], [240], [252], [253], [248], [274], [339], [340], [341]. For local Dirichlet forms in relation to problems of classical continuum mechanics see [153], [155], [387], [388], [345], [399]. For questions of infinite dimensional diffusion processes and Dirichlet forms see [354], [361], [362], [365], [366], [370], [371], [418].

Dirichlet forms and processes on fractals, discrete structures and metric measure spaces Important work has been done for constructing and studying processes on fractals in [338], [387], [389], [336], [332], [331], [141], [249], [275], [304], [315], [414]. For the study of Dirichlet forms and processes on p-adic spaces with relation to certain trees, see [74],[75]. A theory of hyperfinite Dirichlet forms (in the sense of non standard analysis) has been developed in [39],[36]. The construction of local Dirichlet forms and diffusion processes on metric measure spaces was carried out in [482]. The important particular case of Alexandrov spaces was studied in great detail in [342]. See also [445], [454], [487], [488], [489], [490], [491], [504], [508], [509].

Harmonic mappings, non linear Dirichlet forms Dirichlet form techniques turned out to be a powerful tool in the study of generalized harmonic mappings with values in metric spaces. Jost [307] pointed out how to define the energy of mappings from the state space of a Dirichlet form into a metric space. This leads to the concept of nonlinear or generalized Dirichlet forms, [308], [481]. The stochastic counterparts are nonlinear Markov operators and martingales in metric spaces [485], [484], [480]. For work on nonlinear Dirichlet forms see [264], [297], [381], [151].

Non commutative and supersymmetric Dirichlet forms and processes The study of non commutative Dirichlet forms has been initiated in [63] (see also [64], [66]) in the symmetric case. This was extended to the nonsymmetric case in [206], [358], [273]. Associated processes have also been studied in [69], [67], [45]. For recent further work, also connected to non commutative geometry, see [365], [195], [190], [191], [192], [194], [306], [363], [273] (see also [50], [402], [410], [411]). Supersymmetric Dirichlet forms have been considered in [138], [77].

Aknowledgements

I am most grateful to the organizers of the "Ecole d'Eté de St. Flour, 2000", for giving me the opportunity to present an exciting area of mathematics and probability. To them and all participants I am very grateful for creating a most stimulating atmosphere. My special thanks go to Pierre Bernard, without whose patience and understanding the finishing of the writing up of these notes would never have been possible. Sylvie Paycha should also be warmly thanked for her encouragement and support before and during the lectures and in the preparation of the notes. To her and Benedetta Ferrario, Yuri Kondratiev, Zhi Ming Ma, Michael Röckner, Barbara Rüdiger, Karl-Theodor Sturm and Minoru Yoshida I am very grateful for having read my notes and making many suggestions for improvements.

The work presented here would not have been possible without the help over many years by many coworkers and friends, let me express to all of them my hearty thanks.

Finally the skillfull help by Markus Theis in the setting of the typed version of the manuscript is gratefully acknowledged.

References

[1]   Accardi L., Bogachev V.I. (1997), "The Ornstein- Uhlenbeck process associated with the Lévy Laplacian and its Dirichlet forms", Probab. Math. Statist. **17**, no. 1, 95-114

[2]   Accardi L., Heyde C.C. (1998), "Probability Towards 2000", Springer Verlag

[3]   Accardi L., Lu Y.G., Volovich I.V. (2000), "A white-noise approach to stochastic calculus", Acta Appl. Math. **63**, no. 1-3, 3-25

[4]   Acosta E. (1994), "On the essential self-adjointness of Dirichlet operators on group-valued path space", Proc. AMS **122**, 581-590

[5]   Aida S. (1998), "Differential Calculus on Path and Loop Spaces II. Irreducibility of Dirichlet Forms on Loop Spaces", Bull. Sci. math. **122**, 635-666

[6]   Aida S. (2000), "Logarithmic Derivatives of Heat Kernels and Logarithmic Sobolev Inequalities with Unbounded Diffusion Coefficients on Loop Spaces", Journal of Functional Analysis **174**, 430-477

[7]   Aida S. (2000), "On the irreducibility of Dirichlet forms on domains in infinite-dimensional spaces", Osaka J. Math. **37**, no. 4, 953-966

[8]   Aida S., Shigekawa I. (1994), "Logarithmic Sobolev inequalities and spectral gaps: perturbation theory", Journal of Functional Analysis **126**, no. 2, 448-475

[9]   Albeverio S. (1985), "Some points of interaction between stochastic analysis and quantum theory", Springer Verlag, Lect. Notes Control Inform. Sciences **78**, 1-26

[10]   Albeverio S. (1989), "Some new developments concerning Dirichlet forms, Markov fields and quantum fields", B. Simon et al, eds., IX. Intern. congr. Math. Phys., Adam Hilger, Bristol

[11]   Albeverio S. (1993), "Mathematical physics and stochastical analysis - a round table report", Proc. Round Table, St. Chéron, Bull Sci. Math. **117**, 125-152

[12]   Albeverio S. (1996), "Loop groups, random gauge fields, Chern-Simons models, strings: some recent mathematical developments", Espace des lacets Proc. Conf. Loop spaces '94, Eds. R.Léandre, S.Paycha, T.Wurzbacher, Strasbourg, 5-34

[13]   Albeverio S. (1997), "A survey of some developments in loop spaces: associated stochastic processes, statistical mechanics, infinite dimensional Lie groups, topological quantum fields", Proc. Steklov Inst. Math. **217**, no. 2, 203-229

[14]   Albeverio S. (1997), "A survey of some developments in loop spaces: associated stochastic processes, statistical mechanics, infinite dimensional Lie groups, topological quantum fields", Proc. Ste. Inst. Maths. **217**, 203-229

[15]   Albeverio S. (1997), "Some applications of infinite dimensional analysis in mathematical physics", Helv. Phys. Acta **70**, 479-506

[16]   Albeverio S. (1997), "Wiener and Feynman-Path integrals and their Applications", AMS, Proceedings of Symposia in Applied Mathematics **52**, 163-194

[17]   Albeverio S., Belopolskaya Ya., Feller M. (2002), "Lévy Dirichlet forms", in preparation

[18]   Albeverio S., Bernabei M. S. (2002), "Homogenization in random Dirichlet forms", Preprint SFB 611 no. 7, submitted to Forum Mathematicum

[19]   Albeverio S., Blanchard Ph., Kusuoka S., Streit L. (1989), "An inverse problem for stochastic differential equations", J. Stat. Phys 57, 347-356

[20]   Albeverio S., Blanchard Ph., Ma Z.M. (1991), "Feynman-Kac semigroups in terms of signed smooth measures", Birkhäuser, Random Partial Differential Equations, 1-31

[21]   Albeverio S., Bogachev V., Röckner M. (1999), "On uniqueness of invariant measures for finite- and infinite-dimensional diffusions", Communications on Pure and Applied Mathematics 52,325-362

[22]   Albeverio S., Brasche J., Röckner M. (1989), "Dirichlet forms and generalized Schrödinger operators", Springer Verlag, Lect. Notes in Phys. 345, 1-42

[23]   Albeverio S., Brzeźniak Z. (1993), "Finite dimensional approximations approach to oscillatory integrals in infinite dimensions", J. Funct. Anal. 113, 177-244

[24]   Albeverio S., Brzeźniak Z. (1995), "Oscillatory integrals on Hilbert spaces and Schrödinger equation with magnetic fields", J. Math. Phys. 36, 2135-2156

[25]   Albeverio S., Brzeźniak Z., Boutet de Monvel-Berthier A.M. (1995), "Stationary phase in infinite dimensions by finite dimensional approximations: applications to the Schrödinger equation", Pot. Anal. 4, 469-502

[26]   Albeverio S., Brzeźniak Z., Boutet de Monvel-Berthier A.M. (1996), "The trace formula for Schrödinger operators from infinite dimensional oscillatory integrals", Math. Nachr. 182, 21-65

[27]   Albeverio S., Brzeźniak Z., Daletskii A. (2002), "Stochastic differential equations on product loop manifolds", Bonn preprint

[28]   Albeverio S., Cruzeiro A.B. (1990), "Global flows with invariant (Gibbs) measures for Euler and Navier-Stokes two dimensional fluids", Comm. Math. Phys. 129, 431-444

[29]   Albeverio S., Daletskii A. (1998), "Stochastic equations and quasi-invariance on infinite product groups", Infinite Dimensional Analysis, Quantum Probability and Related Topics, Vol. 2, No. 2, 283-288

[30]   Albeverio S., Daletskii A., Kondratiev Y. (1998), "Some examples of Dirichlet operators associated with the actions of infinite dimensional Lie groups", Meth. Funct. Anal. and Top. 2, 1-15

[31]   Albeverio S., Daletskii A., Kondratiev Y. (2000), " DeRham complex over product manifolds: Dirichlet forms and stochastic dynamics", in Mathematical physics and stochastic analysis (Lisbon, 1998), 37-53

[32]   Albeverio S., Daletskii A., Kondratiev Y., Lytvynov E. (2001), "Laplace operators in deRham complexes associated with measures on configuration spaces", Bonn preprint

[33]   Albeverio S., Daletskii A., Lytvynov E. (2000), "Laplace operator and diffusions in tangent bundles over Poisson spaces", in [CHNP], 1-25

[34]   Albeverio S., Daletskii A., Lytvynov E. (2001), "DeRham cohomology of configuration spaces with Poisson measure", Journal of Functional Analysis 185, 240-273

[35]   Albeverio S., De Faria M., Høegh-Krohn R. (1979), "Stationary measures for the periodic Euler flow in two dimensions", J. Stat. Phys. **20**, 585-595

[36]   Albeverio S., Fan R.Z. (2002), "Hyperfinite Dirichlet forms and stochastic processes", preprint Bonn in preparation

[37]   Albeverio S., Ferrario B. (2002), "Uniqueness results for the generators of the two-dimensional Euler and Navier-Stokes flows. The case of Gaussian invariant measures", J. Funct. Anal. **193**, No.1, 77-93

[38]   Albeverio S., Ferrario B. (2002), "2D vortex motion of an incompressible ideal fluid: the Koopman-von Neumann approach", Infinite Dimens. Anal. Quantum Prob., in press

[39]   Albeverio S., Fenstad J.E., Høegh-Krohn R., Lindstrøm T. (1986), "Nonstandard methods in stochastic analysis and mathematical physics", Academic Press, New York

[40]   Albeverio S., Fukushima M., Karwowski W., Streit L. (1981), "Capacity and quantum mechanical tunneling", Comm. Math. Phys. **80**, 301-342

[41]   Albeverio S., Fukushima M., Hansen W., Ma Z.M., Röckner M. (1992), "An invariance result for capacities on Wiener space", J. Funct. Anal. **106**, 35-49

[42]   Albeverio S., Gallavotti G., Høegh-Krohn R. (1979), "Some results for the exponential interaction in 2 or more dimensions", Comm. Math. Phys. **70**, 187-192

[43]   Albeverio S., Gesztesy F., Karwowski W., Streit L. (1985), "On the connection between Schrödinger and Dirichlet forms", J. Math. Phys. **26**(10), 2546-2553

[44]   Albeverio S., Gesztesy F., Høegh-Krohn R., Holden H. (1998), "Solvable Models in Quantum Mechanics", Springer Verlag, Berlin

[45]   Albeverio S., Goswami D. (2002), "A Remark on the structure of symmetric quantum dynamical semigroups on von Neumann algebras", to appear in IDAQP (2002)

[46]   Albeverio S., Gottschalk H., Wu J.L. (1997), "Models of local relativistic quantum fields with indefinite metric (in all dimensions)", Commun. Math. Phys. **184**, 509-531

[47]   Albeverio S., Gottschalk H., Yosida M. (2001), "Representing Euclidean quantum fields as scaling limit of particle systems", Bonn preprint, to appear in J. Stat. Phys.

[48]   Albeverio S., Grothaus M., Kondratiev Y., Röckner M. (2001), "Stochastic dynamics of fluctuations in classical continuous systems", Journal of Functional Analysis **185**, 129-154

[49]   Albeverio S., Guatteri G., Mazzucchi S. (2002), "Phase Space Feynman Path Integrals", J. Math. Phys. **43**, no. 6, 2847-2857

[50]   Albeverio S., Guido D., Ponosov A., Scarlatti S. (1996), "Singular traces and compact operators", J. Funct. Anal. **137**, no. 2, 281-302

[51]   Albeverio S., Haba Z., Russo F. (1996), "On non-linear two-space-dimensional wave equation perturbed by space-time white noise", in Stochastic analysis: random fields and measure-valued processes (Ramat Gan, 93/95), Israel Math. Conf. Proc. **10**, Bar-Ilan Univ., Ramat Gar, 1-25

[52]   Albeverio S., Haba Z., Russo F. (2001), "A two-space dimensional semi-linear heat equation perturbed by (Gaussian) white noise", Probab. Theory Relat. Fields **121**, 319-366

[53]   Albeverio S., Hahn A., Sengupta A. (2002), "Chern-Simons theory, Hida distributions, and state models", Bonn preprint, to appear in Special Issue on "Geometry and Stochastic Analysis", IDAQP

[54]   Albeverio S., Hida T., Potthof J., Streit L. (1989), "The vacuum of the Høegh-Krohn model as a generalized white noise functional", Phys. Lett. B **217**, no. 4, 511-514

[55]   Albeverio S., Hida T., Potthof J., Röckner M., Streit L. (1990), "Dirichlet forms in terms of white noise analysis I - Construction and QFT examples", Rev. Math. Phys **1**, 291-312

[56]   Albeverio S., Hida T., Potthof J., Röckner M., Streit L. (1990), "Dirichlet forms in terms of white noise analysis II - Closability and Diffusion Processes", Rev. Math. Phys **1**, 313-323

[57]   Albeverio S., Høegh-Krohn R. (1974), "A remark on the connection between stochastic mechanics and the heat equation", Journal of Mathematical Physics **15**, 1745-1747

[58]   Albeverio S., Høegh-Krohn R. (1975), "Homogeneous random fields and statistical mechanics", Journal of Functional Analysis **19**, 242-272

[59]   Albeverio S., Høegh-Krohn R. (1976), "Quasi invariant measures, symmetric diffusion processes and quantum fields", Editions du CNRS, Proceedings of the International Colloquium on Mathematical Methods of Quantum Field Theory, Colloques Internationaux du Centre National de la Recherche Scientifique, No. **248**, 11-59

[60]   Albeverio S., Høegh-Krohn R. (1976), "Mathematical Theory of Feynman Path Integrals", Lect. Notes Maths. **523**, Springer Berlin

[61]   Albeverio S., Høegh-Krohn R. (1977), "Dirichlet forms and diffusion processes on rigged Hilbert spaces", Zeitschrift für Wahrscheinlichkeitstheorie und verwandte Gebiete **40**, 1-57

[62]   Albeverio S., Høegh-Krohn R. (1977), "Hunt processes and analytic potential theory on rigged Hilbert spaces", Ann. Inst. H. Poincaré (Probability Theory) **13**, 269-291

[63]   Albeverio S., Høegh-Krohn R. (1977), "Dirichlet forms and Markov semigroups on C^*-algebras", Commun. Math. Phys. **56**, 173-187

[64]   Albeverio S., Høegh-Krohn R. (1979), "The method of Dirichlet forms", Springer Verlag, Lecture Notes in Physics **93**, 250-258

[65]   Albeverio S., Høegh-Krohn R. (1981), "Stochastic Methods in Quantum Field Theory and Hydrodynamics", Physic Reports (Review Section of Physics Letters) **77**, no. 3, 193-214

[66]   Albeverio S., Høegh-Krohn R. (1982), "Some remarks on Dirichlet forms and their applications to quantum mechanics", Springer Verlag, Lect. Notes in Math. **923**, 120-132

[67]   Albeverio S., Høegh-Krohn R. (1985), "A remark on dynamical semigroups in terms of diffusion processes", Springer Verlag, Lect. Notes in Math. **1136**, 40-45

[68]   Albeverio S., Høegh-Krohn R., Marion J., Testard D., Torresani B. (1993), "Non commutative distributions - Unitary representation of gauge groups and algebras", M. Dekker, New York

[69] Albeverio S., Høegh-Krohn R., Olsen G. (1980), "Dynamical semigroups and Markov processes on C^*-algebras", Crelle's J. Reine und Ang. Math.**319**, 25-37

[70] Albeverio S., Høegh-Krohn R., Streit L. (1977), "Energy forms, Hamiltonians, and distorted Brownian paths", J. Math. Phys. **18**, 907–917

[71] Albeverio S., Høegh-Krohn R., Streit L. (1979), "Regularization of Hamiltonians and processes", J. Math. Phys. **21**(7), 1636-1642

[72] Albeverio S., Høegh-Krohn R., Zegarlinski B. (1989), "Uniqueness of Gibbs states for general $P(\varphi)_2$ – weak coupling models by cluster expansion", Commun. Math. Phys. **121**, 683-697

[73] Albeverio S., Johnson G.W., Ma Z.M. (1996), "The Analytic Operator-Valued Feynman Integral via Additive Functionals of Brownian Motion", Acta Applicandae Mathematicae **42**, 267-295

[74] Albeverio S., Karwowski W. (1991), "Diffusion on p-adic numbers", Proc. Third Nagoya Lévy Seminar, Gaussian Random Fields, Ed. Ito, Hida

[75] Albeverio S., Karwowski W., Yasuda K. (2002), "Trace formula for p-adics", Acta Appl. Math. **71**, no. 1, 31-48

[76] Albeverio S., Karwowski W., Zhao X.L. (1999), "Asymptotics and spectral results for random walks on p-adics", Stoch. Proc. Appl. **83**, 39-59

[77] Albeverio S., Kondratiev Y. (1995), "Supersymmetric Dirichlet operators", Ukrainian Math. J. **47**, 583-592

[78] Albeverio S., Kondratiev Y., Kozitsky Y., Röckner M. (2001), "Euclidean Gibbs States of Quantum Lattice Systems", BiBoS preprint, Universität Bielefeld, No. 01-03-03, to appear in Rev. in Math. Phys

[79] Albeverio S., Kondratiev Y., Minlos R., Shchepan'uk G. (2000), "Uniqueness Problem for Quantum Lattice Systems with Compact Spins", Letters in Mathematical Physics **52**, 185-195

[80] Albeverio S., Kondratiev Y., Minlos R., Shchepan'uk G. (2000), "Ground State Euclidean Measures for Quantum Lattice Systems on Compact Manifolds", Rep. Math. Phys. **45**, 419-429

[81] Albeverio S., Kondratiev Y., Pasurek T., Röckner M. (2001), "Euclidean Gibbs States of Quantum Crystals", Moscow Mathematical Journal 1, No. 3, 307-313

[82] Albeverio S., Kondratiev Y., Pasurek T., Röckner M. (2002), "Euclidean Gibbs measures on loop lattices: Existence and a priori estimates", BiBoS Preprint, Universität Bielefeld, No. 02-05-086

[83] Albeverio S., Kondratiev Y., Pasurek T., Röckner M. (2002), "A priori estimates and existence for Euclidean Gibbs measures", BiBos-Preprint N02-06-089, (submitted to Trans. Moscow Math. Soc.)

[84] Albeverio S., Kondratiev Y., Pasurek T., Röckner M. (2002), "A priori estimates and existence for quantum Gibbs states: approach via Lyapunov functions"

[85] Albeverio S., Kondratiev Y. Röckner M. (1992), "An approximative criterium of essential selfadjointness of Dirichlet operators", Potential Anal. 1, no. 3, 307-317

[86] Albeverio S., Kondratiev Y. Röckner M. (1993), "Addendum to the paper 'An approximative criterium of essential self-adjointness of Dirichlet operators'", Potential Anal. 2, 195-198

[87]   Albeverio S., Kondratiev Y., Röckner M. (1994), "Infinite dimensional diffusions, Markov fields, quantum fields and stochastic quantization", Kluwer, Dordrecht, Stochastic Analysis and Applications, 1-94

[88]   Albeverio S., Kondratiev Y., Röckner M. (1995), "Dirichlet Operators via Stochastic Analysis", Journal of Functional Analysis **128**, 102-138

[89]   Albeverio S., Kondratiev Y., Röckner M. (1995), "Uniqueness of the stochastic dynamics for continuous spin systems on a lattice", J. Funct. Anal. **133**, 10-20

[90]   Albeverio S., Kondratiev Y., Röckner M. (1996), "Ergodicity of L^2-Semigroups and Extremality of Gibbs States", Journal of Functional Analysis **144**, No. 2, 394-423

[91]   Albeverio S., Kondratiev Y., Röckner M. (1997), "Ergodicity for the Stochastic Dynamics of Quasi-invariant Measures with Applications to Gibbs States", Journal of Functional Analysis **149**,No. 2, 415-469

[92]   Albeverio S., Kondratiev Y., Röckner M. (1998), "Analysis and geometry on configuration spaces", J. Funct. Anal. **154**, 444-500

[93]   Albeverio S., Kondratiev Y., Röckner M. (1998), "Analysis and geometry on configuration spaces: The Gibbsian case", J. Funct. Anal. **157**, 242-291

[94]   Albeverio S., Kondratiev Y., Röckner M. (1999), "Diffeomorphism groups and current algebras: Configuration spaces analysis in quantum theory", Rev. Math. Phys. **11**, 1-23

[95]   Albeverio S., Kondratiev Y., Röckner M. (2002), "Symmetryzing Measures for Infinite Dimensional Diffusions: An Analytic Approach", BiBoS Preprint, Universität Bielefeld, No. 02-05-087, to appear in Lect. Notes Math., "Geometric Analysis and unlinear partial differential equations", S. Hildebrandt et. al., eds, Springer, Berlin

[96]   Albeverio S., Kondratiev Y., Röckner M. (2002), "Strong Feller Properties for Distorted Brownian Motion and Applications to Finite Particle Systems with Singular Interactions", BiBoS preprint, Universität Bielefeld, No. 02-05-084, to appear in AMS-Volume dedicated to L. Gross

[97]   Albeverio S., Kondratiev Y., Röckner M., Tsikalenko T. (1997), "Dobrushin's Uniqueness for Quantum Lattice Systems with Nonlocal Interaction", Commun. Math. Phys. **189**, 621-630

[98]   Albeverio S., Kondratiev Y., Röckner M., Tsikalenko T. (2000), "A Priori Estimates for Symmetrizing Measures and Their Applications to Gibbs States", Journal of Functional Analysis **171**, 366-400

[99]   Albeverio S., Kusuoka S. (1992), "Maximality of infinite-dimensional Dirichlet forms and Høegh-Krohn's model of quantum fields", Cambridge Univ. Press, in "Ideas and methods in quantum statistical physics",301-330

[100]   Albeverio S., Kusuoka S. (1995), "A basic estimate for two-dimensional stochastic holonomy along Brownian bridges", J. Funct. Anal. **127**, no. 1, 132-154

[101]   Albeverio S., Kusuoka S., Röckner M. (1990), "On partial integration in infinite dimensional space and applications to Dirichlet forms", J. London Math. Soc. **42**, no. 1, 122-136

[102]   Albeverio S., Kusuoka S., Streit L. (1986), "Convergence of Dirichlet forms and associated Schrödinger operators", J. Funct. Anal. **68**, 130-148

[103]   Albeverio S., Léandre R., Röckner M. (1993), "Construction of a rotational invariant diffusion on the free loop spaces", C.R. Acad. Sci **316**, Ser. I, 1-6

[104]   Albeverio S., Ma Z.M. (1991), "Perturbation of Dirichlet forms - lower semiboundedness, closability and form cones", J. Funct. Anal. **99**, 332-356

[105]   Albeverio S., Ma Z.M. (1991), "Necessary and sufficient conditions for the existence of m-perfect processes associated with Dirichlet forms", Springer Verlag, Lect. Notes Math. **1485**, 374-406

[106]   Albeverio S., Ma Z.M. (1992), "Additive Functionals, nowhere Radon and Kato Class Smooth Measures Associated with Dirichlet Forms", Osaka J. Math. **29**, 247-265

[107]   Albeverio S., Ma Z.M. (1992), "Characterization of Dirichlet forms associated with Hunt processes", World Scientific Singapore, Proc. Swansea Conf. Stochastic Analysis, 1-25

[108]   Albeverio S., Ma Z.M. (1992), "A general correspondence between Dirichlet forms and right processes", Bull. Am. Math. Soc. **26**, 245-252

[109]   Albeverio S., Ma Z.M., Röckner M. (1992), "Non-symmetric Dirichlet forms and Markov processes on general state space", CRA Sci., Paris **314**, 77-82

[110]   Albeverio S., Ma Z.M., Röckner M. (1992), "Regularization of Dirichlet spaces and applications", CRAS, Ser. I, Paris **314**, 859-864

[111]   Albeverio S., Ma Z.M., Röckner M. (1992), "A Beurling-Deny structure theorem for Dirichlet forms on general state space", Cambridge University Press, R. Høegh-Krohn's Memorial Volume, Vol 1, 115-123

[112]   Albeverio S., Ma Z.M., Röckner M. (1993), "Local property of Dirichlet forms and diffusions on general state spaces", Math. Annalen **296**, 677-686

[113]   Albeverio S., Ma Z.M., Röckner M. (1993), "Quasi-regular Dirichlet forms and Markov processes", J. Funct. Anal. **111**, 118-154

[114]   Albeverio S., Ma Z.M., Röckner M. (1995), "Potential theory of positivity preserving forms without capacity", Walter de Gruyter & Co., Dirichlet Forms and Stochastic Processes, 47-53

[115]   Albeverio S., Ma Z.M., Röckner M. (1995), "Characterization of (non-symmetric) Dirichlet forms associated with Hunt processes", Random Operators and Stochastic Equations **3**, 161-179

[116]   Albeverio S., Ma Z.M., Röckner M. (1997), "Partition of unity for Sobolev spaces in infinite dimensions", J. Functional Anal. **143**, 247-268

[117]   Albeverio S., Morato L.M., Ugolini S. (1998), "Non-Symmetric Diffusions and Related Hamiltonians", Potential Analysis **8**, 195-204

[118]   Albeverio S., Röckner M. (1989), "Classical Dirichlet forms on topological vector spaces - the construction of the associated diffusion process", Prob. Theory and Rel. Fields **83**, 405-434

[119]   Albeverio S., Röckner M. (1990), "Classical Dirichlet forms on topological vector spaces – closability and a Cameron-Martin formula.", J. Funct. Anal. **88**, 395-436

[120]   Albeverio S., Röckner M. (1991), "Stochastic differential equations in infinite dimension: solution via Dirichlet forms", Prob. Th. Rel. Fields **89**, 347-386

[121]   Albeverio S., Röckner M. (1995), "Dirichlet form methods for uniqueness of martingale problems and applications",in "Stochastic Analysis",Edts. M.C. Cranston et al., AMS, Providence,513-528

[122]   Albeverio S., Röckner M., Zhang T.S. (1993), "Girsanov transform of symmetric diffusion processes in infinite dimensional space", Ann. of Prob. **21**, 961-978

[123]   Albeverio S., Röckner M., Zhang T.S. (1993), "Markov uniqueness and its applications to martingale problems, stochastic differential equation and stochastic quantization", C.R. Math. Rep. Acad. Sci. Canada XV, no. 1, 1-6

[124]   Albeverio S., Röckner M., Zhou X.Y. (1999), "Stochastic quantization of the two dimensional polymer measure", Appl. Math. Optimiz. **40**, 341-354

[125]   Albeverio S., Röckner M., Zhou X.Y. (2002), "Stochastic quantization of the three-dimensional polymer measure", in preparation

[126]   Albeverio S., Rüdiger B. (2002), "The Lévy-Ito decomposition theorem and stochastic integrals, on separable Banach spaces", BiBoS preprint, Universität Bielefeld, No. 02-01-071

[127]   Albeverio S., Rüdiger B. (2002), "Infinite dimensional Stochastic Differential Equations obtained by subordination and related Dirichlet forms", BiBoS preprint, Universität Bielefeld, No. 02-03-077

[128]   Albeverio S., Rüdiger B., Wu J.L. (2000), "Invariant Measures and Symmetry Property of Lévy Type Operators", Potential Analysis **13**, 147-168

[129]   Albeverio S., Rüdiger B., Wu J.L. (2001), "Analytic and Probabilistic Aspects of Lévy Processes and Fields in Quantum Theory", in "Lévy Processes: Theory and Applications", Eds. O.E. Barndorff-Nielsen et al., Birkhäuser, Basel, 187-224

[130]   Albeverio S., Song S. (1993), "Closability and resolvent of Dirichlet forms perturbed by jumps", Pot. Anal. **2**, 115-130

[131]   Albeverio S., Ugolini S. (2000), "Complex Dirichlet Forms: Non Symmetric Diffusion Processes and Schrödinger Operators", Potential Analysis **12**, 403-417

[132]   Albeverio S., Zegarlinski B. (1992), "Global Markov property in quantum field theory and statistical mechanics: a review on results and problems", in R.Høegh-Krohn's Memorial Volume **2**, Edts. S.Albeverio, J.E.Fenstad, H.Holden, T.Lindstrøm, Cambridge University Press, 331-369

[133]   Albeverio S., Zegarlinski B. (2002), in preparation

[134]   Albeverio S., Zhang T.S. (1998), "Approximations of Ornstein-Uhlenbeck processes with unbounded linear drifts", Stoch. and Stoch. Repts. **63**, no. 3-4, 303-312

[135]   Albeverio S., Zhao X.L. (2002), "Stochastic processes on p-adics", book in preparation

[136]   Albeverio S., Ru-Zong F., Röckner M., Stannat W. (1995), "A remark on coercive forms and associated semigroups", Operator Theory: Adv. and Appl. **78**, 1-8

[137]   Aldous D., Evans S.N. (1999), "Dirichlet forms on totally disconnected spaces and bipartite Markov chains", J. Theoret. Probab. **12**, no. 3, 839-857

[138]   Arai A., Mitoma I. (1991), "De Rham-Hodge-Kodaira decomposition in ∞-dimensions", Math. Ann. **291**, 51-73

[139]   Bakry D., Qian Z. (2000), "Some new results on eigenvectors via dimension, diameter, and Ricci curvature", Adv. Math. **155**, no. 1, 98-153

[140]   Bally V., Pardoux E., Stoica L. (2002), "Backward stochastic differential equations associated to a symmetric Markov process", preprint

[141]   Barlow M.T., Kumagai T. (2001), "Transition density asymptotics for some diffusion processes with multifractal structures", Electron. J. Probab. **6**, no. 9, 23pp.

[142]   Bauer H. (2002), "Wahrscheinlichkeitstheorie", Walter de Gruyter & Co., Berlin

[143]   Baxter J., Dal Maso G., Mosco U. (1987), "Stopping times and Γ-convergence", Trans. Amer. Math. Soc. **303**, no. 1, 1-38

[144]   Bendikov A. (1995), "Potential Theory on Infinite- Dimensional Abelian Groups", de Gruyter, Studies in Mathematics **21**

[145]   Bendikov A., Léandre R. (1999), "Regularized Euler-Poincaré number of the infinite dimensional torus", Infinite Dimensional Analysis, Quantum Probability and Related Topics **2**, 617-626

[146]   Bendikov A., Saloff-Coste L. (2000), "On- and Off-Diagonal Heat Kernel Behaviours on Certain Infinite Dimensional Local Dirichlet Spaces", American Journal of Mathematics **122**, 1205-1263

[147]   Berezansky Y.M., Kondratiev Y.G. (1995), "Spectral Methods in Infinite-Dimensional Analysis", Kluwer, Vol. **1,2**, first issued in URSS, 1988

[148]   Berg C., Forst G. (1973), "Non-symmetric translation invariant Dirichlet forms", Invent. Math. **21**, 199-212

[149]   Bertoin J. (1986), "Les processus de Dirichlet en tant qu'espace de Banach", Stochastics **18**, 155-168

[150]   Beurling A., Deny J. (1958), "Espaces de Dirichlet", Acta Math. **99**, 203-224

[151]   van Beusekom P. (1994), "On nonlinear Dirichlet forms", Ph. D. Thesis, Utrecht

[152]   Biroli M. (2000), "Weak Kato measures and Schrödinger problems for a Dirichlet form", Rend. Accad. Naz. Sci. XL Mem. Mat. Appl. (5) **24**, 197-217

[153]   Biroli M., Mosco U. (1992), "Discontinuous media and Dirichlet forms of diffusion type", Developments in partial differential equations and applications to mathmatical physics (Ferrara, 1991), 15-25, Plenum, New York

[154]   Biroli M., Mosco U. (1993), "Sobolev inequalities for Dirichlet forms on homogeneous spaces", RMA Res. Notes Appl. Math. **29**, 305-311

[155]   Biroli M., Mosco U. (1995), "A Saint-Venant principle for Dirichlet forms on discontinuous media", Ann. Mat. Pura Appl. **169**, 125-181

[156]   Biroli M., Mosco U. (1995), "Sobolev and isoperimetric inequalities for Dirichlet forms on homogeneous spaces", Atti. Accad. Naz. Lincei Cl. Sci. Fis. Mat. Natur. Rend. Lincei(9) Mat. Appl. **6**, no. 1, 37-44

[157]   Biroli M., Mosco U. (1999), "Kato space for Dirichlet forms", Potential Anal. **10**, no. 4, 327-345

[158]   Biroli M., Picard C., Tchou N. (2001), " Error estimates for relaxed Dirichlet problems involving a Dirichlet form", Adv. Math. Sci. Appl. 11, no. 2, 673-684

[159]   Biroli M., Tersian S. (1997), "On the existence of nontrivial solutions to a semilinear equation relative to a Dirichlet form", Istit. Lombardo Accad. Sci. Lett. Rend. A 131, no. 1-2, 151-168

[160]   Bliedtner J., Hansen W. (1986), "Potential Theory", Springer Verlag

[161]   Bloom W.R. (1987), "Non-symmetric translation invariant Dirichlet forms on hypergroups", Bull. Austral. Math. Soc. 36, no.1, 61-72

[162]   Blumenthal R.M., Getoor R.K. (1968), "Markov processes and Potential theory", Academic Press, New York

[163]   Boboc N., Bucur G. (1998), "Characterization of Dirichlet forms by their excessive and co-excessive elements", Potential Anal. 8, no. 4, 345-357

[164]   Bogachev V., Krylov N., Röckner M. (1997), "Elliptic regularity and essential self-adjointness of Dirichlet Operators on $\mathbb{R}^n$", Annali Scuola Norm. Super. di Pisa Cl. Sci. (4) 24, no. 3, 451-461

[165]   Bogachev V., Röckner M. (1995), "Mehler Formula and Capacities for Infinite Dimensional Ornstein-Uhlenbeck Processes with General Linear Drift", Osaka J. Math. 32, 237-274

[166]   Bogachev V., Röckner M., Stannat W. (1999), "Uniqueness of invariant measures and maximal dissipativity of diffusion operators on L^1", Preprint SFB 343, Bielefeld, 99-119, 19pp.

[167]   Bogachev V., Röckner M., Wang F.Y. (2001), "Elliptic Equations for Invariant Measures on Finite and Infinite Dimensional Manifolds", J. Math. Pures Appl. 80 No. 2, 177-221

[168]   Bogachev V., Röckner M., Zhang T.S. (2000), "Existence and uniqueness of invariant measures: an approach via sectorial forms", Appl. Math. Optim. 41, no. 1, 87-109

[169]   Borkar V.S., Chari R.T., Mitter S.K. (1988), "Stochastic quantization of field theory in finite and infinite volume", J. Funct. Anal. 81, no. 1, 184-206

[170]   Bouleau N. (1984), "Quelques resultats probabilistes sur la subordination au sens de Bochner", Lect. Notes in Math. 1061, 54-81

[171]   Bouleau N. (2001), "Calcul d'erreur complet lipschitzien et formes de Dirichlet", J. Math. Pures Appl. (9) 80, no. 9, 961-976

[172]   Bouleau N., Hirsch F. (1991), "Dirichlet Forms and Analysis on Wiener Space", deGruyter, Studies in Mathematics 14

[173]   Bouslimi M. (2000), "Subordination operators of Dirichlet forms", Bi-BoS preprint, Universität Bielefeld, No. 00-07-21

[174]   Brasche J.F. (1989), "Dirichlet forms and nonstandard Schrödinger operators, standard and nonstandard", World Sci. Publishing, Teaneck, NJ, 42-57

[175]   Brasche J.F. (2002), "On singular perturbations of order $s, s \leq 2$, of the free dynamics: Existence and completeness of the wave operators", Göteborg Preprint

[176]   Brasche J.F. (2002), "On eigenvalues and eigensolutions of the Schödinger equation on the complement of a set with classical capacity zero", in preparation

[177]   Brasche J.F., Karwowski W. (1990), "On boundary theory for Schrödinger operators and stochastic processes. Order, disorder and chaos in quantum systems", Birkhäuser, Basel, Oper. Theory Adv. Appl. 46

[178]   Briane M., Tchou N. (2001), "Fibered microstructures for some nonlocal Dirichlet forms", Ann. Scuola Norm. Sup. Pisa Cl. Sci. (4) 30, no. 3-4, 681-711

[179]   Carlen E.A., Pardoux E. (1990), "Differential calculus and integration by parts on Poisson space", Kluwer Academic Publishers, Stochastics,Algebra and Analysis in Classical and Quantum Dynamics, 63-73

[180]   Carlen E.A., Kusuoka S., Stroock D.W. (1987), "Upper bounds for symmetric Markov transition functions", Ann. Inst. H. Poincaré Probab. Statist. 23, no. 2, 245-287

[181]   Carmona R., Masters W.C. (1990), "Relativistic Schrödinger Operators: Asymptotic Behaviour of the Eigenfunctions", Journal of Functional Analysis 91, 117-142

[182]   van Casteren J.A. (1999), "Feynman-Kac Semigroups, Martingales and Wave Operators", J. Korean Math. Soc. 38, no. 2, 227-274

[183]   van Casteren J.A. (2000), "Some problems in stochastic analysis and semigroup theory", Progress in Nonlinear Differential Equations and their Applications 42, 43-60

[184]   van Casteren J.A., Demuth M. (2000), "Stochastic spectral theory for selfadjoint Feller operators. A functional integration approach", Probability and its Applications, Birkhäuser, Basel

[185]   Cattiaux P., Fradon M. (1997), "Entropy, reversible diffusion processes, and Markov uniqueness", J. Funct. Anal. 134, 243-272

[186]   Changsoo B., Chul K.K., Yong M.P. (2002), "Dirichlet forms and symmetric Markovian semigroups on CCR algebras with respect to quasi-free states", to appear in J. Math. Phys.

[187]   Chen, M.F. (1989), "Probability metrics and coupling methods", Pitman Res. Notes Math. Ser. 200

[188]   Chen Z.Q., Fitzsimmons P.J., Takeda M., Ying J., Zhang T.S. (2002), "Absolute continuity of symmetric Markov processes", University of Washington, Seattle - preprint

[188a]   Chen Z.Q., Zhao Z. (1994), "Switched diffusion processes and systems of elliptic equations: a Dirichlet space approach", Proc. Roy. Soc. Edinburgh Sect. A 124, no. 4, 673-701

[188b]   Chen Z.Q., Zhang T.S. (2002), "Girsanov and Feynman-Kac type transformations for symmetric Markov processes", Ann. I.H. Poincaré PR 38, 475-505

[189]   Choi V., Park Y.M., Yoo H.J. (1998), " Dirichlet forms and Dirichlet operators for infinite particle systems: essential self-adjointness", J. Math. Phys. 39, no. 12, 6509-6536

[190]   Cipriani F. (1997), "Dirichlet forms and Markovian semigroups on standard forms of von Neumann algebras", J. Funct. Anal. 147, no. 2, 259-300

[191]   Cipriani F. (1998), "The Variational Approach to the Dirichlet Problem in C*-Algebras", Quantum Probability Banach Centre Publications, Vol. 43, 135-146

[192]   Cipriani F. (1999), "Perron Theory for Positive Maps and Semigroups on von Neumann Algebras", Stochastic processes, physics and geometry: new interplays, II, CMS COnf. Proc., 29, 115-123

[193]   Cipriani F. (2000), "Estimates for capacities of nodal sets and polarity criteria in recurrent Dirichlet spaces", Forum Math. 12, 1-21

[194]   Cipriani F. (2002), "Noncommutative potential theory and the sign of the curvature operator in Riemannian geometry", preprint 494/P, Politecnico di Milano

[195]   Cipriani F., Sauvageot J.L. (2000), "Derivations as Square Roots of Dirichlet Forms", Preprint, Politecnico di Milano

[196]   Clément Ph., den Hollander F., van Neerven J., de Pagter B. (2000), "Infinite Dimensional Stochastic Analysis", Roy. Neth. Ac. of Arts & Sci. Amsterdam, ISBN: 90-6984-296-3 60-06

[197]   Courrège P., Renouard P. (1975), "Oscillateur anharmonique, mesures quasi-invariantes sur $C(\mathbb{R}, \mathbb{R})$ et theéorie quantique des champs en dimension $d = 1$", Astérique, no. 22-23, 3-245

[198]   Da Prato G., Debussche A. (2000), "Maximal dissipativity of the Dirichlet operators corresponding to the Burgers equation", CMS Conf. Proc 28, 85-98

[199]   Da Prato G., Debussche A. (2002), "Strong solutions to the stochastic quantization equations", Prepublication de l'IRMAR 02-18

[200]   Da Prato G., Tubaro L. (2000), "Self-adjointness of some infinite dimensional elliptic operators and applications to stochastic quantization", Prob. Theory Related Fields 118, No. 1, 131-145

[201]   Da Prato G., Zabczyk J. (1992), "Stochastic equations in infinite dimensions", Cambridge University Press, Encyclopedia of Mathematics and Applications 44

[202]   Da Silva J., Kondratiev Y. Röckner M. (2001), " On a relation between intrinsic ans extrinsic Dirichlet forms for interacting particle systems", Math. Nachr. 222, 141-157

[203]   Dai Pra P., Roelly S., Zessin H. (2002), " A Gibbs variational principle in space-time for infinite-dimensional diffusions", Prob. Theory Related Fields 122, no. 2, 289-315

[204]   Dal Maso G., De Cicco V., Notarantonio L, Tchou N. (1998), "Limits of variational problems for Dirichlet forms in varying domains", J. Math. Pures Appl. (9) 77, no. 1, 89-116

[205]   Daletsky Yu., Fomin S.V. (1991), "Measures and Differential Equations in Infinite Dimensional Space", Kluwer, Dordrecht

[206]   Davies B.E., Lindsay M.J. (1992), "Noncommutative symmetric Markov semigroups", Math. Z. 210, no. 3, 379-411

[207]   Dellacherie C., Meyer P.A. (1975), "Probabilités et Potentiel", Hermann Paris, Chapitres I à IV

[208]   Dellacherie C., Meyer P.A. (1975), "Probabilités et Potentiel", Hermann Paris, Chapitres V à VIII

[209]   Dellacherie C., Meyer P.A. (1983), "Probabilités et Potentiel", Hermann Paris, Théorie discrète du potentiel, Chapitres IX à XI

[210]   Dellacherie C., Meyer P.A. (1975), "Probabilités et Potentiel", Hermann Paris, Théorie du potentiel associée à une résolvante, Théorie des processus de Markov, Ch. XII-XVI

[211]   Deny J. (1970), "Méthodes Hilbertiennes et théorie du potentiel", Potential Theory, Centro Internazionale Matematico Estivo, Edizioni Cremonesi, Roma

[212]   Deuschel J.D., Stroock D.W. (1989), "A function space large deviation principle for certain stochatic integrals", Probab. Theory Related Fields **83**, no. 1-2, 279-307

[213]   Deuschel J.D., Stroock D.W. (1990), "Hypercontractivity and spectral gap of symmetric diffusions with applications to the stochastic Ising models", J. Funct. Anal. **92**, no. 1, 30-48

[214]   Dohmann J.M.N. (2001), "Feller-type Properties and Path Regularities of Markov Processes", BiBos Preprint E02-04-081

[215]   Dong Z. (1998), "Quasi-regular topologies for Dirichlet forms on arbitrary Hausdorff spaces", Acta Math. Sinica **14**, 683-690

[216]   Dong Z., Gong F.Z. (1997), "A necessary and sufficient condition for quasiregularity of Dirichlet forms corresponding to resolvent kernels", Acta Math. Appl. Sinica **20**, no. 3, 378-385

[217]   Doob J.L. (1984), "Classical Potential Theory and its Probabilistic Counterpart", Springer-Verlag, Grundl. d. math. Wiss. **262**

[218]   Driver B., Röckner M. (1992), "Construction of diffusions on path and loop spaces of compact Riemmanian manifolds", C.R. Acad. Sci. Paris **315**, 603-608

[219]   Dynkin E.B. (1980), "Markov processes and random fields", Bull. Amer. Math. Soc. **3**, no. 3, 975-999

[220]   Dynkin E.B. (1982), "Green's and Dirichlet spaces associated with fine Markov processes", J. Funct. Anal. **47**, 381-418

[221]   Dynkin E.B. (1984), "Gaussian and non-Gaussian random fields associated with Markov processes", J. Funct. Anal. **55**, no. 3, 344-376

[222]   Eberle A. (1997), "Diffusions on path and loop spaces: existence, finite-dimensional approximation and Hölder continuity", Probab. Theory Related Fields **109**, no. 1, 77-99

[223]   Eberle A. (1999), "Uniqueness and Non-Uniqueness of Semigroups generated by Singular Diffusion Operators", Springer Verlag, Lect. Notes Maths. **1718**

[224]   Eberle A. (2001), "Poincaré inequalities on loop spaces", BiBos Preprint 01-07-049

[225]   Elworthy K.D., Le Jan Y., Li X.M. (1999), "On the geometry of diffusion operators and stochastic flows", Lect. Notes in Math. **1720**

[226]   Elworty K.D., Ma Z.M. (1997), "Vector fields on mapping spaces and related Dirichlet forms and diffusions", Osaka J. Math. **34**, no. 3, 629-651

[227]   Ethier S., Kurtz Th.G., "Markov processes – characterization and convergence", J. Wiley, New York (1985)

[228]   Fabes E., Fukushima M., Gross L., Kenig C., Röckner M., Stroock D.W. (1993), " Dirichlet forms", Eds. C. Dell'Antonio, U. Mosco Springer Verlag, Lecture Notes in Math. **1563**

[229]   Farkas W., Jacob N. (2001), "Sobolev spaces on non-smooth domains and Dirichlet forms related to subordinate reflecting diffusions", Math. Nachr. **224**, 75-104

[230]   Feyel D. (1979), "Propriétés de permanence du domaine d'un générateur infinitésimal", Springer Verlag, Lecture N. in Math. **713**, No. 4, 38-50

[231]   Feyel D., de La Pradelle A. (1978), "Processus associés a une classe d'espaces de Banach fonctionnels", Z. Wahrsch. verw. Geb. **43**, 339-351

[232]   Feyel D., de La Pradelle A. (1979), "Dualité des quasi-résolvantes de Ray", Springer Berlin, Lect. Notes in Math. **713**, 67-88

[233]   Finkelstein D.L., Kondratiev Y.G., Röckner M., Konstantinov A.Y. (2001), "Gauss formula and symmetric extensions of the Laplacian on configuration spaces", Inf. Dim. Anal. Quant. Prob. **4**, 489-510

[234]   Fitzsimmons P.J. (1989), "Markov processes and nonsymmetric Dirichlet forms without regularity", J. Funct. Anal. **85**, 287-306

[235]   Fitzsimmons P.J. (1997), "Absolute continuity of symmetric diffusions", Ann. Prob. **25**, 230-258

[236]   Fitzsimmons P.J. (2000), "Hardy's inequality for Dirichlet forms", J. Math. Anal. Appl. **250**, no. 2, 548-560

[237]   Fitzsimmons P.J. (2001), "On the quasi-regularity of semi-Dirichlet forms", Potential Anal. **15**, no. 3, 151-185

[238]   Fitzsimmons P.J., Getoor R.K. (1988), "On the potential theory of symmetric Markov processes", Math. Ann. **281**, 495-512

[239]   Fitzsimmons P.J., Getoor R.K. (2002), "Additive Functionals and Characteristic Measures", Stochastic Analysis and Related Topics, RIMS International Project 2002, 23-24

[240]   Föllmer H. (1981), "Dirichlet processes", Lect. Notes in Math. **851**, 476-478

[241]   Föllmer H. (1984), "Von der Brownschen Bewegung zum Brownschen Blatt: Einige neuere Richtungen in der Theorie der stochastischen Prozesse", Birkhäuser Verlag, Perspectives in Mathematics Anniversary of Oberwolfenbach, 159-189

[242]   Fritz J. (1987), "Gradient dynamics of infinite point systems", Ann. Probab. **15**, no. 2, 478-514

[243]   Fukushima M. (1971), "Dirichlet spaces and strong Markov processes", Trans. Amer. Math. Soc. **162**, 185-224

[244]   Fukushima M. (1980), "Dirichlet Forms and Markov Processes", North Holland, Amsterdam

[245]   Fukushima M. (1981), "On a stochastic calculus related to Dirichlet forms and distorted Brownian motion", Phys. Rep. **77**, 255-262

[246]   Fukushima M. (1982), "On absolute continuity of multidimensional symmetrizable diffusions", Lect. Notes in Math. **923**, 146-176

[247]   Fukushima M. (1984), "A Dirichlet form on the Wiener space and properties of Brownian motion", Springer Verlag, Lecture N. in Math. **1096**, 290-300

[248]   Fukushima M. (1987), "Energy forms and diffusion processes", World Scientific Publishing, Mathematics and Physics **1**, 65-97

[249]   Fukushima M. (1992), "Dirichlet forms, diffusion processes and spectral dimensions for nested fractals", Eds. S. Albeverio et al., Ideas and Methods in Quantum and Statistical Physics, Cambridge University Press

[250]   Fukushima M. (1997), "Dirichlet forms, Caccioppoli sets and Skorohod equation", Progr. Systems Control Theory **23**, 59-66

[251]   Fukushima M. (2000), "BV Functions and Distorted Ornstein Uhlenbeck Processes over the Abstract Wiener Space", Journal of Functional Analysis Vol. **174**, No. 1, 227-249

[252]   Fukushima M. (2001), "Decompositions of Symmetric Diffusion Processes and Related Topics in Analysis", Sugaku Expositions Vol. **14**,No. 1, 1-13

[253]   Fukushima M. (2002), "Function spaces and symmetric Markov processes", Stochastic Analysis and Related Topics, RIMS International Project 2002, 21-22

[254]   Fukushima M., Hino M. (2001), "On the space of BV functions and a related stochastic calculus in infinite dimensions", J. Funct. Anal. **183**, 245-268

[255]   Fukushima M., LeJan Y. (1991), "On quasi-support of smooth measures and closability of pre-Dirichlet forms", Osaka J. Math. **28**, 837-845

[256]   Fukushima M., Ying J. (2002), "A note on regular Dirichlet spaces", Kansai Preprint

[257]   Fukushima M., Nakao S., Takeda M. (1987), "On Dirichlet forms with random data-recurrence and homogenization", Springer Verlag, Lecture Notes in Math. **1250**

[258]   Fukushima M., Oshima Y., Takeda M. (1994), "Dirichlet Forms and Symmetric Markov Processes", Walter de Gruyter

[259]   Fukushima M., Uemura (2002), "On Sobolev and capacitary inequalities for Besov spaces over d-sets", preprint

[260]   Funaki T. (1992), "On the stochastic partial differential equations of Ginzburg-Landau type", Lecture Notes in Control and Inform. Sci. **176**

[261]   Funaki T., Nagai H. (1993), "Degenerative convergence of diffusion process toward a submanifold by strong drift", Stoch. Stoch. Rep. **44**, no. 1-2, 1-25

[262]   Galakhov E.I. (1996), "Sufficient Conditions for the Existence of Feller Semigroups", Mathematical Notes **60**(3), 328-330

[263]   Georgii H.O. (1979), "Canonical Gibbs Measures", Springer Verlag, Lecture Notes in Math. **760**

[264]   Gianazza U. (1997), "Existence for a nonlinear problem relative to Dirichlet forms", Rend. Accad. Naz. Sci. XL Mem. Mat. Appl. (5) **21**, 209-234

[265]   Glimm J., Jaffe A. (1981), "Quantum Physics", Springer, New-York

[266]   Glover J., Rao M., Šikić H., Song R. (1994), "Quadratic Forms Corresponding to the Generalized Schrödinger Semigroups", Journal of Functional Analysis **125**, 358-378

[267]   Gong F.Z., Ma Z.M. (1998), "The Log-Sobolev inequality on loop space over a compact Riemannian manifold", J. Funct. Anal. **157**, 599-623

[268]   Gong F., Röckner M., Liming W. (2001), "Poincaré Inequality for Weighted First Order Sobolev Spaces on Loop Spaces", Journal of Functional Analysis **185**, 527-563

[269]   Gross L. (1976), "Logarithmic Sobolev inequalities", Amer. J. Math. **97**, 1061-1083

[270]   Gross L. (1992), "Logarithmic Sobolev Inequalities and Contractivity Properties of Semigroups", Springer Verlag, Lecture Notes in Math. **1563**

[271]   Grothaus M. (1998), "New Results in Gaussian Analysis and their Applications in Mathematical Physics", Ph. D. thesis, University Bonn

[272]   Guerra F., Rosen L., Simon B. (1975), "The $P(\phi)_2$ euclidean field theory as classical statistical mechanics II", Ann. of Math. (2) **110**, 111-189

[273]   Guido D., Isola T., Scarlatti S. (1996), "Non-symmetric Dirichlet forms on semifinite von Neumann algebras", J. Funct. Anal. **135**, no. 1, 50-75

[274]   Hagemann B.U. (1997), "Eine Klasse von Pseudodifferentialoperatoren die als Erzeuger regulärer Dirichlet-Formen auftreten", Diplomarbeit Bochum

[275]   Hambly B.M., Kigami J., Kumagai T. (2002), "Multifractal formalisms for the local spectral and walk dimensions", Math. proc. Cambridge Philos. Soc. **132**, no. 3, 555-571

[276]   Heyer H. (1979), "Einführung in die Theorie Markoffscher Prozesse", BI Mannheim

[277]   Hida T., Hitsuda M. (1993), "Gaussian processes", Translations of Mathematical Monographs **120**, xvi+183pp

[278]   Hida T., Kuo H.H., Potthoff J., Streit L. (1993), "White Noise, An Infinite Dimensional Calculus", Kluwer Academic Publishers

[279]   Hida T., Potthoff J., Streit L. (1988), "Dirichlet forms and white noise analysis", Comm. Math. Phys. **116**, 235-245

[280]   Hida T., Potthoff J., Streit L. (1989), "Energy forms and white noise analysis. New Methods and results in nonlinear field equations", Lect. Notes in Phys. **347**, 115-125

[281]   Hino M. (2002), "On Dirichlet spaces over convex sets in infinite dimensions", Contemp. Math.

[282]   Hino M. (2002), "A weak version of the extension theorem for infinite dimensional Dirichlet spaces", Stochastic Analysis and Related Topics, RIMS International Project 2002, 27-29

[283]   Hino M., Ramirez J.A. (2002), "Smalltime Gaussian behaviour of symmetric diffusion semigroups", Kyoto Preprint

[284]   Hirsch F. (1984), "Générateurs étendus et subordination au sens de Bochner", Seminar on potential theory, paris, no. 7, Lect. Notes in Math. **1061**, 134-156

[285]   Hirsch F. (2002), "Intrinsic metrics and Lipschitz functions", preprint

[286]   Hirsch F., Song S. (2001), "Markovian uniqueness on Bessel space", Forum Math. **13**, no. 2, 287-322

[287]   Hoh W. (1993), "Feller semigroups generated by pseudo differential operators", de Gruyter, Dirichlet Forms and Stochastic Processes, 199-206

[288]   Hoh W. (2002), "Perturbation of pseudodifferential operators with negative definite symbol", Appl. Math. Optim. **45**, no. 3, 269-281

[289]   Hoh W., Jacob N. (1992), "Pseudo Differential Operators, Feller Semigroups and the Martingale Problem", Stochastic Monographs **7**,95-103

[290]   Huang Z.Y., Yan J.A (2000), "Introduction to infinite dimensional stochastic analysis", Science Press, Kluwer, Beijing, Dordrecht

[291]   Ikeda N. (1990), "Probabilistic methods in the study of asymptotics", Lecture Notes in Math. **1427**, 195-325

[292]   Ismagilov R.S. (1996), "Representations of Infinite-Dimensional Groups", Amer. Math. Soc., Providence

[293]   Iwata K. (1985), "Reversible measures of a $P(\phi)_1$-time evolution", Probabilistic methods in mathematical physics, 195-209

[294]   Jacob N. (1996), "Pseudo-Differential Operators and Markov Processes", Akademie Verlag, Mathematical Research **94**

[295]   Jacob N. (2001), "Pseudo-Differential Operators and Markov Processes. Vol. I. Fourier analysis and semigroups", Imperial College Press, London

[296]   Jacob N. (2002), "Pseudo-Differential Operators and Markov Processes.
Vol. II. Generators and their potential theory", Imperial College Press,
London

[297]   Jacob N., Moroz V. (2000), "On the semilinear Dirichlet problem for a
class of nonlocal operators generating Dirichlet forms", Progr. Nonlinear
Differential Equations Appl. **40**, 191-204

[298]   Jacob N., Moroz V. (2002), "On the log-Laplace equation for nonlocal
operators generating sub-Markovian semigroups", Appl. Math. Optim.
45, no. 2, 237-250

[299]   Jacob N., Schilling R.L. (1999), "Some Dirichlet spaces obtained by
subordinate reflected diffusions", Revista Matemática Iberoamericana
15,No. 1, 59-91

[300]   Jacob N., Schilling R.L. (2000), "Fractional derivatives, non-symmetric
and time-dependent Dirichlet forms and the drift form", Z. Anal. Anwendungen **19**, no. 3, 801-830

[301]   Janson S. (1997), "Gaussian Hilbert spaces", Cambridge University
Press

[302]   Jona-Lasinio G., Mitter P.K. (1985), "On the stochastic quantization of
field theory", Comm. Math. Phys. **101**, 409-436

[303]   Jona-Lasinio G., Sénéor R. (1996), "Study of stochastic differential equations by constructive methods I.", J. of Statistical Physics **83**, 1109-1148

[304]   Jonsson A. (2000), "Dirichlet forms and Brownian motion penetrating
fractals", Pot. Anal. **13**, 69-80

[305]   Jørgensen P.T. (1985), "Monotone convergence of operator semigroups
and the dynamics of infinite particle systems", J. Approx. Theory **43**,
no. 3, 205-230

[306]   Jørgensen P.T. (1993), "Spectral theory for self-adjoint operator extensions associated with Clifford algebras", Contemp. Math. **143**, 131-150

[307]   Jost J. (1997), "Generalized Dirichlet forms and harmonic maps", Calc.
Var. Partial Differ. Equ. **5**, no. 1, 1-19

[308]   Jost J. (1998), "Nonlinear Dirichlet forms", Stud. Adv. Math. **8**, 1-47

[309]   Jost J., Kendall W., Mosco U., Röckner M., Sturm K.T. (1998), "New
Directions in Dirichlet Forms", American Mathematical Society

[310]   Kallianpur G., Perez-Abreu V. (1988), "Stochastic evolution equations
driven by nuclear-space-valued martingales", Appl. Math. Optim. **17**,
237-272

[311]   Kassmann M. (2001), "On Regularity for Beurling-Deny Type Dirichlet
Forms", preprint SFB 256, University Bonn, No. 729

[312]   Kato T. (1980), "Perturbation Theory for Linear Operators",
Grundlehren der mathematischen Wissenschaften **132**, Springer-Verlag,
Berlin, Heidelberg, New York,

[313]   Kendall W.S. (1994), "Probability, Convexity, and Harmonic Maps II.
Smoothness via Probabilistic Gradient Inequalities", Journal of Functional Analysis **126**, 228-257

[314]   Kiefer S. (1995), "Beispiele für Nicht-Symmetrische Dirichlet Formen",
Diplomarbeit, Ruhr Universität Bochum

[315]   Kigami J. (2000), "Markov property of Kusuoka-Zhou's Dirichlet forms
on self-similar sets", J. Math. Sci. Univ. Tokyo **7**, no. 1, 27-33

[316]   Kim D.H., Kim J.H, Yun Y.S. (1998), "Non- symmetric Dirichlet forms
for oblique reflecting diffusions", Math. Japon. **47**, no. 2, 323-331

[317]   Kolokoltsov V.N. (2000), "Semiclassical Analysis for Diffusions and Stochastic Processes", LN Math. **1724**, Springer, Berlin

[318]   Kolsrud T. (1988), "Gaussian random fields, infinite dimensional Ornstein-Uhlenbeck processes, and symmetric Markov processes", Acta Appl. Math. **12**, 237-263

[319]   Kondratiev Y. (1985), "Dirichlet operators and the smoothness of solutions of infinite -dimensional elliptic equations", Soviet Math. Dokl. **31**, no. 3, 269-273

[320]   Kondratiev Y. (1986), "Infinite-dimensional hyperbolic equations corresponding to Dirichlet operators", Soviet Math. Dokl. **33**, no. 3, 82-85

[321]   Kondratiev Y., Barbuljak V. (1991), "Functional integrals and quantum lattice systems: III. Phase transitions", ibid., no. 10, 19-22

[322]   Kondratiev Y., Globa S.A. (1990), "The construction of Gibbs states of quantum lattice systems", Selecta Math. Sovietica **9**, no. 3, 297-307

[323]   Kondratiev Y., Konstantinov A.Y. (1988), "The scattering problem for special perturbations of harmonic systems", in "Limit problems for differential equations", Kiev, Institute of Mathematics. English translation (1994): Selecta Math. Sov. **13**, 217-224

[324]   Kondratiev Y., Koshmanenko V.D. (1982), "Scattering problem for Dirichlet operators", Soviet Math. Dokl. **267**, no. 2, 285-288

[325]   Kondratiev Y., Kuna T. (1999), "Harmonic analysis on configuration spaces I. General theory", Preprint SFB 256, University Bonn, submitted to Infinite Dimensional Analysis, Quantum Theory and Related Topics

[326]   Kondratiev Y., Roelly S., Zessin H. (1996), "Stochastic dynamics for an infinite system of random closed strings: a Gibbsian point of view", Stoch. Processes Appl. **61**, no. 2, 223-248

[327]   Kondratiev Y., Tsycalenko T.V. (1986), "Infinite dimensional hyperbolic equations for Dirichlet operators", Soviet Math. Dokl. **228**, no. 4, 814-817

[328]   Kondratiev Y., Tsycalenko T.V. (1991), "Dirichlet operators and associated differential equations", Selecta Math. Sovietica **10**, no. 4, 345-397

[329]   Krée M., Krée P. (1983), "Continuité de la divergence dans les espaces de Sobolev relatifs á l'espace de Wiener ", CRAS, Paris, **296**, no. 20, 833-836

[330]   Krylov N., Röckner M., Zabczyk J. (1999),"Stochastic PDE's and Kolmogorov equations in infinite dimensions", Springer Verlag, Lecture Notes in Mathematics **1715**

[331]   Kumagai T. (1993), "Regularity, closedness and spectral dimensions of the Dirichlet forms on P.C.F. self-similar sets", J. Math. Kyoto Univ. **33**, no. 3, 765-786

[332]   Kumagai T. (2002), "Sub-Gaussian estimates of heat kernels on a class of fractal-like graphs and the stability under rough isometries", RIMS, Kyoto University

[333]   Kuna T. (1999), "Studies in Configuration Space Analysis and Applications", Ph.D. thesis, University Bonn

[334]   Kuo H. (1975), "Gaussian measures on Banach space", Springer Verlag, Lecture Notes in Math. **463**

[335]   Kusuoka S. (1982), "Dirichlet forms and diffusion processes on Banach spaces", J. Fac. Sc. Univ. Tokyo **29**, 79-95

[336]   Kusuoka S. (1989), "Dirichlet forms on fractals and products of random matrices", Publ. RIMS, Kyoto Univ. **25**, 659-680

[337]   Kusuoka S. (2000), "Term structure and stochastic partial differential equations. Advances in mathematical economics", Adv. Math. Eco. **2**,67-85

[338]   Kusuoka S., Zhou X.Y. (1992), "Dirichlet form on fractals: Poincaré constant and resistance", Prob. Theo. Rel. Fields **93**, 169-196

[339]   Kuwae K. (1994), "Permanent sets of measures charging no exceptional sets and the Feynman-Kac formula", preprint 1994, to appear in Forum Math.

[340]   Kuwae K. (2000), "On a strong maximum principle for Dirichlet forms", CMS Conf. Proc. **29**, 423-429

[341]   Kuwae K. (2002), "Reflected Dirichlet forms and the uniqueness of Silverstein's extension", Potential Anal. **16**, no. 3, 221-247

[342]   Kuwae K., Machigashira Y., Shioya T. (2001), "Sobolev spaces, Laplacian, and heat kernel on Alexandrov spaces", Math. Z. **238**, no. 2, 269-316

[343]   Kuwae K., Shioya T. (2002), "Convergence of spectral structures: a functional analytic theory and its applications to spectral geometry", submitted to Communications in Analysis & Geometry

[344]   Lang R. (1977), "Unendlich-dimensionale Wienerprozesse mit Wechselwirkung II. Die reversiblen Masse sind kanonische Gibbs-Masse", Z. Wahrscheinlichkeitstheorie und Verw. Gebiete **39**, no. 4, 277-299

[345]   Laptev A., Weidl T. (1999), "Hardy inequalities for magnetic Dirichlet forms", Oper. Theory Adv. Appl. **108**, 299-305

[346]   Le Jan Y. (1978), "Mesures associées à une forme de Dirichlet", Bull. Soc. Math. France **106**, 61-112

[347]   Le Jan Y. (1983), "Quasicontinuous functions and Hunt processes", J. Math. Soc. Japan **35**, no. 1, 37-42

[348]   Le Jan Y. (1995), "New examples of Dirichlet spaces", Dirichlet forms and stochastic processes, de Gruyter, Berlin, 253-256

[349]   Léandre R. (2001), "Stochastic diffeology and homotopy", Progr. Probab. **50**, 51-57

[350]   Léandre R. (2002), "An example of a brownian non linear string theory", preprint

[351]   Léandre R. (2002), "Full stochastic Dirac-Ramond operator over the free loop space", preprint Institut Elie Cartan

[352]   Léandre R. (2002), "Super Brownian motion on a loop group", preprint Institut Elie Cartan

[353]   Léandre R., Roan S.S. (1995), "A stochastic approach to the Euler-Poincaré number of the loop space of developable orbifold", J. Geom. Phys. **16**, 71-98

[354]   Leha G., Ritter G. (1993), "Lyapunov-type conditions for stationary distributions of diffusion processes on Hilbert spaces", Stoch. and Stoch. Rep. **48**, 195-225

[355]   Lescot, Röckner M. (2002), "Pertubations of generalized Mehler semigroups and applications to stochastic heat equations with Lévy noise and singular drift", BiBos Preprint 02-03-076

[356]   Lim H.Y., Park Y.M., Yoo H.J. (1997), " Dirichlet forms, Dirichlet operators and log-Sobolev inequalities for Gibbs measures of classical unbounded spin systems", J. Korean Math. Soc. **34**, no. 3, 731-770

[357]   Lim H.Y., Park Y.M., Yoo H.J. (1998), " Dirichlet forms and Dirichlet operators for Gibbs measures of quantum unbouded spin systems: essential self-adjointness and log-Sobolev inequality", J. Statist. Phys. **90**, no. 3-4, 949-1002

[358]   Lindsay J.M. (1993), "Non-commutative Dirichlet forms", de Gruyter, Dirichlet Forms and Stochastic Processes, 257-270

[359]   Liskevich V., Röckner M. (1998), "Strong Uniqueness for Certain Infinite Dimensional Dirichlet Operators and Applications to Stochastic Quantization", Ann. Scuola Norm. Pisa **27**

[360]   Löbus J.U. (1993), "Generalized diffusion operators", Akademie Verlag, Berlin

[361]   Lyons T.J., Röckner M., Zhang T.S. (1996),"Martingale decomposition of Dirichlet processes on the Banach space $C_0[0,1]$", Stochastic Process. Appl. **64**, no. 1, 31-38

[362]   Lyons T.J., Zhang T.S. (1994), "Decomposition of Dirichlet processes and its application", Ann. Prob. **22**, no. 1, 494-524

[363]   Lyons T.J., Zheng W.A. (1988), "A crossing estimate for the canonical process on a Dirichlet space and a tightness result", Société Math. de France, Astérisque No. 157-158, 249-271

[364]   Ma Z.M. (1990), "Some new results concerning Dirichlet forms, Feynman-Kac semigroups and Schrödinger equations", in Probability Theory in China (Contemp. Math.), Amer. Math. Soc., Providence, RI

[365]   Ma Z.M. (1995), "Quasi-regular Dirichlet forms and Applications", Birkhäuser Basel, Proceedings of the International Congress of Mathematicans Vol **1,2**, 1006-1016

[366]   Ma Z.M., Overbeck L., Röckner M. (1995), "Markov processes associated with Semi-Dirichlet forms", Osaka J. of Math. **32**, 97-119

[367]   Ma Z.M., Röckner M. (1992), "An Introduction on the Theory of (Nonsymmetric) Dirichlet Forms", Springer-Verlag, Berlin

[368]   Ma Z.M., Röckner M. (1995), "Markov Processes Associated with Positivity Preserving Coercive Forms", Canadian J. Math. **47**, 817-840

[369]   Ma Z.M., Röckner M. (2000), "Construction of diffusions on configuration spaces", Osaka J. Math. **37**, No. 2, 273-314

[370]   Ma Z.M., Röckner M, Sun W. (2000), "Approximation of Hunt processes by multivariate Poisson processes", Acta Appl. Math. **63**, 233-243

[371]   Ma Z.M., Röckner M., Zhang T.S. (1998), "Approximations of arbitrary Dirichlet processes by Markov chains", Ann. Inst. Henri Poincaré **34**, No. 1, 1-22

[372]   Malliavin P. (1997), "Stochastic Analysis", Springer Verlag

[373]   Malliavin M.P., Malliavin P. (1990), "Integration on loop groups. I. Quasi invariant measures", J. Funct. Anal. **93**, 207-237

[374]   Malý J., Mosco U. (1999), "Remarks on measure-valued Lagrangians on homogeneous spaces", Ricerche Mat. **48**, 217-231

[375]   Manavi A., Voigt J. (2002), "Maximal operators associated with Dirichlet forms perturbed by measures", Potential Anal. **16**, no. 4, 341-346

[376]   Marchi S. (1997), "Influence of the nonlocal terms on the regularity of equations involving Dirichlet forms", Istit. Lombardo Accad. Sci. Lett. Rend. A **131**, no. 1-2, 189-199

[377]   Mataloni S. (1999), "Representation formulas for non-symmetric Dirichlet forms", Z. Anal. Anwendungen **18**, no. 4, 1039-1064

[378]   Mataloni S. (1999), "On a type of convergence for non-symmetric Dirichlet forms", Adv. Math. Sci. Appl. **9**, no.2, 749-773

[379]   Mataloni S. (2001), "Quasi-linear relaxed Dirichlet problems involving a Dirichlet form", Rend. Accad. Naz. Sci. XL Mem. Mat. Appl.(5) **25**, 67-96

[380]   Mataloni S., Tchou N. (2001), "Limits of relaxed Dirichlet problems involving a nonsymmetric Dirichlet form", Ann. Mat. Pura Appl. (4) **179**, 65-93

[381]   Matzeu M. (2000), "Mountain pass and linking type solutions for semilinear Dirichlet forms", Progr. Nonlinear Differential Equations Appl. **40**, 217-231

[382]   McGilivray I. (1997), "A recurrence condition for some subordinated strongly local Dirichlet forms", Forum Math. **9**, no. 2, 229-246

[383]   Menendez S.C. (1975), "Processus de Markov associé à une forme de Dirichlet non symétrique", Z. Wahrscheinlichkeitstheorie verw. Gebiete **33**, 139-154

[384]   Mikulievikius R., Rozovskii B.L. (1999), "Martingale problems for stochastic PDEs", Stochastic Partial Differential Equations. Six Perspectives, AMS Rhode Island, 251-333

[385]   Mitter S. (1989), "Markov random fields, stochastic quantization and image analysis", Math. Appl. **56**, 101-109

[386]   Morato L. (2002), in preparation

[387]   Mosco U. (1994), "Composite Media and Asymptotic Dirichlet Forms", Journal of Functional Analysis **123**, no. 2, 368-421

[388]   Mosco U. (1998), "Dirichlet forms and self-similarity. New directions in Dirichlet forms", AMS/IP Stud. Adv. Math. **8**, 117-155

[389]   Mosco U. (2000), "Self-similar measures in quasi-metric spaces", Birkhäuser, Basel, Progr. Nonlinear Differential Equations Appl. **40**, 233-248

[390]   Mück S. (1993), "Large deviations w.r.t. quasi-every starting point for symmetric right processes on general state spaces", Prob. Theor. Rela. Fields **99**, 527-548

[391]   Nagai H. (1986), "Stochatic control of symmetric Markov processes and nonlinear variational inequalities", Stochastics **19**, no. 1-2, 83-110

[392]   Nagai H. (1992), "Ergodic control problems on the whole Euclidean space and convergence of symmetric diffusions", Forum Math. **4**, no. 2, 159-173

[393]   Nelson E., "Analytic vectors", Ann. Math. **70**, 572-615 (1959)

[394]   Nelson E., "Dynamical theories of Brownian motion", Princeton Univ. Press

[395]   Nelson E. (1973), "The free Markov field", J. Funct. Anal. **12**, 217-227

[396]   Nualart D. (1996), "The Malliavin Calculus and Related Topics", Springer Berlin, Probability and its Applications

[397]   Ogura Y., Tomisaki M., Tsuchiya M. (2001), "Esitence of a strong solution for an integro-differential equation and superposition of diffusion
processes", Birkhäuser, Boston, Stochastics in finite and infinite dimensions, 341-359

[398]   Ogura Y., Tomisaki M., Tsuchiya M. (2002), "Convergence of local
Dirichlet forms to a non-local type one", Ann. I. H. Poincaré 30, no. 4,
507-556

[399]   Oksendal B. (1988), "Dirichlet forms, quasiregular functions and Brownian motion", Invent. math. 91, 273-297

[400]   Ôkura H. (2002), "Recurrence and transience criteria for subordinated
symmetric Markov processes", Forum Math. 14, 121-146

[401]   Oliveira M.J. (2002), "Configuration Space Analysis and Poissonian
White Noise Analysis", Ph. D. thesis, Universidade de Lisboa

[402]   Olkiewicz R., Zegarlinski B. (1999), "Hypercontractivity in Noncommutative L_p Spaces", Journal of Functional Analysis 161, 246-285

[403]   Osada H. (1996), "Dirichlet form approach to infinite-dimensional
Wiener process with singular interactions", Comm. Math. Phys. 176,
117-131

[404]   Oshima Y. (1992), "On a construction of Markov Processes associated
with time dependent Dirichlet spaces", Forum Mathematicum 4, 395-
415

[405]   Oshima Y. (1992), "Some properties of Markov processes associated
with time dependent Dirichlet forms", Osaka J. Math. 29, 103-127

[406]   Oshima Y. (1993), "Time dependent Dirichlet forms and its application to a transformation of space-time Markov processes", de Gruyter,
Dirichlet Forms and Stochastic Processes, 305-320

[407]   Ouhabaz E.M. (1992), "Propriétés d'ordre et de contractivité des semi-
groupes avec applications aux opérateures elliptiques", Ph.D. thesis, Be-
sançon

[408]   Paclet Ph. (1978), "Espaces de Dirichlet et capacités fonctionelles sur
les triplets de Hilbert-Schmidt", Séminare Equations aux Dérivées Par-
tielles en Dimension Infinite, No. 5

[409]   Pardoux E., Veretennikov A.Y. (2001), "On the Poisson equation and
diffusion approximation I.", Ann. Probab. 29, no. 3, 1061-1085

[410]   Park Y.M. (2000), "Construction of Dirichlet forms on standard forms
of von Neumann algebras", Infin. Dim. Anal. Quantum Probab. Theory
Related Top. 3, no. 1, 1-14

[411]   Park Y.M. (2002), "Dirichlet forms and symmetric Markovian semigroups on von Neumann algebras", Proceedings of the Third Asian
Mathematical Conference, 2000 (Diliman), 427-443

[412]   Park Y.M., Yoo H.J. (1997), "Dirichlet operators on loop spaces: Essential self-adjointness and log-Sobolev inequality", J. Math. Phys. 38,
3321-3346

[413]   Pazy A. (1983), "Semigroups of linear operators and applications to
partial diffential equations", Applied Mathematical Sciences, 44. New
York, Springer

[414]   Peirone R. (2000), "Convergence and uniqueness problems for Dirichlet
forms on fractals", Boll. Unione Mat. Ital. Sez. B Artic. Ric. Mat. (8)
3, no. 2, 431-460

[415]   Pontier M. (2000), "A Dirichlet form on a Poisson space", Potential Anal. **13**, no. 4, 329-344

[416]   Popovici A. (2001), "Forward interest rates: a Hilbert space SDE approach", preprint BiBoS,

[417]   Posilicano A. (1996), "Convergence of Distorted Brownian Motions and Singular Hamiltonians", Potential Anal. **5**,no. 3, 241-271

[418]   Potthoff J., Röckner M. (1990), "On the contraction property of energy forms for infinite-dimensional space", J. of Funct. Anal. **92**,No. 1, 155-165

[419]   Priouret P., Yor M. (1975), "Processus de diffusion à valeurs dans $\mathbb{R}$ et mesures quasi-invariantes sur $C(\mathbb{R},\mathbb{R})$", Astérique 22-23, 247-290

[420]   Privault N. (1999), "Equivalence of gradients on configuration spaces", Random Oper. Stoch. Eq. **7**, 241-262

[421]   Pugachëv O.V. (2001), "On the closability of classical Dirichlet forms in the plane", Dokl. Akad. Nauk **380**, no. 3, 315-318

[422]   Ramirez J.A. (2001), "Short time symptotics in Dirichlet spaces", Comm. Appl. Math. **54**, 259-293

[423]   Reed M., Simon B. (1980), "Methods of Modern Mathematical Physics. I. Functional Analysis", Academic Press, London

[424]   Reed M., Simon B. (1975), "Methods of Modern Mathematical Physics. II. Fourier Analysis, Self-Adjointness", Academic Press, London

[425]   Reed M., Simon B. (1979), "Methods of modern mathematical physics. III. Scattering Theory", Academic Press, New York

[426]   Reed M., Simon B. (1978), "Methods of modern mathematical physics. IV. Analysis of operators", Academic Press, New York

[427]   Rellich F. (1969), "Perturbation Theory of Eigenvalue Problem", Gordon and Breach, New York

[428]   Röckle H. (1997), "Banach Space-Valued Ornstein-Uhlenbeck Processes with General Drift Coefficients", Acta Applicandae Mathematicae **47**, 323-349

[429]   Röckner M. (1985), "A Dirichlet problem for distributions and specifications for random fields", Memoirs of the American Mathematical Society **324**

[430]   Röckner M. (1988), "Traces of harmonic functions and a new path space for the free quantum field", J. Funct. Anal. **79**, 211-249

[431]   Röckner M. (1992), "Dirichlet Forms on infinite dimensional state space and applications", pp. 131-186 in H.Korezlioglu, A.S. Üstümel, eds., Birkhäuser, Boston

[432]   Röckner M. (1993), "General theory of Dirichlet forms and applications", Lecture Notes in Math. **1563**, 129-193

[433]   Röckner M. (1995), "Dirichlet forms on infinite-dimensional 'manifold-like' state spaces: a survey of recent results and some prospects for the future", Springer, New York, Lecture Notes in Statis. **128**, 287-306

[434]   Röckner M. (1998), "Stochastic analysis on configuration spaces: Basic ideas and recent results", New Directions in Dirichlet Forms (eds. J. Jost et al.), Studies in Advanced Mathematics, Vol. 8, American Math. Soc., 157-232

[435]   Röckner M. (1998), "L^p-Analysis of Finite and Infinite Dimensional Diffusion Operators", Lect. Notes in Math. **1715**, 65-116

[436]   Röckner M., Schied A. (1998), "Rademacher's Theorem on Configuration Spaces and Applications", J. Funct. Anal. **169**, 325-356

[437]   Röckner M., Schmuland B. (1995), "Quasi-regular Dirichlet forms: examples and counter examples", Canad. J. math. **47**, 165-200

[438]   Röckner M., Schmuland B. (1998), "A support property for infinite-dimensional interacting diffusion processes", C.R. Acad. Sci. Paris Sér. I Math. **326**, no. 3, 359-364

[439]   Röckner M., Zhang T.S. (1991), "Decomposition of Dirichlet processes on Hilbert space", London Math. Soc. Lecture Note Ser. **167**, 321-332

[440]   Röckner M., Zhang T.S. (1992), "Uniqueness of generalized Schrödinger operators and applications", J. Funct. Anal. **105**, no. 1, 187-231

[441]   Röckner M., Zhang T.S. (1994), "Uniqueness of generalized Schrödinger operators", J. Funct. Anal. **119**, no. 2, 455-467

[442]   Röckner M., Zhang T.S. (1996), "Finite-dimensional approximation of diffusion processes of infinite-dimensional spaces", Stochastics Stochastics Rep. **57**, no. 1-2, 37-55

[443]   Röckner M., Zhang T.S. (1997), "Convergence of operators semigroups generated by elliptic operators", Osaka J. Math. **34**, no. 4, 923-932

[444]   Röckner M., Zhang T.S. (1998), "On the strong Feller property of the semigroups generated by non-divergence operators with L^p-drift", Progr. Probab. **42**, 401-408, Birkhäuser, Boston

[445]   Röckner M., Zhang T.S. (1999), "Probabilistic representations and hyperbound estimates for semigroups", Infin. Dimens. Anal. Quantum Probab. Relat. Top. **2**, no. 3, 337-358

[446]   Roelly S., Zessin H. (1993), "Une caractérisation des mesures de Gibbs sur $C((0,1),\mathbb{Z}^d)$ par le calcul des variations stochastiques", Ann. Inst. H. Poincaré Prob. Stat. **29**, 327-338

[447]   Roelly S., Zessin H. (1995), "Une caractérization des champs gibbsiens sur un espace de trajectories", C.R. Acad. Sci. Paris Ser I. Math. **321**, 1377-1382

[448]   Roman L.J., Zhang X., Weian Z. (2002), "Rate of convergence in homogenization of parabolic PDEs", preprint

[449]   Roth J.P. (1976), "Les Operateurs Elliptiques comme Générateurs Infinitésimaux de Semi-Groupes de Feller", Lecture Notes in Math **681**, 234-251

[450]   Royer G., Yor M. (1976), "Représentation intégrale de certaines mesures quasi-invariantes sur $C(R)$ mesures extrémales et propriété de Markov", Ann. Inst. Fourier (Grenoble) **26**, no. 2, ix, 7-24

[451]   Rüdiger B. (2002), "Processes with jumps properly associated to non local quasi-regular (non symmetric) Dirichlet forms obtained by subordination", in preparation

[452]   Rüdiger B. (2002), "Stochastic integration w.r.t. compensated Poisson random measures on separable Banach spaces", in preparation

[453]   Rüdiger B., Wu J.L. (2000), "Construction by subordination of processes with jumps on infinite-dimensional state spaces and corresponding non local Dirichlet forms", CMS Conf. Proc. **29**, 559-571

[454]   Saloff-Coste (2002), "Aspects of Sobolev-Type Inequalities", Cambridge University Press

[455]   Schachermayer W. (2002), , St. Flour Lectures 2000, this volume

[456]   Scheutzow M., v. Weizsäcker H. (1998), "Which moments of a logarith-
        mic derivative imply quasiinvariance", Doc. Math. **3**, 261-272
[457]   Schmuland B. (1990), "An alternative compactification for classical
        Dirichlet forms on topological vector spaces", Stochastics **33**, 75-90
[458]   Schmuland B. (1993), "Non-symmetric Ornstein-Uhlenbeck processes in
        Banach space via Dirichlet forms", Canad. J. Statis. **45**,no. 6, 1324-1338
[459]   Schmuland B. (1994), "A Dirichlet Form Primer", American Mathemat-
        ical Society, CRM Proceedings and Lecture Notes **5**, 187-197
[460]   Schmuland B. (1998), "Some Regularity Results on Infinite Dimensional
        Diffusions via Dirichlet Forms", Stochastic Analysis and Applications
        6(3), 327-348
[461]   Schmuland B. (1999), "Dirichlet forms: some infinite dimensional exam-
        ples", Canad. J. Statist. **27**, No. 4, 683-700
[462]   Schwartz L. (1973), "Random measures on arbitrary topological spaces
        and cylindrical measures", Oxford University Press
[463]   Sharpe M. (1988), "General Theory of Markov Processes", Academic
        Press, New York
[464]   Shigekawa I. (2002), "Square root of a Schrödinger operator and its L^p
        norms", Stochastic Analysis and Related Topics, RIMS International
        Project 2002, 25-26
[465]   Shigekawa I., Taniguchi S. (1992), "Dirichlet forms on separable metric
        spaces", Ed. A.N. Shizyaev et al., World Scient. Singapore, Probability
        Theory and Mathematical Statistics, 324-353
[466]   Simon B. (1974), "The $P(\varphi)_2$ Euclidean quantum field theory", Prince-
        ton Univ. Press
[467]   Simon B. (1979), "Functional Integration", Academic Press, New York
[468]   Silverstein M.L. (1974), "Symmetric Markov Processes", Springer Ver-
        lag, Lect. Notes in Math. **426**
[469]   Silverstein M.L. (1976), "Boundary theory for symmetric Markov pro-
        cesses", Springer Verlag, Lecture Notes in Math. **516**
[470]   Smolyanov O.G., von Weizsäcker H. (1999), "Smooth probability mea-
        sures and associated differential operators", Infin. Dimens. Anal. Quan-
        tum Probab. Relat. Top. **2**, no. 1, 51-78
[471]   Song S. (1994), "A study on markovian maximality, change of probabil-
        ity and regularity", Potential Anal. **3**, 391-422
[472]   Spönemann U. (1997), "An existence and a structural result for Dirichlet
        forms", Ph. D. thesis, University Bielefeld
[473]   Srimurthy V.K. (2000), "On the equivalence of measures on loop space",
        Prob. Th. Rel. Fields **18**, 522-546
[474]   Stannat W. (1999), "The Theory of Generalized Dirichlet Forms and
        Its Applications in Analysis and Stochastics", Memoirs of the American
        Mathematical Society **678**
[475]   Stannat W. (1999), "(Nonsymmetric) Dirichlet Operators on L^1: Ex-
        istence, Uniqueness and Associated Markov Processes", Ann. Scuola
        Norm. Sup. Pisa **28**, 99-140
[476]   Stannat W. (2000), "Long-Time Behaviour and Regularity Properties
        of Transition Semigroups of Fleming-Viot Processes", Probab. Theory
        Related Fields **122**,no. 3, 431-469
[477]   Stannat W. (2000), "On the validity of the log- Sobolev inequality for
        symmetric Fleming-Viot operators", Ann. Prob. **28**, 667-684

[478]   Streit L. (1981), "Energy Forms: Schrödinger Theory, Processes", Physics Reports **77**, no. 3, 363-375

[479]   Streit L. (1985), "Quantum theory and stochastic processes - some contact points", Lect. Notes Math. **1203**, 197-213

[480]   Sturm K.T. (1995), "On the geometry defined by Dirichlet forms", Birkhäuser Verlag, Seminar on stochastic analysis, random fields and applications (Ascona 1993)

[481]   Sturm K.T. (1997), "Monotone approximation of energy functionals for mappings into metric spaces", J. Reine Angew. Math. **468**, 129-151

[482]   Sturm K.T. (1998), "Diffusion processes and heat kernels on metric spaces", Ann. Probab. **26**, no. 1, 1-55

[483]   Sturm K.T. (1998), "The geometric aspect of Dirichlet forms", AMS/IP Stud. Adv. Math. **8**, 233-277

[484]   Sturm K.T. (2002), "Nonlinear martingale theory for processes with values in metric spaces of nonpositive curvature", Annals of Probab.

[485]   Sturm K.T. (2002), "Nonlinear Markov operators, discrete heat flow, and harmonic maps between singular spaces", Potential Anal. **16**, no.4, 305-340

[486]   Sun W. (1999), "Mosco convergence of quasi-regular Dirichlet forms", Acta Math. Appl. Sinica **15**, no. 3, 225-232

[487]   Takeda M. (1992), "The maximum Markovian self-adjoint extensions of generalized Schrödinger operators", J. Math. Soc. Jap. **44**, 113-130

[488]   Takeda M. (1994), "Transformations of local Dirichlet forms by supermartingale multiplicative functionals", preprint 94, to appear in Proc. ICDFSP (Z.M. Ma et al., eds.) Walter de Gruyter, Berlin and Hawthorne, NY

[489]   Takeda M. (1999), "Topics on Dirichlet forms and symmetric Markov processes", Sugaku Expositions **12**, no. 2, 201-222

[490]   Takeda M. (2002), "Conditional gaugeability of generalized Schrödinger operators", Stochastic Analysis and Related Topics, RIMS International Project 2002, 15-16

[491]   Tanemura H. (1997), "Uniqueness of Dirichlet forms associated with systems of infinitely many Brownian balls in $\mathbb{R}^{d}$", Probab. Theory Relat. Fields **109**, 275-299

[492]   Trutnau G. (2000), "Stochastic Calculus of Generalized Dirichlet Forms and Applications", Osaka J. Math. **37**, no. 2, 315-343

[493]   Uemura T. (1995), "On Weak Convergence of Diffusion Processes Generated by Energy Forms", Osaka J. Math. **32**, 861-868

[494]   Üstünel A.S. (1995), "An Introduction to Analysis on Wiener Space", LN Math. **1610**, Springer Berlin

[495]   Weidmann J. (1976), "Lineare Operatoren in Hilberträumen", Teubner, Stuttgart

[496]   Wu L. (1997), "Uniqueness of Schrödinger Operators Restricted in a Domain", Journal of Functional Analysis **153**, No. 2, 276-319

[497]   Wu L. (1999), "Uniqueness of Nelson's diffusions", Probab. Theory Relat. Fields **114**, 549-585

[498]   Wu L. (1999), "Uniqueness of Nelson's diffusions II: Infinite Dimensional Setting and Applications", Potential Analysis **13**, 269-301

[499]   Wu L. (2001), "L^p-uniqueness of Schrödinger operators and capacitary positive improving property", J. Funct. Anal **182**, 51-80

[500]   Wu L. (2001), "Martingale and Markov uniqueness of infinite dimensional Nelson diffusions", Birkhäuser, Boston, Progr. Probab. **50**, 139-150

[501]   Wu L., in preparation

[502]   Yalovenko I. (1998), "Modellierung des Finanzmarktes und unendlich dimensionale stochastische Prozesse", Diplomarbeit, Bochum

[503]   Ying J. (1997), "Dirichlet forms perturbated by additive functionals of extended Kato class", Osaka J. Math. **34**, no. 4, 933-952

[504]   Yor M. (1976), "Une remarque sur les formes de Dirichlet et les semi-martingales", Lect. Notes. Math. **563**, 283-292

[505]   Yoshida M. (1996), "Construction of infinite dimensional interacting diffusion processes through Dirichlet forms", Prob. Theory related Fields **106**, 265-297

[506]   K.Yosida (1980), "Functional analysis", Springer, Berlin (VI ed.)

[507]   Zambotti L.(2000), "A reflected heat equation as symmetric dynamics with respect to 3-d Bessel Bridge", J. Funct. Anal. **180**, no. 1, 195-209

[508]   Zhang T.S. (2000), "On the small time asymptotics of diffusion processes on Hilbert spaces", Ann. Prob. **28**, 537-557

[509]   Zheng W. (1995), "Conditional propagation of chaos and a class of quasilinear PDE", Ann. Probab. **23**, no. 3, 1389-1413

Index

Part II

Walter Schachermayer: Introduction to the
Mathematics of Financial Markets

Table of Contents

Summary. In this introductory course we review some of the basic concepts of Mathematical Finance. We start with an account on the thesis of L. Bachelier, which was defended as "Théorie de la Spéculation" in Paris in 1900. We hope that this historic approach gives a good motivation for a critical appreciation of the modern theory.

In section 2 we then present the basic framework of the modern no-arbitrage theory in the simple setting of finite probability spaces Ω.

The celebrated Black-Scholes model, based on geometric Brownian motion, is presented in section 3. It is compared to Bachelier's model, which is based on (arithmetic) Brownian motion.

The first three sections are kept on a relatively low level of technical sophistication. In section 4 we pass to a higher level of technicality and review the general theory of semi-martingale models of financial markets. We discuss in some detail the "fundamental theorem of asset pricing", which establishes the relation between the no-arbitrage theory on the one hand, and martingale theory on the other.

Finally, in section 5 we briefly discuss some of the applications of the fundamental theorem.

1 Introduction: Bachelier's Thesis from 1900

The fact that this course is given in the year 2000 at the école d'été in Saint Flour makes it particularly appealing to start this course with a review of the seminal thesis of Louis Bachelier: "Théorie de la Spéculation" [B 00]. This historical account will provide a good motivation for the general theory. We note, however, that readers only interested in a presentation of the theory in modern terms, can immediately pass to section 2.

Bachelier's thesis was defended in Paris on March 29, 1900, and H. Poincaré was a member of the thesis committee. He wrote a very positive and insightfull report on this thesis (this opinion as well as many other value judgements below only reflect my personal point of view). One may consult this report in Courtault et al. [CK 00], where one can also find a copy of the handwritten manuscript of Poincaré's report. We also refer to the interview of M. Taqqu with B. Bru [T 00] for an account on the personal life of Bachelier, who — in spite of his brilliant and original work, and the fame and support of his thesis adviser — remained an outsider to the French mathematical establishment during all of his life.

L. Bachelier was born in 1870 and became an orphan at the age of 19. In order to make a living, he had to work at the Bourse de Paris where he was exposed on a daily basis with the erratic movement of prices of financial securities.

In these days there was massive trade at the Bourse de Paris in the so-called "rentes", which were perpetual bonds paying an annual interest rate, typically 3 % (paid in 4 quarterly coupons of 75 centimes per 100 francs par value). The reason why these instruments had such importance in France goes back to the French revolution, when many wealthy aristocrats left the country and lost their property. When they returned after the restauration, they wanted their property back, but this turned out to be impossible after 25 years. The solution adopted by the government in order to recompensate them, was to issue "rentes", and to distribute an appropriate amount of them among the expropriated noble-men. While the quaterly coupons would provide them with an adequate income, the capital was never paid. These rentes were passed on in the families and they were also traded massively at the Bourse de Paris (for more information see [T 00]). Of course, they were not necessarily traded at par value but rather at changing prices similarly as in today's bond markets.

We spoke in some detail about the "rentes", because their special properties are important to understand the choice of Bachelier's model for the stock price process.

(i) There was a (very) liquid market, and price fluctuations happened "in continuous time", similarly as in the major stock and bond markets of today.

(ii) The price of a "rente" would typically not deviate too much from its par value, e.g. 100 francs; hence the absolute price changes (expressed in francs) and the relative price changes (expressed in percent) would roughly be the same.

(iii) The price fluctuations were relatively mild, if compared, e.g., with today's price fluctuations of stocks: one may deduce from the data provided by L. Bachelier that the standard deviation of the price change of a "rente" with par value of 100 francs over a year was about 2.4 francs (which roughly corresponds to a yearly volatility of 2.4 percent in the Black-Scholes model analyzed in section 3 below).

L. Bachelier was interested in designing a rational theory for the prices of term contracts. The two forms which were traded at the Bourse at that time also play a basic role today as forward contracts and options.

Definition 1. *A forward contract on an underlying security S consists of the* right and the obligation *to buy a fixed quantity (which we normalize to be one) of the underlying security, at a fixed price K and a fixed time T in the future.*

The underlying was in Bachelier's case typically a "rente", but it may just as well be any risky security such as a share, a foreign currency, a commodity etc.

Depending on the value of the "strike price" K, the present day value (i.e., at time $t = 0$) of such a forward contract could be positive or negative. The price $K = F$ at which a forward is contracted today at price zero is called the *forward price* of the underlying S (see, e.g., [H 99] or any introductory text on Mathematical Finance for more explanation).

We shall show now — in a similar way as Bachelier did in 1900 — that the forward price F is determinated by some very elementary *no-arbitrage arguments*. For the sake of clarity, we provide an example in a slightly different economic context than the one considered by Bachelier.

Example 1. Let $X = (X_t)_{0 \leq t \leq T}$ model the exchange rate of the US \$ vs. the €, i.e., the price of 1 US \$ in terms of €. The present day rate X_0 (the "spot price") can be looked up in the newspaper, hence this is just a positive number, say $X_0 = 1.1$. On the other hand, for $t > 0$, we do not know the exchange rate X_t. Later on we shall model X as a stochastic process, but presently it is not even necessary to speak about probability at all. X_t simply is some quantity which will be known at time t.

To compare cash-flows at different times we assume that there are "cash-accounts" B_t^d and B_t^f for € and US \$ respectively, which are given by

$$B_t^d = e^{r_d t}, \quad B_t^f = e^{r_f t}, \tag{1}$$

where d stands for "domestic", i.e. €, while f stands for "foreign", i.e. US \$. The idea behind the notion of "cash account" is that an investor has the possibility of investing in a "riskless" way, which means that the value of her investment in the "cash-account" will develop deterministically in a way which is known in advance. The reader should think of a bank account (in €

or US $ resp.) yielding an interest rate of r_d or r_f respectively. We in advance shall also assume that these investments may be either positive or negative ("long" or "short" in the financial lingo), in other words we may invest or borrow at the same conditions. Of course, this is an unrealistic assumption for small investors, but we should think of large investors (banks, investment funds, broker houses etc.) for which this assumption is a close approximation to reality.

Claim. Given the time horizon T, there is a unique forward price F which does not allow arbitrage opportunites, namely

$$F = X_0 e^{(r_d - r_f)T}. \tag{2}$$

We have not yet defined the notion of arbitrage — and we shall give a formal mathematical definition only much later. But the best way to grasp this — very primitive and economically convincing — concept is to consider the subsequent argument.

Consider two portfolios which can be established on the market today.

Portfolio A: Invest $e^{-r_f T}$ US $ into a US $ cash account. This investment will be worth one US $ at time T, and we can buy it today at a price of $e^{-r_f T} X_0 \in$.

Portfolio B: Invest $e^{-r_d T} F \in$ into a $\in$ cash account and buy one forward contract with maturity T and strike-price $K = F$. A moment's reflection reveals that this investment will also be worth one US $ at time T and that we can buy it today at a price of $e^{-r_d T} F \in$ (recall that F is defined in such a way that we can "buy" (i.e., enter into) a forward contract today at cost zero).

Hence the portfolios A and B are worth the same at time T (independent of how the exchange rate $(X_t)_{0 \le t \le T}$ develops!). We therefore claim that they also must have the same value today which results in equation (2).

Indeed, suppose for example that $F > X_0 e^{(r_d - r_f)T}$. In this case an "arbitrageur" would profit of the situation by buying portfolio A and selling portfolio B, thus obtaining the strictly positive difference $e^{-r_d T} F - e^{r_f T} X_0$, as a *riskless profit*: at time T the two positions will cancel out surely.

If the inequality is of the form $F < X_0 e^{(r_d - r_f)T}$ just reverse the roles of portfolio A and B. Also note that we have given our example in terms of the rather symbolic quantity of one US $. But of course there is no normalizing factor in front of the above argument and — if the market circumstances permit — you are free to multiply it with your favourite power of 10. Hence it is economically quite obvious that a market, where equation (2) is violated, cannot be in equilibrium as such an arbitrage opportunity would quickly be exploited by economic agents; a moment's reflection reveals that the market forces triggered by an arbitrageur behaving according to the above recipe will act towards making a possible violation of the identity (2) become smaller, and that people would continue to exploit such a violation of the "no-arbitrage principle" up to the point where (2) is satisfied to a sufficient degree as to make this arbitrage opportunity unattractive, even for a large investor.

The reader also should note that it is not necessary that all market participants behave rationally (and that they are aware of the identity (2)). It suffices that some of them (in theory even one would suffice!), who are ready and able to act with large sums on the market, are aware of (2) and eager to exploit arbitrage opportunities, whenever they come up.

Let us recapitulate the assumptions on the financial market which we have made above (more or less tacitly) in order for the no-arbitrage argument — and therefore the formula (2) — to be valid: we assumed that we can go long and short in the cash accounts (1) as well as in the forward contract at prices, which do not depend on the sign of the investment, without any transaction costs and with arbitrarily large quantities. As mentioned above, these assumptions are not fully satisfied in practice, but the economic situation of the "big players" in the market is quite close to these assumptions.

The attentive reader has noticed that we did not fully rely on the assumption (1) that there exist "riskless" cash accounts behaving according to (1), for all $0 \leq t \leq T$; all we needed was, that the relation holds true for $t = T$. In other words and using the financial lingo, we had to assume the existence of "riskless" (in practice this means that the government guarantees for the payment) "zero coupon bonds" maturing at time T, i.e., a contract, which pays $1 \in$ (or $1\,US\,\$$) at time T. Such contracts — or close approximations to them — are indeed traded in massive volume in financial markets.

At this point the reader is advised to convince herself — by consulting the financial section of a standard newspaper — that the above arguments are not merely theoretical but confirm very well to reality: the forward price of a currency depends on the difference of the interest rates in the corresponding currencies, pertaining to the maturity T, via (2) — and it only depends on this difference. Also observe that, in the case $r_d = r_f$, (2) reduces to $F = X_0$, i.e. the forward price then simply equals the spot price.

Let us turn back to L. Bachelier and the "rentes" traded at the Bourse de Paris. There was a liquid market in forward contracts on these "rentes" and Bachelier noticed the above relation between the spot price and the forward price. To link to our US $\$/\in$ example, the role of the accumulated interest of a "rente" plays a similar role as the interest rate r_f for the foreign currency, at least for periods $[0, T]$ which contain no coupon payment (in the case of coupon payments one has to make some rather straightforward adaptations). On the other hand, there was a complicated system of partial recompensation of the buyer of a forward contract with respect to this accumulated interest, called — "contangoes" (in french: "reports") — which — roughly speaking — plays a similar role as r_d above. The details are quite complicated, only of historical interest, and not relevant for our purposes. We shall therefore assume that the system of contangoes would fully recompensate the accumulated interest of the "rentes"; while this was not the case in reality, it was explicitly mentioned as a theoretical case by Bachelier. This corresponds to the case $r_d = r_f$ for the case of foreign exchange considered above, and implies — by similar no-arbitrage arguments — that the forward price (called the "true price" by Bachelier) coincides with the spot price.

Assumption 1 *We assume for the rest of this section that, for every maturity T, the spot price S_0 of the underlying security, and the forward price F with respect to T, coincide.*

We shall see later that this convenient assumption does not restrict the generality of the argument. What it does in practice: it dispenses us of making boring calculations of upcounting and discounting as reflected by the identity (2).

One final comment on whether Bachelier used the same no-arbitrage arguments as we did above: Bachelier does not argue by no-arbitrage but simply states that bond prices must "logically increase" by the accumulated interest which is tantamount to (2). He would simply appeal to common sense without explicitly mentioning the rather obvious no-arbitrage arguments. He saves this for more complicated securities where the argument becomes less obvious, as we shall presently see.

After this elementary treatment of forward contracts and forward prices in the first pages of his thesis, Bachelier passes to the case of options, which — in today's terminology — were European options.

Definition 2. *A European call (resp. put) option on an underlying security S consists in the right (but not the obligation) to buy (resp. to sell) a fixed quantity (which we normalize to be one) of the underlying security, at a fixed price K and a fixed time T in the future.*

In fact, there is a slight — but for the mathematical modelling rather crucial — difference between the way options are traded today and the way they were traded in Bachelier's days, at least in France. Nowadays the option premium C, i.e., the price, the buyer of an option has to pay, in order to enter into the contract, (the letter C standing for call option) is paid up front, i.e., at $t = 0$. In 1900 it was paid at the exercise time $t = T$ of the option. We denote the latter premium by $\widehat{C}$ to indicate that it corresponds to the upcounted premium C (from $t = 0$ to $t = T$) with respect to the risk free rate of interest (more precisely and in modern terminology: with respect to the zero coupon rate with maturity T). Fixing the letter K for the strike price of the option one arrives — after a moment's reflection — at the usual "hockey-stick" shape for the pay-off function of the option at time T. We draw the value of the option as a function of the price S_T of the underlying at time T:

This famous picture appears explicitly (with different letters for notation) in Bachelier's thesis. In fact, Bachelier compares the pay-off function of an option to the pay-off function of a forward contract with forward price F:

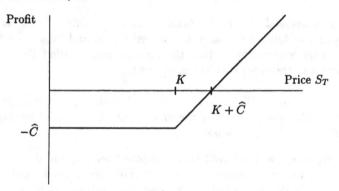

Fig. 1. Pay-off function of a call option at time T.

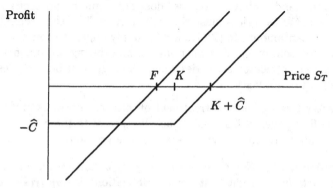

Fig. 2. Pay-off function of a call option and a forward contract at time T.

Today the quotation system in option markets usually fixes the strike price K while the premium C is variable and fluctuates (the reader might look up the financial section of any standard newspaper); in Bachelier's times it was done the other way round (at least for the "rentes"): the (upcounted) premium $\widehat{C}$ was fixed, typically $\widehat{C} = 50$, 25, or 10 centimes, and the strike price K would fluctuate according to demand and supply. In fact, the way people were quoting options was in terms of the "spread" $K + \widehat{C} - F$, which is very natural as we now shall see.

Bachelier gives the following numerical example: Suppose that the forward price F (for fixed horizon T which at these times was in the order of one or two months) for a "rente" equals 104 francs. He then continues: "*If we buy a forward contract on 3000 francs par value, we expose ourselves to a potential loss which may become considerable if a fall in the market occurs. To avoid this risk, we could buy an option at 50 centimes paying no longer 104 francs but 104.15 francs, for example.*" In our notation this amounts to $F = 104$,

$K = 103.65$, $\widehat{C} = 0.50$, $K + \widehat{C} = 104.15$, and the spread $K + \widehat{C} - F = 0.15$. The idea is that one agrees to pay $K + \widehat{C} - F$ (the "spread") more than in the case of a forward contract when exercising the contract at time T; en revanche, one has limited the maximal loss to $\widehat{C} = 0.50$.

He then remarks the "obvious fact" that the spread is a decreasing function of the premium $\widehat{C}$ and again he does not bother to give the — rather trivial — corresponding no-arbitrage argument (which we leave to the attentive reader). But then he also observes the concavity of this function by a less trivial combination of investments: this combination of options is known today under the name of "butterfly" in finance. We don't give the details here; the interested reader may look it up in Bachelier's thesis [B 00, p. 24] and compare it to the "butterfly" argument as explained, e.g., in [H 99]. Bachelier does not use the word arbitrage, which is today's terminology, but refers to *"operations in which one of the traders would profit regardless of eventual prices"*, which amounts to the same, and is in fact a very pretty description of the notion of arbitrage. Working at the bourse he was very aware of the no-arbitrage principle [B 00, p. 24]: *"We will see that such spreads are never found in practice"*.

After these preparations, L. Bachelier passes to the central topic, *Probabilities in Operations on the Exchange.* He had already addressed the basic difficulty of introducing probability in the context of the stock exchange in the introduction to the thesis in a very sceptical way: *"The calculus of probabilities, doubtless, could never be applied to fluctuations in security quotations, and the dynamics of the Exchange will never be an exact science."*

Nevertheless he now proceeds to model the price change of securities by a probability distribution distinguishing

"two kinds of probabilities:
(i) The probability which might be called "mathematical", which can be determined a priori and which is studied in games of chance.
(ii) The probability dependent on future events and, consequently impossible to predict in a mathematical manner.
This last is the probability that the speculator tries to predict."

My personal interpretation of this — somewhat confusing — definition is the following: sitting daily at the Bourse and watching the movement of prices, Bachelier got the same impression that we get today when observing price movements in financial markets, e.g., on the internet. The development of the charts of prices of stocks, indices etc. on the screen or blackboard resembles a "game of chance". On the other hand, the second kind of probability seems to refer to the expectations of a speculator who has a personal opinion on the development of prices. Bachelier continues: *"His (the speculator's) inductions are absolutely personal, since his counterpart in a transaction necessarily has the opposite opinion."*

Here he is led to the remarkable conclusion, which in today's terminology is called the "efficient market hypothesis":

"*It seems that the market, the aggregate of speculators, at a given instant can believe in neither a market rise nor a market fall since, for each quoted price, there are as many buyers as sellers.*"

He then makes clear that this principle should be understood in terms of "true prices", i.e., forward prices (compare the up- and discounting arguments as well as assumption 1 above). Finally he ends up with his famous dictum:

"*In sum, the consideration of true prices permits the statement* of this fundamental *principle:*
The mathematical expectation of the speculator is zero."

This is a truly fundamental principle and the reader's admiration for Bachelier's pathbreaking work will increase even more when continuing to the subsequent paragraph of Bachelier's thesis:

It is necessary to evaluate the generality of this principle carefully: It means that the market, at a given instant, considers not only currently negotiable transactions, but even those which will be based on a subsequent fluctuation in prices as having a zero expectation.
For example, I buy a bond with the intention of selling it when it will have appreciated by 50 centimes. The expectation of this complex transaction is zero exactly as if I intended to sell my bond on the liquidation date, or at any time whatever.

In my opinion, in these two paragraphs, the basic ideas underlying the concepts of martingales, stopping times, trading strategies, and Doob's stopping theorem already appear in a very intuitive way. It also sets the basic theme of the modern approach to option pricing which is based on the notion of a martingale.

In the remainder of this introductory review of Bachelier's thesis we shall discuss the implications of this **fundamental principle** and we shall address the following natural basic question:

Is the fundamental principle of L. Bachelier true?

There are, at least, two aspects to this question:

(i) Is it true, from a practical point of view, i.e., does it agree with data from financial markets?
(ii) Is it true, from a mathematical point of view, i.e., are there theorems that support his claim?

But let us first look at the implications of the *fundamental principle*: In order to draw conclusions from it, Bachelier had to determine the probability distribution of the random variable S_T (the price of the underlying security at expiration time T), or, more generally, on the entire stochastic process

$(S_t)_{0 \leq t \leq T}$. It is important to note that Bachelier had the approach of considering this object as a *process*, i.e., by thinking of the *pathwise behaviour* of the trajectories $(S_t(\omega))_{0 \leq t \leq T}$; this was very natural for him, as he was constantly exposed to observing the behaviour of the prices, as t "varies in continuous time".

To fix the process S, Bachelier fixes the maturity time T and chooses the coordinates such that the forward price F which — according to our assumption 1 — coincides with the current price S_0 of the underlying security, is at the origin. Then Bachelier assumes — more or less tacitly — that, for $0 \leq t \leq T$, the probability $p_{x,t}dx$, that the price S_t of the underlying security, starting at time $t_0 = 0$ from the point $x = 0$, lies at time t in the infinitesimal interval $(x, x + dx)$, is *symmetric around the origin and homogeneous in time t as well as in space x*.

Of course, Bachelier notices that this creates a problem, as it gives positive probabilities to negative values of the underlying security, which is absurd. But one should keep in mind the proportions mentioned above: a typical yearly standard deviation σ of the prices of the bonds considered by L. Bachelier was of the order of 2.4 %. Hence the region where the bond price after a year becomes negative is roughly 40 standard deviations away from the mean; anticipating that Bachelier uses the normal distribution this is — in his words — *"considered completely negligible"*, as the time horizons for the options were just fractions of a year. On the other hand, we should be warned when considering Bachelier's results asymptotically for $t \to \infty$ (or $\sigma \to \infty$ which roughly amounts to the same), as in these circumstances the effect of assigning positive probabilities to negative values of S_t is not *"completely negligible"* any more.

After these specifications, Bachelier argues that *"by the principle of joint probabilities"* (apparently he means the independence of the increments), we obtain

$$p_{z,t_1+t_2} = \int_{-\infty}^{+\infty} p_{x,t_1} p_{z-x,t_2} dx. \tag{3}$$

In other words, he obtains what we call today the Chapman-Kolmogoroff equation. Then he observes that *"this equation is confirmed by the function"*

$$p_{x,z} = \frac{1}{\sigma\sqrt{2\pi t}} \exp\left(-\frac{x^2}{2\sigma^2 t}\right), \tag{4}$$

concluding that *"evidently the probability is governed by the Gaussian law already famous in the calculus of probabilities"*.

Some remarks seem in order here: firstly, for the convenience of the reader who looks up Bachelier's original text, we mention that Bachelier did not use the quantity σ for the parametrisation but rather the quantity $H = \frac{\sigma}{\sqrt{2\pi}}$. Secondly, he obviously did not bother about the uniqueness of the solution to (3). Thirdly, he was well aware — and explicitly mentions — that he models the price movements in absolute terms and not in relative terms (w.r.t. the stock prices). As already mentioned, this distinction is not very important in the case of the "rentes", where the current price is typically close to the par value of 100 francs.

Summing up, Bachelier derived from some basic principles the transition law of Brownian Motion and it's relation to the Chapman-Kolmogoroff equation.

Bachelier then gives an *"Alternative Determination of the Law of Probability"*. He approximates the continuous time model $(S_t)_{t \geq 0}$ by a random walk, i.e., a process which during the time interval Δt moves up or down with probability $\frac{1}{2}$ by Δx. He clearly works out that Δx must be of the order $(\Delta t)^{\frac{1}{2}}$ and — using only Stirling's formula — he obtains the convergence of the one-dimensional marginal distributions of the random walk to those of Brownian motion.

Now a section follows, which is not directly needed for the subsequent applications in finance, but which — retrospectively — is of utmost mathematical importance: *"Radiation of probability"*. Consider the random walk model and suppose that the grid in space is given by $\ldots, x_{n-2}, x_{n-1}, x_n, x_{n+1}, x_{n+2}, \ldots$ having the same distance $\Delta x = x_n - x_{n-1}$, for all n, and such that at time t these points have probabilities $\ldots, p_{n-2}^t, p_{n-1}^t, p_n^t, p_{n+1}^t, p_{n+2}^t, \ldots$ under the random walk under consideration. What are the probabilities $\ldots, p_{n-2}^{t+\Delta t}, p_{n-1}^{t+\Delta t}, p_n^{t+\Delta t}, p_{n+1}^{t+\Delta t}, p_{n+2}^{t+\Delta t}, \ldots$ of these points at time $t + \Delta t$? A moment's reflection reveals the rule which is so nicely described by Bachelier in the subsequent phrases:

> *"Each price x during an element of time radiates towards its neighboring price an amount of probability proportional to the difference of their probabilities.*
>
> *I say proportional because it is necessary to account for the relation of Δx to Δt.*
>
> *The above law can, by analogy with certain physical theories, be called the law of radiation or diffusion of probability."*

Passing formally to the continuous limit and denoting by $P_{x,t}$ the distribution function associated to the density function (4)

$$P_{x,t} = \int_{-\infty}^{x} p_{z,t} dz \qquad (5)$$

Bachelier deduces in an intuitive and purely formal way the relation

$$\frac{\partial P}{\partial t} = \frac{1}{c^2} \frac{\partial p}{\partial x} = \frac{1}{c^2} \frac{\partial^2 P}{\partial x^2} \qquad (6)$$

where $c > 0$ is a constant. Of course, the heat equation was known to Bachelier: he notes that *"this is a Fourier equation"*.

Hence Bachelier in 1900 very explicitly discovered the fundamental relation between Brownian motion and the heat equation; this fact was rediscovered five years later by A. Einstein and resulted in a goldmine of mathematical investigation through the work of Kolmogoroff, Kakutani, Feynman, Kac, and many others up to recent research. It is worth noting that H. Poincaré in his report on Bachelier's thesis apparently saw the seminal importance of this idea when he wrote *"On peut regretter que M. Bachelier n'ait pas developpé d'avantage cette partie de sa thèse"* (One may regret that M. Bachelier did not develop further this part of his thesis).

With all these considerations L. Bachelier has fixed the model for the price changes of the underlying security — namely the normal distribution — up to the crucial parameter σ, which he calls the *"coefficient of instability or of nervousness of a security"* (strictly speaking he considers $\frac{\sigma}{\sqrt{2\pi}}$ rather than σ, which is just a matter of normalization). Fixing the parameter σ and applying the "fundamental principle" to the pay-off function in figure 2 one obtains — using the identity $F = S_0$ from assumption 1 — the equation

$$-\widehat{C} + \int_{K-S_0}^{\infty} (x - (K - S_0))\, f(x)dx = 0, \qquad (7)$$

where

$$f(x) = \frac{1}{\sigma\sqrt{2\pi T}} e^{-\frac{x^2}{2\sigma^2 T}}, \qquad (8)$$

which clearly determines the relation between the premium $\widehat{C}$ of the option and $K - S_0$ and therefore also the relation between $\widehat{C}$ and the "spread" $K + \widehat{C} - S_0$. In other words, equation (7) determines the price for the option and therefore solves the basic problem considered by Bachelier.

It is straightforward to derive from (7) an "option pricing formula" by calculating the integral in (7) (compare, e.g., [Sh 99]): denoting by $\varphi(x)$ the standard normal density function, i.e., $\varphi(x)$ equals (8) for $\sigma^2 T = 1$, by $\Phi(x)$ the corresponding distribution function, and using the relation $\varphi'(x) = -x\varphi(x)$, an elementary calculation reveals that

$$\widehat{C} = \int_{\frac{K-F}{\sigma\sqrt{T}}}^{\infty} \left(x\sigma\sqrt{T} - (K - F)\right) \varphi(x)dx \qquad (9)$$

$$= (F - K)\Phi\left(\frac{F - K}{\sigma\sqrt{T}}\right) + \sigma\sqrt{T}\varphi\left(\frac{F - K}{\sigma\sqrt{T}}\right).$$

Interestingly, Bachelier does not bother to write up this easy formula which gives $\widehat{C}$ as a function of K (the way which is useful in determining option prices today). As mentioned above, he was rather interested in expressing inversely the "spread" $K + \widehat{C} - F$ as a function of $\widehat{C}$, and apparently there is no explict way of writing down this relation.

Instead, he does something much more interesting: he first specializes to the case of *simple options* (this is terminology from 1900), when $K = F$, which at his time were the usual options on commodities; in modern terms they correspond to so called at-the-money options where the strike price K equals the forward price F (which in our setting is equal to the spot price S_0 by assumption 1). In this case the solution to (7) obviously results in

$$\widehat{C} = \frac{\sigma}{\sqrt{2\pi}}\sqrt{T}, \tag{10}$$

which is a remarkably simple formula. Bachelier also uses this formula to turn the point of view upside down, or — in modern terms — to determine the "implied volatility", thus discovering yet one more basic idea of modern mathematical finance: if we can observe the (upcounted) premium $\widehat{C}$ of an at-the-money option on the market, formula (10) determines very directly the "nervousness" parameter $\frac{\sigma}{\sqrt{2\pi}}$ and therefore specifies the probability distribution $p_{x,t}$.

Still, formula (10) depends on the parameter σ and Bachelier — following the reflexes of a true mathematician — wanted to find quantities invariant under variations of the parameter σ and the expiration date T: For example, he determines the probability that the buyer of an at-the-money option (i.e., $K = F$) makes a profit. Glancing at figure 2 this probability p equals

$$p = \int_{\widehat{C}}^{\infty} f(x)dx = \int_{\frac{\sigma\sqrt{T}}{\sqrt{2\pi}}}^{\infty} f(x)dx, \tag{11}$$

where $f(x)$ is given by (8). Calculating this expression, the term $\sigma\sqrt{T}$ cancels out, and we obtain

$$p = \int_{\frac{1}{\sqrt{2\pi}}}^{\infty} \frac{1}{\sqrt{2\pi}}e^{-\frac{x^2}{2}}dx = 1 - \Phi\left(\frac{1}{\sqrt{2\pi}}\right) \approx 0.345. \tag{12}$$

In other words, according to Bachelier's model, the buyer of an at-the-money option makes a profit in 34,5 % of the cases, and a loss in 65,5 % of the cases. Isn't it a remarkable and surprising result that this number does not depend on any parameter? Bachelier also derives explicit numbers (not depending on any parameter) for the probability of success in a number of similar situations.

Then he treats the case of options where the strike price K is not necessarily equal to the forward price F, i.e., options which are not necessarily at the money. He uses a quadratic approximation of the behaviour of the relation between K and $\widehat{C}$ determined by (7) in a neighbourhood of $K = F$ which again yields very explicit and practical formulae, displaying a good fit for the values appearing in practice, i.e.,when $|K - F|$ is small as compared to F.

After all these derivations Bachelier compares his theoretical results with the financial data observed for the "rentes" in the period of 1894-1898.

He just considers averages over these five years and in particular the "nervousness parameter" $\frac{\sigma}{\sqrt{2\pi}}$ is an average estimate, while it becomes clear from the remarks of Bachelier, that the "nervousness" $\frac{\sigma}{\sqrt{2\pi}}$ of the market was varying in time (just as it does today).

He estimates the yearly standard deviation of the price movement of a rente to be approximately equal to 240 centimes, which corresponds to the above mentioned 2.4 % of the par value of 100 francs.

Then he compares the quantities derived from his model (using this parameter) to the empirical financial data (taking again averages over these five years).

This comparison of calculated figures with observed data does not live up to the standards of a modern statistical analysis; also the match is not overwhelmingly good — the difference sometimes being in the range of 10 or 20 percent — but it shows that the qualitative features are well captured.

To sum up the issue of the match of his theory with empirical financial data Bachelier makes the remarkable comment:

> "If, with respect to several questions treated in this study, I have compared the results of observation with those of theory, it was not to verify formulae established by mathematical methods, but only to show that the market, unwittingly, obeys a law which governs it, the law of probability."

It is interesting to have a look into Poincaré's report on Bachelier's thesis where he gives an argument in favor of Bachelier's *fundamental principle* (which, of course, is the basis of the above methodology) relying on the law of large numbers; Poincaré also clearly stresses the relative weakness of this argument (the reader should compare the argument below involving the law of large numbers to the much more convincing no-arbitrage arguments encountered above):

> "One should not expect a very exact verification. The principle of the mathematical expectation holds in the sense that, if it were violated, there would always be people who would act so as to re-establish it and they would eventually notice this. But they would only notice it, if the deviations were considerable. The verification, then, can only be gross. The author of the thesis gives statistics where this happens in a very satisfactory manner."

In the final part of his thesis L. Bachelier makes another seminal discovery: the law of the maximum of a Brownian path. Here we again see clearly that Bachelier had a *pathwise* approach to stochastic processes. The fact that the density function of the maximum of the Brownian path equals twice the density of the corresponding Gaussian density function on the positive axis (while it is of course zero on the negative axis) is today the standard example for the use of the "reflection principle", which reduces this fact almost to a triviality.

Interestingly, Bachelier does not derive it in this way, but rather by approximation with a discrete random walk and using a combinatorial result obtained in 1888 by D. André, called the solution to the ballot problem ("problème du scrutin").

Using this theorem and passing in an appropriate way to the limit, Bachelier obtains the result on the distribution of the maximum of Brownian motion. It is only then that he uses the reflection principle to *interpret* this result:

> "The interpretation of this result is very simple: The price cannot be exceeded at the moment t without having been attained previously. The probability $\mathcal{P}$ is therefore equal to the probability P multiplied by the probability that, given that the price was quoted previously, it will be exceeded at the moment t, i.e., multiplied by $\frac{1}{2}$. Thus
>
> $$\mathcal{P} = \tfrac{P}{2} \ . "$$ (13)

To explain the notation: letting $c > 0$ denote "the price" referred to above, and $(W_t)_{t \geq 0}$ Brownian motion, the letter $\mathcal{P}$ denotes the probability $\mathbb{P}\{W_t \geq c\}$ while P denotes $\mathbb{P}\{\sup_{0 \leq u \leq t} W_u \geq c\}$, so that (13) describes the well known law of the maximum of a Brownian path.

Bachelier was led to consider this problem by a very interesting idea from the financial point of view, which may be considered as a precursor of the idea of dynamical hedging, which in turn is the central idea of modern mathematical finance.

Consider again the buyer of an at-the-money option with $K = F = S_0$ and a premium $\widehat{C}$. We have seen above that the probability of success of the buyer of such an option is

$$\mathbb{P}[S_T \geq S_0 + \widehat{C}] = 1 - \Phi\left(\frac{1}{\sqrt{2\pi}}\right) \approx 0.345.$$ (14)

Now suppose that the buyer of this option follows the subsequent strategy: at the first moment when S_t reaches the level $S_0 + \widehat{C}$ (if this happens before T), she "locks in" her profit by going short (i.e., selling) one unit of the underlying security. Of course, the "first moment when ..." is a stopping time τ in modern terminology. A moment's reflection reveals that in the case $\tau \leq T$ the speculator cannot end up with a negative result and will have a strictly positive gain, when, in addition to $\tau \leq T$, the price S_T at expiration time happens to be less than K. But, of course, this operation of "locking in" the profit only happens if S_t attains the level $S_0 + \widehat{C}$ for some $0 \leq t \leq T$, while in the other case the speculator will end up with a loss.

What is the probability of success (i.e., a non-negative result) of a speculator pursuing this strategy? Clearly it equals

$$\mathbb{P}\left[\max_{0 \leq t \leq T} S_t \geq S_0 + \widehat{C}\right]$$ (15)

for which we obtain, using the law of the maximum of Brownian motion,

$$\mathbb{P}\left[\max_{0 \le t \le T} S_t \ge S_0 + \widehat{C}\right] = 2\left(1 - \varPhi\left(\frac{1}{\sqrt{2\pi}}\right)\right) \approx 0.69. \qquad (16)$$

In other words, the probability of a non-negative result of this strategy is about 69 %. Again we find it remarkable that this result does not depend on any parameter.

Let us try to give a résumé of this review of Bachelier's remarkable thesis and to compare it with the modern theory, in particular with the Black-Scholes model considered below.

The usual argument against Bachelier and in favor of Black-Scholes is the fact that Bachelier's model of Brownian motion assigns positive probability to negative prices of the underlying stock, while the Black-Scholes model (using geometric Brownian motion) does not. (For the remainder of this section we assume that the reader is already sufficiently familiar with the basic features of the Black-Scholes model as discussed in section 3 below.)

In my opinion this argument is to a large extent a myth: in basic applications of statistics (say, quality control) there are good reasons to model the quantities under consideration (say, the length of a screw) by a normal distribution. Apparently nobody worries that this model also assigns positive probability to a negative length of the screw, although this is at least as absurd as a negative stock price. The reason is, that — if expressed numerically — these probabilities are *"completely negligible"*, as was so nicely phrased by Bachelier.

One might compare the relation of modelling price processes with Brownian motion as opposed to geometric Brownian motion, to that of using linear interest as opposed to continuously compounded interest for cash accounts. Of course, the latter one is logically more appealing, but we all know, that the difference between these two procedures is very minor for short periods (say, less than a year, in the case of reasonable values of the interest rate). On the other hand, in the long run the difference is spectacular.

Similarly, the differences between the Bachelier and the Black-Scholes option pricing formulae are very minor, as long as σ and T remain in a reasonable range, which certainly was the case for the options on the rentes considered by Bachelier (compare the more quantitative discussion at the end of section 3). On the other hand it is worth noting that, for $T \to \infty$, Bachelier's formula $\widehat{C} = \frac{\sigma}{\sqrt{2\pi}}\sqrt{T}$ (see (10) above) for the option price assigns arbitrarily large values to the premium of an option, while an obvious no-arbitrage argument (using assumption 1 and the non-negativity of the underlying security) reveals that $\widehat{C}$ is certainly less than S_0.

In my opinion, L. Bachelier has obtained an option pricing model which, for practical purposes, is just as satisfactory as the model obtained by Black and Scholes some 70 years later, with the shortcomings of these models being very similar (e.g., underestimation of extreme movements of the underlying by using normal or lognormal distributions). But there is one crucial idea which

L. Bachelier has missed and which is of central importance for the modern theory: the concept of *dynamic hedging* which allows to reduce Bachelier's "fundamental principle" to no-arbitrage considerations. The use of this idea to determine option prices is due to R. Merton (in a footnote of the original Black-Scholes paper [BS 73] this is explicitly acknowledged) and plays a truly fundamental role.

On the other hand, we have seen above that L. Bachelier was already close to this idea when considering trading strategies where the selling of a security would happen at a *stopping time*. But for a full-fledged theory of dynamic hedging, Bachelier would have had to make quite a number of additional pioneering steps in his lonely endeavour of investigating Brownian motion. His situation was in sharp contrast to the situation encountered by the researchers in Mathematical Finance in the last third of the 20th century, who could build on a well-established theory of stochastic integration, as notably developed by K. Itô and by the school of P.A. Meyer in Strasbourg.

In any case, let us stop here with the review of Bachelier's seminal achievements and turn to a systematic development of the modern theory of option pricing, which is based on the notion of *no-arbitrage*.

2 Models of Financial Markets on Finite Probability Spaces

In order to reduce the technical difficulties of the theory of option pricing to a minimum, we assume throughout this section that the probability space Ω underlying our model will be finite, say, $\Omega = \{\omega_1, \omega_2, \ldots, \omega_N\}$. This assumption implies that all functional-analytic delicacies pertaining to different topologies on $L^\infty(\Omega, \mathcal{F}, \mathbb{P})$, $L^1(\Omega, \mathcal{F}, \mathbb{P})$ and $L^0(\Omega, \mathcal{F}, \mathbb{P})$ evaporate, as all these spaces are simply $\mathbb{R}^N$ (we assume w.l.o.g. that the sigma-algebra $\mathcal{F}$ is the power set of Ω and that $\mathbb{P}$ assigns strictly positive value to each $\omega \in \Omega$). Hence all the functional analysis, which we shall need in section 4 for the case of more general processes, reduces to simple linear algebra in the setting of the present chapter.

Nevertheless we shall write $L^\infty(\Omega, \mathcal{F}, \mathbb{P})$, $L^1(\Omega, \mathcal{F}, \mathbb{P})$ etc. below (knowing very well that these spaces are isomorphic in the present setting) to indicate, what we shall encounter in the setting of the general theory.

Definition 3. *A model of a financial market is an $\mathbb{R}^{d+1}$-valued stochastic process $S = (S)_{t=0}^T = (S_t^0, S_t^1, \ldots, S_t^d)_{t=0}^T$, based on and adapted to the filtered stochastic base $(\Omega, \mathcal{F}, (\mathcal{F})_{t=0}^T, \mathbb{P})$. We shall assume that the zero coordinate S^0, which we call the cash account, satisfies $S_t^0 \equiv 1$, for $t = 0, 1, \ldots, T$. The letter ΔS_t denotes the increment $S_t - S_{t-1}$.*

Definition 4. *$\mathcal{H}$ denotes the set of trading strategies for the financial market S. An element $H \in \mathcal{H}$ is an $\mathbb{R}^d$-valued process $(H_t)_{t=1}^T = (H_t^1, H_t^2, \ldots, H_t^d)_{t=1}^T$ which is predictable, i.e. each H_t is $\mathcal{F}_{t-1}$-measurable.*

We then define the stochastic integral $(H \cdot S)$ as the $\mathbb{R}$-valued process $((H \cdot S)_t)_{t=0}^T$ given by

$$(H \cdot S)_t = \sum_{j=1}^t (H_j, \Delta S_j), \quad t = 0, \ldots, T, \tag{17}$$

where $(.,.)$ denotes the inner product in $\mathbb{R}^d$.

The reader might be puzzeled why we chose S to be $\mathbb{R}^{d+1}$-valued, while we chose H to be $\mathbb{R}^d$-valued. The reason is that we defined the zero coordinate S^0 of S to be identically equal to 1 so that $\Delta S_t^0 \equiv 0$ and this coordinate can not contribute to the stochastic integral (17). We note that this assumption does not restrict the generality of the model as we always may choose the cash account as numéraire. This means, that the stock prices are expressed in units of the cash account, or — in more practical terms — we have expressed stock prices in discounted terms.

On the other hand we want to stress for later use (the change of numéraire theorem 3 below) the role of the cash account — which we choose as numéraire — in the definition of a financial market, although the coordinate S^0 presently only serves as a dummy.

Definition 5. *We call the subspace K of $L^0(\Omega, \mathcal{F}, \mathbb{P})$ defined by*

$$K = \{(H \cdot S)_T : H \in \mathcal{H}\} \tag{18}$$

the set of contingent claims attainable at price 0.

The economic interpretation is the following: the random variables $f = (H \cdot S)_T$, for some $H \in \mathcal{H}$, are precisely those contingent claims, i.e., the pay-off functions at time T depending on $\omega \in \Omega$, that an economic agent may replicate with zero initial investment, by pursuing some predictable trading strategy H.

For $a \in \mathbb{R}$, we call the *set of contingent claims attainable at price a* the affine space K_a obtained by shifting K by the constant function $a\mathbf{1}$, in other words the random variables of the form $a + (H \cdot S)_T$, for some trading strategy H. Again the economic interpretation is that these are precisely the contingent claims that an economic agent may replicate with an initial investment of a by pursuing some predictable trading strategy H.

Definition 6. *We call the convex cone C in $L^\infty(\Omega, \mathcal{F}, \mathbb{P})$ defined by*

$$C = \{g \in L^\infty(\Omega, \mathcal{F}, \mathbb{P}) \text{ s.t. there is } f \in K, f \geq g\} \tag{19}$$

the set of contingent claims super-replicable at price 0.

Economically speaking, a contingent claim $g \in L^\infty(\Omega, \mathcal{F}, \mathbb{P})$ is *super-replicable at price 0*, if we can achieve it with zero net investment, subsequently pursuing some predictable trading strategy H — thus arriving at some contingent claim f — and then, possibly, "throwing away money" to arrive at g. This operation of "throwing away money" may seem awkward at this stage, but we shall see later that the set C plays an important role in the development of the theory. Observe that C is a convex cone containing the negative orthant $L^\infty_-(\Omega, \mathcal{F}, \mathbb{P})$. Again we may define C_a as the *contingent claims super-replicable at price a* if we shift C by the constant function $a\mathbf{1}$.

Definition 7. *A financial market S satifies the* no-arbitrage condition *(NA) if*

$$K \cap L^0_+(\Omega, \mathcal{F}, \mathbb{P}) = \{0\} \tag{20}$$

or, equivalently,

$$C \cap L^\infty_+(\Omega, \mathcal{F}, \mathbb{P}) = \{0\}, \tag{21}$$

where 0 denotes the function identically equal to zero.

In other words we now have formalized the concept of an arbitrage possibility: it consists of the existence of a trading strategy H such that — starting from an initial investment zero — the resulting contingent claim $f = (H \cdot S)_T$ is non-negative and not identically equal to zero. If a financial market does not allow for arbitrage we say it satisfies the *no-arbitrage condition (NA)*.

Definition 8. *A probability measure Q on $(\Omega, \mathcal{F})$ is called an* equivalent martingale measure *for S, if $Q \sim \mathbb{P}$ and S is a martingale under Q.*

We denote by $\mathcal{M}^e(S)$ the set of equivalent martingale probability measures and by $\mathcal{M}^a(S)$ the set of all (not necessarily equivalent) martingale probability measures. The letter a stands for "absolutely continuous with respect to $\mathbb{P}$" which in the present setting (finite Ω and $\mathbb{P}$ having full support) automatically holds true, but which will be of relevance for general probability spaces $(\Omega, \mathcal{F}, \mathbb{P})$ later. We shall often identify a measure Q on $(\Omega, \mathcal{F})$ with its Radon-Nikodym derivative $\frac{dQ}{d\mathbb{P}} \in L^1(\Omega, \mathcal{F}, \mathbb{P})$.

Lemma 1. *For a probability measure Q on $(\Omega, \mathcal{F})$ the following are equivalent:*

(i) $Q \in \mathcal{M}^a(S)$,
(ii) $\mathbf{E}_Q[f] = 0$, for all $f \in K$,
(iii) $\mathbf{E}_Q[g] \leq 0$, for all $g \in C$.

Proof The equivalences are rather trivial, as (ii) is tantamount to the very definition of S being a martingale under Q, and the equivalence of (ii) and (iii) is straightforward. $\square$

After having fixed these formalities we may formulate and prove the central result of the theory of pricing and hedging by no-arbitrage, sometimes called the "fundamental theorem of asset pricing", which in its present form (i.e., finite Ω) is due to Harrison and Pliska [HP 81].

Theorem 1 (Fundamental Theorem of Asset Pricing). *For a financial market S modeled on a finite stochastic base $(\Omega, \mathcal{F}, (\mathcal{F}_t)_{t=0}^T, \mathbb{P})$ the following are equivalent:*

(i) S satisfies (NA).
(ii) $\mathcal{M}^e(S) \neq \emptyset$.

Proof (ii) $\Rightarrow$ (i): This is the obvious implication. If there is some $Q \in \mathcal{M}^e(S)$ then by lemma 1 we have that

$$\mathbf{E}_Q[g] \leq 0, \quad \text{for } g \in C. \tag{22}$$

On the other hand, if there were $g \in C \cap L_+^\infty$, $g \neq 0$, then, using the assumption that Q is equivalent to $\mathbb{P}$, we would have

$$\mathbf{E}_Q[g] > 0, \tag{23}$$

a contradiction.

(i) $\Rightarrow$ (ii) This implication is the important message of the theorem which will allow us to link the no-arbitrage arguments with martingale theory. We give a functional analytic existence proof, which will be generalizable — in spirit — to more general situations.

By assumption the space K intersects L_+^∞ only at 0. We want to separate the disjoint convex sets $L_+^\infty \backslash \{0\}$ and K by a hyperplane induced by a linear functional $Q \in L^1(\Omega, \mathcal{F}, \mathbb{P})$. Unfortunately this is a situation, where the usual versions of the separation theorem (i.e., the Hahn-Banach Theorem) do not apply (even in finite dimensions!).

One way to overcome this difficulty (in finite dimension) is to consider the convex hull of the unit vectors $(1_{\{\omega_n\}})_{n=1}^N$ in $L^\infty(\Omega, \mathcal{F}, \mathbb{P})$ i.e.

$$P := \left\{ \sum_{n=1}^N \mu_n 1_{\{\omega_n\}} : \mu_n \geq 0, \sum_{n=1}^N \mu_n = 1 \right\}. \tag{24}$$

This is a convex, compact subset of $L_+^\infty(\Omega, \mathcal{F}, \mathbb{P})$ and, by the *(NA)* assumption, disjoint from K. Hence we may strictly separate the sets P and K by a linear functional $Q \in L^\infty(\Omega, \mathcal{F}, \mathbb{P})^* = L^1(\Omega, \mathcal{F}, \mathbb{P})$, i.e., find $\alpha < \beta$ such that

$$\begin{aligned} \langle Q, f \rangle &\leq \alpha \quad \text{for} \quad f \in K, \\ \langle Q, h \rangle &\geq \beta \quad \text{for} \quad h \in P. \end{aligned} \tag{25}$$

As K is a linear space, we have $\alpha \geq 0$ and may, in fact, replace α by 0. Hence $\beta > 0$. Therefore $\langle Q, 1 \rangle > 0$, and we may normalize Q such that $\langle Q, 1 \rangle = 1$. As Q is strictly positive on each $1_{\{\omega_n\}}$, we therefore have found a probability measure Q on $(\Omega, \mathcal{F})$ equivalent to $\mathbb{P}$ such that condition (ii) of lemma 1 holds true. In other words, we found an equivalent martingale measure Q for the process S.   □

Corollary 1. *Let S satisfy (NA) and $f \in L^\infty(\Omega, \mathcal{F}, \mathbb{P})$ be an attainable contingent claim so that*

$$f = a + (H \cdot S)_T, \tag{26}$$

for some $a \in \mathbb{R}$ and some trading strategy H.

Then the constant a and the process $(H \cdot S)$ are uniquely determined by (26) and satisfy, for every $Q \in \mathcal{M}^e(S)$,

$$a = \mathbf{E}_Q[f], \quad and \quad a + (H \cdot S)_t = \mathbf{E}_Q[f|\mathcal{F}_t] \text{ for } 0 \leq t \leq T. \tag{27}$$

Proof As regards the uniqueness of the constant $a \in \mathbb{R}$, suppose that there are two representations $f = a^1 + (H^1 \cdot S)_T$ and $f = a^2 + (H^2 \cdot S)_T$ with $a^1 \neq a^2$. Assuming w.l.o.g. that $a^1 > a^2$ we find an obvious arbitrage possibility: we have $a^1 - a^2 = ((H^1 - H^2) \cdot S)_T$, i.e. the trading strategy $H^1 - H^2$ produces a strictly positive result at time T, a contradiction to *(NA)*.

As regards the uniqueness or the process $H \cdot S$ we simply apply a conditional version of the previous argument: assume that $f = a + (H^1 \cdot S)_T$ and $f =$

$a + (H^2 \cdot S)_T$ such that the processes $H^1 \cdot S$ and $H^2 \cdot S$ are not identical. Then there is $0 < t < T$ such that $(H^1 \cdot S)_t \neq (H^2 \cdot S)_t$; w.l.o.g. $A := \{(H^1 \cdot S)_t > (H^2 \cdot S)_t\}$ is a non-empty event, which clearly is in $\mathcal{F}_t$. Hence, using the fact that $(H^1 \cdot S)_T = (H^2 \cdot S)_T$, the trading strategy $H := (H^2 - H^1)\chi_A \cdot \chi_{]t,T]}$ is a predictable process producing an arbitrage, as $(H \cdot S)_T = 0$ outside A, while $(H \cdot S)_T = (H^1 \cdot S)_t - (H^2 \cdot S)_t > 0$ on A, which again contradicts *(NA)*.

Finally, the equations in (27) result from the fact that, for every predictable process H and every $Q \in \mathcal{M}^a(S)$, the process $H \cdot S$ is a Q-martingale. $\square$

Denote by $\text{cone}(\mathcal{M}^e(S))$ and $\text{cone}(\mathcal{M}^a(S))$ the cones generated by the convex sets $\mathcal{M}^e(S)$ and $\mathcal{M}^a(S)$, respectively. The subsequent result clarifies the polar relation between these cones and the cone C. Recall (see, e.g., [Sch 66]) that, for a pair (E, E') of vector spaces in separating duality via the scalar product $\langle ., . \rangle$, the polar C^0 of a set C in E is defined as

$$C^0 = \{g \in E' : \langle f, g \rangle \leq 1, \text{ for all } f \in C\}. \tag{28}$$

In the case when C is closed under multiplication with positive scalars (e.g., if C is a convex cone) the polar C^0 may equivalently be defined by

$$C^0 = \{g \in E' : \langle f, g \rangle \leq 0, \text{ for all } f \in C\}. \tag{29}$$

The *bipolar theorem* (see, e.g., [Sch 66]) states that the bipolar $C^{00} := (C^0)^0$ of a set C in E is the $\sigma(E, E')$-closed convex hull of C.

After these general considerations we pass to the concrete setting of the cone $C \subseteq L^\infty(\Omega, \mathcal{F}, \mathbb{P})$ of contingent claims super-replicable at price 0. Note that in our finite-dimensional setting this convex cone is closed as it is the algebraic sum of the closed linear space K (a linear space in $\mathbb{R}^N$ is always closed) and the closed polyhedral cone $L^\infty_-(\Omega, \mathcal{F}, \mathbb{P})$ (the verification, that the algebraic sum of a space and a polyhedral cone in $\mathbb{R}^N$ is closed, is an easy, but not completely trivial exercise). Hence we deduce from the bipolar theorem, that C equals its bipolar C^{00}.

Proposition 1. *Suppose that S satisfies (NA). Then the polar of C is equal to $\text{cone}(\mathcal{M}^a(S))$ and $\mathcal{M}^e(S)$ is dense in $\mathcal{M}^a(S)$. Hence the following assertions are equivalent for an element $g \in L^\infty(\Omega, \mathcal{F}, \mathbb{P})$*

(i) $g \in C$,
(ii) $\mathbf{E}_Q[g] \leq 0$, for all $Q \in \mathcal{M}^a(S)$,
(iii) $\mathbf{E}_Q[g] \leq 0$, for all $Q \in \mathcal{M}^e(S)$,

Proof The fact that the polar C^0 and $\text{cone}(\mathcal{M}^a(S))$ coincide, follows from lemma 1 and the observation that $C \supseteq L^\infty_-(\Omega, \mathcal{F}, \mathbb{P})$ implies $C^0 \subseteq L^\infty_+(\Omega, \mathcal{F}, \mathbb{P})$. Hence the equivalence of (i) and (ii) follows from the bipolar theorem.

As regards the density of $\mathcal{M}^e(S)$ in $\mathcal{M}^a(S)$ we first deduce from theorem 1 that there is at least one $Q^* \in \mathcal{M}^e(S)$. For any $Q \in \mathcal{M}^a(S)$ and $0 < \mu \leq 1$

we have that $\mu Q^* + (1 - \mu)Q \in \mathcal{M}^e(S)$, which clearly implies the density of $\mathcal{M}^e(S)$ in $\mathcal{M}^a(S)$. The equivalence of (ii) and (iii) now is obvious.   □

The subsequent theorem tells us precisely what the principle of no-arbitrage can tell us about the possible prices for a contingent claim f. It goes back to the work of D. Kreps [K 81] and was subsequently extended by several authors.

For given $f \in L^\infty(\Omega, \mathcal{F}, \mathbb{P})$, we call $a \in \mathbb{R}$ an *arbitrage-free price*, if in addition to the financial market S, the introduction of the contingent claim f at price a does not create an arbitrage possibility. Mathematically speaking, this can be formalized as follows. Let $C^{f,a}$ denote the cone spanned by C and the linear space spanned by $f - a$; then a is an arbitrage-free price for f if $C^{f,a} \cap L^\infty_+(\Omega, \mathcal{F}, \mathbb{P}) = \{0\}$.

Theorem 2 (Pricing by No-Arbitrage). *Assume that S satisfies (NA) and let $f \in L^\infty(\Omega, \mathcal{F}, \mathbb{P})$. Define*

$$\overline{\pi}(f) = \sup\{\mathbf{E}_Q[f] : Q \in \mathcal{M}^e(S)\}, \tag{30}$$

$$\underline{\pi}(f) = \inf\{\mathbf{E}_Q[f] : Q \in \mathcal{M}^e(S)\}, \tag{31}$$

Either $\underline{\pi}(f) = \overline{\pi}(f)$, in which case f is attainable at price $\pi(f) := \underline{\pi}(f) = \overline{\pi}(f)$, i.e. $f = \pi(f) + (H \cdot S)_T$ for some $H \in \mathcal{H}$; therefore $\pi(f)$ is the unique arbitrage-free price for f.

Or $\underline{\pi}(f) < \overline{\pi}(f)$, in which case $\{\mathbf{E}_Q[f] : Q \in \mathcal{M}^e(S)\}$ equals the open interval $]\underline{\pi}(f), \overline{\pi}(f)[$, which in turn equals the set of arbitrage-free prices for the contingent claim f.

Proof First observe that the set $\{\mathbf{E}_Q[f] : Q \in \mathcal{M}^e(S)\}$ forms a bounded non-empty interval in $\mathbb{R}$, which we denote by I.

We claim that a number a is in I, iff a is an arbitrage-free price for f. Indeed, supposing that $a \in I$ we may find $Q \in \mathcal{M}^e(S)$ s.t. $\mathbf{E}_Q[f - a] = 0$ and therefore $C^{f,a} \cap L^\infty_+(\Omega, \mathcal{F}, \mathbb{P}) = \{0\}$.

Conversely suppose that $C^{f,a} \cap L^\infty_+(\Omega, \mathcal{F}, \mathbb{P}) = \{0\}$. Note that $C^{f,a}$ is a closed convex cone (it is the algebraic sum of the linear space $\operatorname{span}(K, f - a)$ and the closed, polyhedral cone $L^\infty_-(\Omega, \mathcal{F}, \mathbb{P})$). Hence by the same argument as in the proof of theorem 1 there exists a probability measure $Q \sim \mathbb{P}$ such that $Q|_{C^{f,a}} \leq 0$. This implies that $\mathbf{E}_Q[f - a] = 0$, i.e., $a \in I$.

Now we deal with the boundary case: suppose that a equals the right boundary of I, i.e., $a = \overline{\pi}(f) \in I$, and consider the contingent claim $f - \overline{\pi}(f)$; by definition we have $\mathbf{E}_Q[f - \overline{\pi}(f)] \leq 0$, for all $Q \in \mathcal{M}^e(S)$, and therefore by proposition 1, that $f - \overline{\pi}(f) \in C$. We may find $g \in K$ such that $g \geq f - \overline{\pi}(f)$. If the sup in (30) is attained, i.e., if there is $Q^* \in \mathcal{M}^e(S)$ such that $\mathbf{E}_{Q^*}[f] = \overline{\pi}(f)$, then we have $0 = \mathbf{E}_{Q^*}[g] \geq \mathbf{E}_{Q^*}[f - \overline{\pi}(f)] = 0$ which in view of $Q^* \sim \mathbb{P}$ implies that $f - \overline{\pi}(f) \equiv g$; in other words f is attainable at price $\overline{\pi}(f)$. This in turn implies that $\mathbf{E}_Q[f] = \overline{\pi}(f)$, for all $Q \in \mathcal{M}^e(S)$, and therefore I is reduced to the singleton $\{\overline{\pi}(f)\}$.

Hence, if $\underline{\pi}(f) < \overline{\pi}(f)$, $\overline{\pi}(f)$ cannot belong to the interval I, which is therefore open on the right hand side. Passing from f to $-f$, we obtain the analogous result for the left hand side of I, which therefore equals $I =]\underline{\pi}(f), \overline{\pi}(f)[$. □

Corollary 2 (complete financial markets). *For a financial market S satisfying the no-arbitrage condition (NA) the following are equivalent:*

(i) $\mathcal{M}^e(S)$ consists of a single element Q.
(ii) Each $f \in L^\infty(\Omega, \mathcal{F}, \mathbb{P})$ may be represented as

$$f = a + (H \cdot S)_T, \quad \text{for some } a \in \mathbb{R}, \text{ and } H \in \mathcal{H}. \tag{32}$$

In this case $a = \mathbf{E}_Q[f]$, the stochastic integral $(H \cdot S)$ is unique, and we have that

$$\mathbf{E}_Q[f|\mathcal{F}_t] = \mathbf{E}_Q[f] + (H \cdot S)_t, \quad t = 0, \ldots, T. \tag{33}$$

Proof The implication (i) $\Rightarrow$ (ii) immediately follows from the preceding theorem; for the implication (ii) $\Rightarrow$ (i), note that, (32) implies that, for elements $Q_1, Q_2 \in \mathcal{M}^a(S)$, we have $\mathbf{E}_{Q_1}[f] = a = \mathbf{E}_{Q_2}[f]$; hence it suffices to note that if $\mathcal{M}^e(S)$ contains two different elements Q_1, Q_2 we may find $f \in L^\infty(\Omega, \mathcal{F}, \mathbb{P})$ s.t. $\mathbf{E}_{Q_1}[f] \neq \mathbf{E}_{Q_2}[f]$. □

Let us pause here for a moment and recapitulate the above results from an economic point of view. In particular we address the question: how does this theory relate to Bachelier's *fundamental principle*?

We consider a model S of a financial market satisfying the assumptions of corollary 2. The reason why these (models of) financial markets are called *complete* in the Mathematical Finance literature is related to assertion (ii) above: in such a market any contingent claim f is already replicable by an initial investment a and a properly chosen trading strategy H. We shall see in the next section that the arch-example of a complete financial market in discrete time is the random walk, also called the binomial model. We have seen that this model was already considered by Bachelier (over a grid in arithmetic progression); some 70 years later, Cox, Ross and Rubinstein [CRR 79] studied this model over a grid in geometric progression.

The basic problem of Bachelier, as well as of modern Mathematical Finance in general, is that of assigning a price a to a contingent claim f; corollary 2 tells us that, in the case of a complete market, we simply have to take the expectation $\mathbf{E}_Q[f]$, similarly as Bachelier proposed in his "fundamental principle". But now the argument in favor of this methodology is based on a no-arbitrage argument, which is more robust from an economic point of view than the equilibrium argument used by Bachelier.

Also, the message of corollary 2 is not quite identical to Bachelier's "fundamental principle". The subtle difference is that in modern Mathematical Finance one takes the expectation with respect to a *risk neutral* probability measure Q, i.e., a measure under which S is a martingale and which does not necessarily coincide with the *physical* measure P. This distinction between P and Q does not show up in Bachelier's work (although he also is speaking about "two kinds of probability", but apparently he has something different in mind in this passage of his thesis). Bachelier argues somehow in the opposite direction as compared to the modern approach: he postulates that the process S has to be a martingale already under the "physical" measure P (this is what his "fundamental principle" amounts to in modern terminology).

The distinction between the measure Q and P is one of the crucial features of the modern approach to Mathematical Finance. It is implicit in the early work of Black and Scholes [BS 73] and Merton [M 73], and has clearly been crystallized in the later work of Harrison, Kreps and Pliska ([HK 79], [HP 81], [K 81]).

In this respect Bachelier's approach really misses something crucial: for example, there is massive empirical evidence that — in the long run — stocks perform better than bonds. At least, this happened in the previous hundred or two hundred years. Many people believe that this will also be the case in the future (but, of course, we don't know that). In any case, Bachelier has no way of modelling such a phenomenon without violating the "fundamental principle".

One might try to argue in favor of Bachelier that such a long term effect is not of crucial importance for short term option prices and may therefore be ignored.

But there are also other obstructions to the somewhat naive application of the "fundamental principle", which involve logical inconsistencies (which is, of course, particularly annoying from a mathematical part of view): let's take up again the foreign exchange example 1 and assume, mainly for notational convenience, that the domestic and foreign interest rates r_d and r_f equal zero. The stochastic process $(X_t)_{0 \le t \le T}$ models the price of one US \$ in terms of €. By applying Bachelier's "fundamental principle" to the situation of a €-investor "speculating" in US \$, we must have

$$X_0 = \mathbf{E}[X_T]. \tag{34}$$

On the other hand, the same principle applied to the situation of a US \$-investor "speculating" in € implies

$$X_0^{-1} = \mathbf{E}[X_T^{-1}]. \tag{35}$$

But Jensen's inequality tells us that (34) and (35) cannot hold simultaneously (except for the trivial case when X_T is constant). Hence we find a logical obstruction to the "fundamental principle" of Bachelier.

At this stage a distiction between the measure P and Q is unavoidable and we also see from the above argument that the "risk-neutral" measure Q apparently depends on the choice of a numéraire.

We therefore pass to a thorough analysis of the role of the numéraire S^0 in our modelling of a financial market. In particular, we investigate what happens under a "change of numéraire", i.e., by passing from one unit of denomination, say €, to another one, say US $.

Let us consider once more the basic example 1 of a financial market consisting of a €-bond and a $-bond (which we now consider in discrete time, to confirm with the setting of this section). We now drop the assumption $r_d = r_f = 0$ and assume that these bonds develop (expressed in terms of € and $ respectively) by

$$B_t^{€} = e^{r_d t}, \quad B_t^{\$} = e^{r_f t}, \quad t = 0, 1, \dots, T. \tag{36}$$

Denoting again by $(X_t)_{0 \le t \le T}$ the stochastic process modelling the exchange rate, the value (in terms of €) of an investment into the $-bond is given by the stochastic process $(e^{r_f t} X_t)_{t=0}^T$. But note that this refers to the € as numéraire, which *is not a traded asset*, unless we have $r_d = 0$. This may seem odd a first glance; but remember our standing assumption that we can go long and short in traded assets at the same conditions. If the Euro were a traded asset, this would imply that we could borrow Euros at nominal value (i.e., without paying interest); combining this operation with an investment into a €-bond paying positive interest, clearly creates an arbitrage.

We have agreed to choose a traded asset as numéraire: from the point of view of a €-investor, the natural choice in our example is the €-bond. Hence from her point of view the financial market is modeled by

$$S_t = (S_t^0, S_t^1) = \left(1, e^{(r_f - r_d)t} X_t \right), \quad t = 0, 1, \dots, T, \tag{37}$$

where S now is expressed in terms of units of the €-bond.

But adopting the point of view of a $-investor it is natural to express everything in terms of the $-bond, i.e.

$$\check{S}_t = \left(\frac{S_t^0}{S_t^1}, \frac{S_t^1}{S_t^1} \right) = \left(e^{(r_d - r_f)t} X_t^{-1}, 1 \right), \quad t = 0, 1, \dots, T. \tag{38}$$

The previous theorem 2 and corollary 2 tell us, how to relate the arbitrage free prices of derivative securities $f \in L^\infty(\Omega, \mathcal{F}, \mathbb{P})$ to the expectations under the "risk-neutral" probabilities $Q \in \mathcal{M}^e(S)$.

How do these things change, when we pass to a new numéraire? Of course, the arbitrage free prices should remain unchanged (after denominating things in the new numéraire), as the notion of arbitrage should not depend on whether we do the book-keeping in € or in $. On the other hand, we shall presently see that the risk-neutral measures Q do depend on the choice of numéraire.

Let us analyze the situation in the proper degree of generality: the model of a financial market $S = (S_t^0, S_t^1, \dots, S_t^d)_{t=0}^T$ is defined as above. Recall that we assumed that the traded asset S^0 serves as numéraire, i.e., the value S_t^j of the j'th asset at time t is expressed in units of S^0. In particular, we have $S_t^0 = 1$, for all $0 \leq t \leq T$.

We also assume that the process $(S_t^1)_{0 \leq t \leq T}$ is strictly positive; choosing this asset as the new numéraire we find the process $\check{S}$ denoting the prices of the assets $S^0, S^1, \ldots, S^d$ in terms of S^1:

$$\check{S} = \left(\frac{S_t^0}{S_t^1}, 1, \frac{S_t^2}{S_t^1}, \ldots, \frac{S_t^d}{S_t^1} \right)_{t=0}^T. \tag{39}$$

To link with the previous example, we might have that S^0 is a cash account in €, S^1 a cash account in US\$, while $S^2, \ldots, S^d$ model some other stocks, commodities etc.

We now have the proper setting to formulate the theorem clarifying the situation:

Theorem 3 (change of numéraire). *Assume that the financial market $S = (S_t^0, S_t^1, \ldots, S_t^d)$ satisfies (NA) and recall that we have assumed $S_t^0 \equiv 1$, i.e., we have chosen the zero coordinate as numéraire.*

We also assume that the first coordinate $(S_t^1)_{t=0}^T$ is a strictly positive process, so that we may define the "process S in terms of the numéraire S^1" by passing to

$$\check{S} = \left(\frac{S_t^0}{S_t^1}, 1, \frac{S_t^2}{S_t^1}, \ldots, \frac{S_t^d}{S_t^1} \right)_{t=0}^T. \tag{40}$$

Then the set $\mathcal{M}^e(\check{S})$ of equivalent martingale measures for $\check{S}$ equals

$$\mathcal{M}^e(\check{S}) = \left\{ \check{Q} : \frac{d\check{Q}}{d\mathbb{P}} = \frac{S_T^1}{S_0^1} \frac{dQ}{d\mathbb{P}}, \ Q \in \mathcal{M}^e(S) \right\}. \tag{41}$$

For a contingent claim $f \in L^\infty(\Omega, \mathcal{F}, \mathbb{P})$ the interval of arbitrage-free prices therefore does not depend on the chosen numéraire, as we have

$$\{ \mathbf{E}_Q[f] : Q \in \mathcal{M}^e(S) \} = \left\{ S_0^1 \mathbf{E}_{\check{Q}} \left[\frac{f}{S_T^1} \right] : \check{Q} \in \mathcal{M}^e(\check{S}) \right\}. \tag{42}$$

Proof Note that the fact that S is a Q-martingale implies that $\mathbf{E}_Q[\frac{S_T^1}{S_0^1}] = 1$, for all $Q \in \mathcal{M}^e(S)$, so that the set defined by the right hand side of (41) consists of probability measures. Also note that, by our assumption on the strict positivity of S^1, these measures are equivalent to $\mathbb{P}$.

We now calculate the space $\check{K} \subseteq L^\infty(\Omega, \mathcal{F}, \mathbb{P})$ of claims attainable at price 0 with respect to $\check{S}$,

$$\check{K} = \{ (L \cdot \check{S})_T : L \in \mathcal{H} \}. \tag{43}$$

We claim that

$$\check{K} = \left\{ \frac{1}{S_T^1} f : f \in K \right\}. \tag{44}$$

In fact, we claim more generally, that the class of processes of the form $S_t^1(L \cdot \check{S})_t$ coincides with the class of processes $(H \cdot S)_t$ where L and H run through the predictable processes.

Economically speaking this means that the possible gains processes $H \cdot S$, obtained by trading with respect to some trading strategy H in terms of the numéraire S^0, coincide with the possible gains processes $S^1(L \cdot \check{S})$, where $L \cdot \check{S}$ run through the possible gains processes in terms of the numéraire S^1, which subsequently are transformed into units of the numéraire S^0 by multiplication with S^1.

To verify this — economically rather obvious — identity in a formal way we use a little stochastic calculus, namely the stochastic version of the product formula (similarly as in [DS 95, theorem 11]). We have

$$S_t = S_t^1 \check{S}_t, \tag{45}$$

the right hand side refering to multiplication of the positive scalar $S_t^1(\omega)$ with the $(d+1)$-vector $\check{S}_t(\omega) = (\frac{S_t^0(\omega)}{S_t^1(\omega)}, 1, \frac{S_t^2(\omega)}{S_t^1(\omega)}, \ldots, \frac{S_t^d(\omega)}{S_t^1(\omega)})$. Hence by elementary algebra we obtain

$$\Delta S_t = S_{t-1}^1 \Delta \check{S}_t + \check{S}_{t-1} \Delta S_t^1 + \Delta S_t^1 \Delta \check{S}_t. \tag{46}$$

Now we fix any predictable process L and calculate the increment of the process $S_t^1(L \cdot \check{S})_t$ in a similar way:

$$\begin{aligned}
\Delta(S_t^1(L \cdot \check{S})_t) &= (L \cdot \check{S})_{t-1} \Delta S_t^1 + S_{t-1}^1 \Delta((L \cdot \check{S})_t) + \Delta S_t^1 \Delta((L \cdot \check{S})_t) \quad (47) \\
&= (L \cdot \check{S})_{t-1} \Delta S_t^1 + L_t(S_{t-1}^1 \Delta \check{S}_t + \Delta S_t^1 \Delta \check{S}_t) \\
&= (L \cdot \check{S})_{t-1} \Delta S_t^1 + L_t(S_{t-1}^1 \Delta \check{S}_t + \check{S}_{t-1} \Delta S_t^1 + \Delta S_t^1 \Delta \check{S}_t) \\
&\quad - L_t \check{S}_{t-1} \Delta S_t^1 \\
&= ((L \cdot \check{S})_{t-1} - L_t \check{S}_{t-1}) \Delta S_t^1 + L_t \Delta S_t,
\end{aligned}$$

where in the last equality we have used (46). In other words, the increment $\Delta(S_t^1(L \cdot \check{S})_t)$ of the process $S^1(L \cdot \check{S})$ is the product of some $\mathcal{F}_{t-1}$-measurable functions with the increments ΔS_t^1 and ΔS_t respectively. Noting that ΔS_t^1 is just one of the coordinates of ΔS_t, we conclude that the process $S_t^1(L \cdot \check{S}_t)$ may be represented as a stochastic integral of the form $(H \cdot S)$ for some predictable process H. Reversing the roles of S and $\check{S}$ and using the strict positivity of the process S^1, we also conclude that each process of the form $(H \cdot S)$ may be presented in the form $S^1(L \cdot \check{S})$, for some predictable process L, which shows in particular (44).

Hence the linear map $M : L^\infty \to L^\infty$ of multiplication by the function $\frac{S_0^1}{S_T^1}$

$$Mf = \frac{S_0^1}{S_T^1} f \tag{48}$$

maps K bijectively onto $\check{K}$. By basic linear algebra the adjoint M^* of M, which is equal to M, maps the polar of $\check{K}$ onto the polar of K and therefore $cone(\mathcal{M}^a(\check{S}))$ and $cone(\mathcal{M}^e(\check{S}))$ onto $cone(\mathcal{M}^a(S))$ and $cone(\mathcal{M}^e(S))$ respectively. Hence we obtain the identity (41). Finally observe that equality (42) is an immediate consequence of equality (41), noting that, when Q runs through $\mathcal{M}^e(S)$, then $M^{-1}(Q)$ runs through $\mathcal{M}^e(\check{S})$.   □

Corollary 3 (change of numéraire in a complete market). *Assume in addition to the assumptions of theorem 3 that $\mathcal{M}^e(S)$ consists of a singleton $\{Q\}$. Then $\mathcal{M}^e(\check{S}) = \{\check{Q}\}$ where $\frac{d\check{Q}}{d\mathbb{P}} = \frac{S_T^1}{S_0^1}\frac{dQ}{d\mathbb{P}}$.*

For $f \in L^\infty(\Omega, \mathcal{F}, \mathbb{P})$, we obtain the unique arbitrage free price as

$$\mathbf{E}_Q[f] = S_0^1 \mathbf{E}_{\check{Q}}\left[\frac{f}{S_T^1}\right]. \quad \square \tag{49}$$

We finish this section by a dynamic version of theorem 2 on pricing by no-arbitrage, due to D. Kramkov (in a much more general setting; see [K 96] and section 5 below).

Theorem 4 (Optional Decomposition). *Assume that S satisfies (NA) and let $V = (V_t)_{t \geq 0}$ be an adapted process.*

The following assertions are equivalent:

(i) V is a super-martingale for each $Q \in \mathcal{M}^e(S)$.
(i') V is a super-martingale for each $Q \in \mathcal{M}^a(S)$
(ii) V may be decomposed into $V = V_0 + H \cdot S - C$, where $H \in \mathcal{H}$ and $C = (C_t)_{t \geq 0}$ is an increasing adapted process starting at $C_0 = 0$.

To explain the terminology "*optional decomposition*" let us compare this theorem with Doob's celebrated decomposition theorem for non-negative super-martingales $(V_t)_{t \geq 0}$ (see, e.g., [P 90]): this theorem asserts that, for a non-negative (adapted, càdlàg) process V, we have equivalence between the following two statements:

(i) V is a super-martingale (with respect to the fixed measure $\mathbb{P}$),
(ii) V may be decomposed in a unique way into $V = V_0 + M - C$, where $M = (M_t)_{t \geq 0}$ is a local martingale (with respect to $\mathbb{P}$) and C an increasing predictable process s.t. $M_0 = C_0 = 0$.

We see the similarity in spirit, but, of course, there are differences. As regards condition (i) the difference is that, in the setting of the optional decomposition theorem, the super-martingale properly pertains to *all* martingale measures Q for the process S. As regards condition (ii) the role of the local martingale M in Doob's theorem is taken by the stochastic integral $H \cdot S$. A decisive difference between the two theorems is that, in theorem 4, the decomposition is not unique any more and one cannot choose, in general, C to be predictable. The process C can only be chosen to be adapted and therefore optional (for finite Ω, a process is adapted iff it is optional).

The economic interpretation of the optional decomposition theorem goes as follows: a process of the form $V = V_0 + H \cdot S - C$ describes the wealth process of an economic agent, starting at an initial wealth V_0, subsequently investing in the financial market according to the trading strategy H, and consuming as described by the process C: the random variable C_t models the accumulated consumption during the time interval $\{1, \ldots, t\}$. The message of the optional decomposition theorem is that these wealth processes are characterised by condition (i) (or, equivalently, (i')).

Proof of theorem 4 First assume that $T = 1$, i.e., we have a one-period model $S = (S_0, S_1)$. In this case the present theorem is an immediate consequence of theorem 2: if V is a super-martingale under each $Q \in \mathcal{M}^e(S)$, then

$$\mathbf{E}_Q[V_1] \leq V_0, \qquad \text{for all } Q \in \mathcal{M}^e(S). \tag{50}$$

Hence there is a predictable trading strategy H such that $V_0 + (H \cdot S)_1 \geq V_1$. Letting $C_0 = 0$ and writing $\Delta C_1 = C_1 = V_1 - (V_0 + (H \cdot S)_1)$ we have obtained the desired decomposition.

Recall our general assumption that $\mathcal{F}_0$ is trivial; it implies that the trading strategy $H = H_1$ simply is a vector in $\mathbb{R}^d$, as an $\mathcal{F}_0$-measurable function is constant. But this assumption is not at all essential for the above argument: if $\mathcal{F}_0$ is not trivial, we simply apply the above argument to each of the atoms of the sigma-algebra $\mathcal{F}_0$ to obtain an $\mathcal{F}_0$-measurable function H_1.

Hence we may apply, for each fixed $t \in \{1, \ldots, T\}$, the same argument as above to the one-period financial market (S_{t-1}, S_t) based on $(\Omega, \mathcal{F}, \mathbb{P})$ and adapted to the filtration $(\mathcal{F}_{t-1}, \mathcal{F}_t)$. We thus obtain an $\mathcal{F}_{t-1}$-measurable $\mathbb{R}^d$-valued function H_t and a non-negative $\mathcal{F}_t$-measurable function ΔC_t such that

$$\Delta V_t = (H_t, \Delta S_t) - \Delta C_t, \tag{51}$$

where $(.,.)$ denotes the inner product in $\mathbb{R}^d$.

This finishes the construction of the optional decomposition: define the predictable process H as $(H_t)_{t=1}^T$, and the adapted increasing process C by $C_t = \Sigma_{j=1}^t \Delta C_j$.

This shows the implication (i) $\Rightarrow$ (ii); the implications (ii) $\Rightarrow$ (i') $\Rightarrow$ (i) are trivial. $\square$

3 The Binomial Model, Bachelier's Model and the Black-Scholes Model

The canonical example of a finite probability space Ω, to which the no-arbitrage theory applies very nicely, is the *binomial* model. Let $\Omega = \{-1, +1\}^T$ be equipped with the filtration $(\mathcal{F}_t)_{t=0}^T$, where $\mathcal{F}_t$ is generated by the first t coordinate maps on Ω. As probability measure $\mathbb{P}$ we chose the uniform measure on $\mathcal{F} = \mathcal{F}_T$, but we remark that the subsequent results do not depend on this special choice of $\mathbb{P}$; the only property of $\mathbb{P}$ which is needed is, that $\mathbb{P}$ assigns positive mass to each point of Ω.

Consider a financial market based on $(\Omega, (\mathcal{F}_t)_{t=0}^T, \mathbb{P})$ consisting of a cash account $B_t := (S_t^0)_{t=0}^T \equiv 1$ and a risky asset (stock) $(S_t^1)_{t=0}^T$ which is an $\mathbb{R}$-valued adapted process defined on $(\Omega, (\mathcal{F}_t)_{t=0}^T, \mathcal{F})$. By abuse of notation we also shall write S for the one-dimensional process S^1.

To avoid trivialities we assume that

$$\mathbb{P}[S_t \neq S_{t-1}|\mathcal{F}_{t-1}] > 0 \quad \text{everywhere, for } t = 1, \ldots, T. \tag{52}$$

It ist rather obvious and very intuitive that S does not allow arbitrage iff

$$\mathbb{P}[S_t > S_{t-1}|\mathcal{F}_{t-1}] > 0 \text{ and } \mathbb{P}[S_t < S_{t-1}|\mathcal{F}_{t-1}] > 0, \quad \text{for } t = 1, \ldots, T. \tag{53}$$

It is just as obvious — using, e.g., backward induction (compare [LL 96]) — that in this case there exists a unique equivalent martingale measure Q. Hence we know that, for any contingent claim $f \in L^\infty(\Omega, \mathcal{F}, \mathbb{P})$, we can find a trading strategy H such that

$$f = \mathbf{E}_Q[f] + (H \cdot S)_T \tag{54}$$

and that we have, for every $t = 0, \ldots, T$,

$$\mathbf{E}_Q[f|\mathcal{F}_t] = \mathbf{E}_Q[f] + (H \cdot S)_t. \tag{55}$$

We now specialize to two concrete cases for the financial market model S: the first example is the simple random walk; this was considered by Bachelier as a discrete approximation to Brownian motion. The second one is the multiplicative version of the random walk — i.e., $(\ln(\frac{S_t}{S_0}))_{t=0}^T$ is a random walk, possibly with drift. In finance the latter model is called the Cox, Ross, and Rubinstein model [CRR 79]. These authors analyzed this model as a discrete analogue to geometric Brownian motion.

In the former case, i.e., the simple random walk, where $(S_t - S_{t-1})_{t=1}^T$ are i.i.d. random variables taking values $\pm \sigma \Delta x$ with probability $\frac{1}{2}$, the original measure $\mathbb{P}$ is already the unique martingale measure for the process $(S_t)_{t=0}^T$. Hence we deduce from corollary 3 that the unique arbitrage-free price of a contingent claim $f \in L^\infty(P)$ is given by $\mathbf{E}_P[f]$, which justifies Bachelier's "fundamental principle" on the basis of no-arbitrage arguments for the model of a simple random walk.

In the Cox-Ross-Rubinstein case the original measure $\mathbb{P}$ — in general — is not a martingale measure, but it is easy to explicitly calculate the density $\frac{dQ}{d\mathbb{P}}$ (which amounts to a discrete version of Girsanov's theorem).

In both cases the pricing formulae for an option reduce to the calculation of the expected value of the hockey-stick function $f(x) = (x - K)_+$ with respect to a binomial distribution, placed on a sequence of points in arithmetic progression in the former and on a sequence in geometric progression in the latter case.

We leave the elementary but somewhat cumbersome calculations of the resulting formulae in the first case to the energetic reader (who may also find the calcuations essentially in Bachelier's thesis) and in the second case we refer to the beautiful book by Lamberton and Lapeyre [LL 96], where these calculations are presented in a clean and transparent way.

We now pass on to the continuous limits of these models (if properly normalised), where — as usual in mathematics — the results and formulae become more elegant and more transparent.

To do so, we recall the *martingale representation theorem* for Brownian motion, which is the continuous analogue to the elementary considerations on the binomial model above.

Theorem 5. *(see, e.g., [RY 91]) Let $(W_t)_{0 \le t \le T}$ be a standard Brownian motion modeled on $(\Omega, (\mathcal{F}_t)_{0 \le t \le T}, \mathbb{P})$, where $(\mathcal{F}_t)_{0 \le t \le T}$ is the natural (saturated) filtration generated by W.*

Then $\mathbb{P}$ is the unique measure on $\mathcal{F}_T$ which is absolutely continuous with respect to $\mathbb{P}$, and under which W is a martingale.

Correspondingly, for every function $f \in L^1(\Omega, \mathcal{F}_T, \mathbb{P})$ there is a unique predictable process $H = (H_t)_{0 \le t \le T}$ such that

$$f = \mathbf{E}[f] + (H \cdot W)_T, \tag{56}$$

and

$$\mathbf{E}[f|\mathcal{F}_t] = \mathbf{E}[f] + (H \cdot W)_t, \quad 0 \le t \le T, \tag{57}$$

which implies in particular that $(H \cdot W)$ is a uniformly integrable martingale.

Bachelier's model revisited:

Let us restate Bachelier's model in the framework of the formalism developed above: let $B_t \equiv 1$ and $S_t = S_0 + \sigma W_t$, $0 \le t \le T$, where S_0 is the current stockprice, $\sigma > 0$ is a fixed constant, and W is standard Brownian motion on its natural base $(\Omega, (\mathcal{F}_t)_{0 \le t \le T}, \mathbb{P})$.

Fixing the strike price K, we want to price and hedge the contingent claim

$$f(\omega) = (S_T(\omega) - K)_+ \in L^1(\Omega, \mathcal{F}_T, \mathbb{P}). \tag{58}$$

Using the martingale representation theorem we may find a trading strategy $\overline{H}$ s.t.

$$f = \mathbf{E}[f] + (\overline{H} \cdot W)_T \tag{59}$$
$$= \mathbf{E}[f] + (H \cdot S)_T,$$

where $H = \frac{\overline{H}}{\sigma}$. Noting that $B_t \equiv 1$ implies that assumption 1 is satisfied, we deduce from (9) above that

$$C(S_0, T) := \mathbf{E}[f] = (S_0 - K)\Phi\left(\frac{S_0 - K}{\sigma\sqrt{T}}\right) + \sigma\sqrt{T}\varphi\left(\frac{S_0 - K}{\sigma\sqrt{T}}\right). \tag{60}$$

By the same token we obtain, for every $0 \le t \le T$, and conditionally on the stock price having the value S_t at time t,

$$C(S_t, T - t) := \mathbf{E}[f|S_t] = (S_t - K)\Phi\left(\frac{S_t - K}{\sigma\sqrt{T-t}}\right) + \sigma\sqrt{T-t}\varphi\left(\frac{S_t - K}{\sigma\sqrt{T-t}}\right). \tag{61}$$

Hence this solves the pricing problem, which now is based on the no-arbitrage considerations rather than on accepting Bachelier's *fundamental principle*, as we now have the "replication formula" (59).

But what is the trading strategy H, in other words, the recipe to replicate the option by trading dynamically? Economic intuition suggests that we have

$$H(S, t) = \frac{\partial}{\partial S}C(S, T - t). \tag{62}$$

Indeed, consider the following heuristic reasoning using infinitesimals: suppose at time t the stock price equals S_t so that the value of the option equals $C(S_t, t)$. During the infinitesimal interval $(t, t + dt)$ the Brownian motion W_t will move by $dW_t = W_{t+dt} - W_t = \epsilon_t\sqrt{dt}$, where $\mathbb{P}[\epsilon_t = 1] = \mathbb{P}[\epsilon_t = -1] = \frac{1}{2}$, so that S_t will move by $dS_t = S_{t+dt} - S_t = \epsilon_t\sigma\sqrt{dt}$. Hence the value of the option $C(S_t, t)$ will move by $dC_t = C(S_{t+dt}, T - (t + dt)) - C(S_t, T - t) \approx \epsilon_t\frac{\partial C}{\partial S}(S_t, T - t)\sigma\sqrt{dt}$, where we neglect terms of smaller order than $\sqrt{dt}$. In other words, the ratio between the up or down movement of the underlying stock S and the option is

$$dC_t : dS_t = \epsilon_t\frac{\partial C}{\partial S}(S_t, T - t)\sigma\sqrt{dt} : \epsilon_t\sigma\sqrt{dt} \tag{63}$$
$$= \frac{\partial C}{\partial S}(S_t, T - t).$$

If we want to replicate the option by investing the proper quantity H of the underlying stock, formula (63) suggests that this quantity should equal $\frac{\partial C}{\partial S}(S_t, T - t)$.

After these motivating remarks, let us deduce the equation

$$H(S_t, t) = \frac{\partial C}{\partial S}(S_t, T - t) \tag{64}$$

more formally. Consider the stochastic process

$$C(S_t, T - t) = C(S_0 + \sigma W_t, T - t), \quad 0 \le t \le T, \tag{65}$$

of the value of the option. By Itô's formula

$$dC(S_t, t) = \frac{\partial C}{\partial S} dS_t + \left(\frac{\partial C}{\partial t} + \frac{1}{2} \frac{\partial^2 C}{\partial S^2} \sigma^2 \right) dt, \tag{66}$$

where we have used $dS_t = \sigma dW_t$. One readily deduces from formula (61) that C verifies the heat equation with parameter $\frac{\sigma^2}{2}$ displayed in (68) below (time is running into the negative direction in the present setting). In particular, for the function C defined in (61), the drift term in (66) vanishes as it must be the case according to the general theory (the option price process is a martingale by (57)). Hence (66) reduces to the formula

$$C(S_t, T - t) = C(S_0, T) + (H \cdot S)_t, \tag{67}$$

where H is given by (64). Rephrasing this result once more we have shown that the trading strategy H, whose existence was guaranteed by the martingale representation, is of the form (64).

One more word on the fact that $C(S, T - t)$ satisfies the heat equation, which may be verified by simply calculating the partial derivatives in (61). Admitting this calculation, we concluded above that the drift term in (66) vanishes. One may also turn the argument around to conclude from (57) that the drift term in (66) must vanish, which then *implies* that $C(S, T - t)$ must satisfy the heat equation (time running inversely)

$$\frac{\partial C}{\partial t}(S, T - t) = -\frac{\sigma^2}{2} \frac{\partial^2 C}{\partial S^2}(S, T - t). \tag{68}$$

Imposing the boundary condition $C(S, T - T) = C(S, 0) = (S - K)_+$ one may *derive* from this p.d.e. by standard methods the solution (61). This is, in fact, how F. Black and M. Scholes originally proceeded (in the framework of their model, which we shall analyse in a moment). Let us also give the heuristic argument to deduce the p.d.e. (C10) from Bachelier's "fundamental principle" and Itô's formula.

Suppose there is a *"formula"* $C(S_t, T - t)$ which gives the value of an option for every $0 \le t \le T$ and $S_t \in \mathbb{R}$. By assumption, at the terminal date $t = T$ we have the boundary condition $C(S_T, T - T) = C(S_T, 0) = (S_T - K)_+$.

Applying Bachelier's fundamental principle (remember this wonderful passage following the formulation of his "fundamental principle", which describes the idea of a martingale!) the stochastic process $(C(S_t, T - t))_{0 \le t \le T}$ should be a martingale. Therefore the drift term in (66) should vanish, which amounts to formula (68).

The Black-Scholes model:
This model of a stock market was proposed by the famous economist P. Samuelson in 1965 ([S 65]), who at this time was aware of Bachelier's work.

In fact, triggered by a question of J. Savage, it was P. Samuelson who had rediscovered Bachelier's work for the economic literature some years before 1965.

The model is usually called the Black-Scholes model today and became the standard reference model in the context of option pricing:

$$B_t = e^{rt},$$
$$S_t = S_0 e^{\sigma W_t + (\mu - \frac{\sigma^2}{2})t}, \quad 0 \le t \le T. \tag{69}$$

Again W is a standard Brownian motion with natural base $(\Omega, (\mathcal{F}_t)_{0 \le t \le T}, \mathbb{P})$.

The paramenter r models the "riskless rate of interest", while the parameter μ models the average increase of the stock price. Indeed using Itô's formula one may describe the model equivalently by the differential equations:

$$\frac{dB_t}{B_t} = rdt, \tag{70}$$

$$\frac{dS_t}{S_t} = \mu dt + \sigma dW_t. \tag{71}$$

The numéraire in this model is just the relevant currency (say €); in order to remain consistent with the above theory, we shall rather follow our usual procedure of taking a traded asset as numéraire, namely the bond. We then have

$$\tilde{B}_t \equiv 1, \tag{72}$$
$$\tilde{S}_t = S_0 e^{\sigma W_t + (\mu - r - \frac{\sigma^2}{2})t}.$$

The tilde indicates that we now have denominated B_t and S_t in terms of the bond B_t, i.e., we have discounted them. We shall write ν for $\mu - r$ which is called the "excess return". The only thing we have to keep in mind by passing to the bond as numéraire, is that now quantities have to be expressed in terms of the bond: in particular, if K denotes the strike price of an option at time T (expressed in € at time T), we have to express it as Ke^{-rT} units of the bond.

Contrary to Bachelier's setting, the process

$$\tilde{S}_t = S_0 e^{\sigma W_t + (\nu - \frac{\sigma^2}{2})t}, \quad 0 \le t \le T, \tag{73}$$

is *not* a martingale under $\mathbb{P}$ (unless $\nu = 0$, which typically is not the case).

The unique martingale measure Q for $\tilde{S}$ (which is absolutely $\mathbb{P}$-continuous) is given by Girsanov's theorem (see [RY 91] or any introductory text to Mathematical Finance)

$$\frac{dQ}{d\mathbb{P}} = \exp\left(-\frac{\nu}{\sigma}W_T - \frac{\nu^2}{2\sigma^2}T\right). \tag{74}$$

Let us price and hedge the contingent claim $f(\omega) = (\tilde{S}_T(\omega) - Ke^{-rT})_+$, which is the pay-off function of the European call option with exercise time T and strike price K (expressed in terms of €).

Noting that $(W_t + \nu t)_{t \geq 0}$ is a standard Brownian motion under Q and applying theorem 5 to the Q-martingale $\tilde{S}$, we may calculate

$$C(S_0, T) = \mathbf{E}_Q[f] = \mathbf{E}_Q \left[\left(S_0 e^{\sigma(W_T + \nu T) - \frac{\sigma^2}{2}T} - Ke^{-rT} \right)_+ \right] \qquad (75)$$

$$= S_0 \mathbf{E}_Q \left[e^{\sigma\sqrt{T}Z - \frac{\sigma^2 T}{2}} \chi_{\{S_T \geq K\}} \right] - Ke^{-rT} Q[S_T \geq K],$$

where Z denotes a $N(0, 1)$-distributed random variable under Q.

After an elementary but tedious calculation (see, e.g., [LL 96]) this yields the famous *Black-Scholes formula*

$$C(S_0, T) = S_0 \Phi \left(\frac{\ln(\frac{S_0}{K}) + (r + \frac{\sigma^2}{2})T}{\sigma\sqrt{T}} \right) \qquad (76)$$

$$- Ke^{-rT} \Phi \left(\frac{\ln(\frac{S_0}{K}) + (r - \frac{\sigma^2}{2})T}{\sigma\sqrt{T}} \right)$$

and, by the same token, for $0 \leq t \leq T$, and $S_t > 0$,

$$C(S_t, T - t) = S_0 \Phi \left(\frac{\ln(\frac{S_t}{K}) + (r + \frac{\sigma^2}{2})(T - t)}{\sigma\sqrt{T - t}} \right) \qquad (77)$$

$$- Ke^{-rT} \Phi \left(\frac{\ln(\frac{S_t}{K}) + (r - \frac{\sigma^2}{2})(T - t)}{\sigma\sqrt{T - t}} \right).$$

Let us take some time to contemplate on this truly remarkable formula (for which R. Merton and M. Scholes received the Nobel prize in economics in 1997; F. Black unfortunately had passed away in 1995).

1.) As a warm-up consider the limits as $\sigma \to \infty$ (which yields $C(S_0, T) = S_0$) and $\sigma \to 0$ (which yields $C(S_0, T) = (S_0 - Ke^{-rT})_+$). The reader should convince herself that this does make sense economically. For an extremely risky underlying S, an option on one unit of S is almost as valuable as one unit of S itself (think, for example, of a call option on a lottery ticket with $K = 100$ and exercise time T, such that T is later than the drawing where it is decided, whether the ticket wins a million or not). On the other hand, if the underlying S is almost riskless a similar consideration reveals that the value of an option is almost equal to its "inner value" $(S_0 - Ke^{-rT})_+$.

This behavior of the Black-Scholes formula should be contrasted to Bachelier's formula (specializing to the case $S_0 = K$ and $r = 0$)

$$C^{\text{Bachelier}}(S_0, T) = \frac{\sigma}{\sqrt{2\pi}} \sqrt{T} \qquad (78)$$

obtained in (10) above, which tends to infinity as $\sigma \to \infty$; this limiting behaviour is economically absurd and contradicts an obvious no-arbitrage argument which — using the fact that S_T is non-negative — shows that the value of a call option always must be less than the value of the underlying stock.

The reason for this difference in the behaviour of the Black-Scholes formula and Bachelier's one, for large values of σ, is that geometric Brownian motion always remains positive, while Brownian motion may also attain negative values, a fact which has strong effects for very large σ or — what amounts to the same, at least in the case $r = 0$ — for very large T. Nevertheless we shall presently see that — for reasonable values of σ and T — the Black-Scholes formula and Bachelier's formula (78) are very close. This seems to be the essential fact, keeping in mind Keynes' dictum telling us, *not* to look at the limit $T \to \infty$: in the long run we all are dead.

2.) Let us compare the Black-Scholes formula (76) and Bachelier's formula (78) more systematically. To do so we specialize in the Black-Scholes formula to $r = 0$ and $S_0 = K$, and we have to let the σ in the Black-Scholes formula, which we now denote by σ^{BS}, correspond to the σ appearing in Bachelier's formula, denoted by σ^B. As the former pertains to the relative standard deviation of stock prices and the latter to the absolute standard deviation, we roughly find the correspondence — at least for small values of T —

$$\sigma^B \approx \sigma^{BS} S_0 \tag{79}$$

Hence, in this special case, the Black-Scholes and Bachelier option prices to be compared are

$$C^{BS} = S_0 \left[\Phi \left(\frac{\sigma^{BS}\sqrt{T}}{2} \right) - \Phi \left(-\frac{\sigma^{BS}\sqrt{T}}{2} \right) \right], \tag{80}$$

while

$$C^B = \frac{\sigma^B}{\sqrt{2\pi}} \sqrt{T} \approx S_0 \frac{\sigma^{BS}}{\sqrt{2\pi}} \sqrt{T}. \tag{81}$$

The difference of the two quantities is best understood by looking at the shaded area in the subsequent graph involving the density $\varphi(x) = \frac{1}{\sqrt{2\pi}} e^{-\frac{x^2}{2}}$ of the standard normal distribution, and noting that $\varphi(0) = \frac{1}{\sqrt{2\pi}}$.

Developping $\varphi(x)$ into a Taylor series around zero and using $\varphi''(0) = -\frac{1}{\sqrt{2\pi}}$ we get the asymptotic expression

$$C^B - C^{BS} = S_0 \left[\frac{1}{24\sqrt{2\pi}} \left(\sigma^{BS}\sqrt{T} \right)^3 \right] + o \left(\left(\sigma^{BS}\sqrt{T} \right)^3 \right), \tag{82}$$

which indicates a very good fit, if $\sigma^{BS}\sqrt{T}$ is small. Evaluating this expression for the empirical data reported by Bachelier, i.e., $\sigma^{BS} \approx 2.4\,\%$ on a yearly basis, and $T \approx 2\,\text{months} = \frac{1}{6}\,\text{year}$ (this is a generous upper bound for the periods considered by Bachelier which were ranging between 10 and 45 days) we find

$$C^B - C^S \approx S_0 \frac{1}{24\sqrt{2\pi}} \left(0.024\sqrt{\tfrac{1}{6}} \right)^3 \approx 1.56 * 10^{-8} S_0. \tag{83}$$

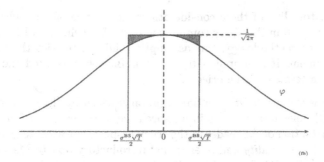

Fig. 3. Comparison of the Bachelier with the Black-Scholes formula.

Hence for this data the difference of the option value obtained from Bachelier's or the Black-Scholes model is of the order 10^{-8} times the value S_0 of the underlying; keeping in mind, that for Bachelier's data, the price of an option was of the order of $S_0/100$, we find that the difference is of the order 10^{-6} of the price of the option.

In view of all the uncertainties involved in option pricing, in particular as regards the estimation of σ, one might be tempted to call this quantity "completely negligible, a priori" (this expression was used by Bachelier when discussing the drawbacks of the normal distribution giving positive probability to negative stock prices).

3.) Let us now comment on the role of the riskless rate of interest r, appearing in the Black-Scholes formula and the reason why this variable does not show up in Bachelier's formula: noting the obvious fact that

$$\ln\left(\frac{S_0}{K}\right) + rT = \ln\left(\frac{S_0}{Ke^{-rT}}\right), \tag{84}$$

one readily observes that this quantity only enters the Black-Scholes formula (76) via the discounting of the strike price, i.e., transforming K units of $\text{€}_{t=T}$ into Ke^{-rT} units of $\text{€}_{t=0}$. When comparing the setting of Black-Scholes to that of Bachelier one should recall that the option premium in Bachelier's days pertained to a payment at time T or, in modern terms, was expressed in terms of a zero coupon bond maturing at time T. Under the assumption of a constant riskless interest — as is the case in the Black-Scholes model — this amounts to considering the present day quantities upcounted by e^{rT}. This was perfectly taken into account by Bachelier, who stressed that the quantities appearing in his formulae have to be understood in terms of "true prices", i.e., forward prices in modern terminology, which amounts to upcounting by e^{rT} in the present setting. In fact, we have seen in section 1 that Bachelier did even more, as he in addition was considering the "contangoes", which — in modern terminology — correspond to a continuous yield on the stock.

The bottom line of these considerations on the role of r is: when we assumed that $r = 0$ in the above comparison of the Bachelier and Black-Scholes option pricing methodology, this assumption did not restrict the generality of the argument: it also applies to $r \neq 0$ as Bachelier denoted the relevant quantities in terms of "true prices".

4.) What is the partial differential equation satisfied by the solution (77) of the Black-Scholes formula? Again we specialize to the case $r = 0$ in order to focus the attention of the reader to the crucial aspect, but we note that now we *do restrict the generality* and refer to any introductory text to Mathematical Finance (e.g., [LL 96]) for the Black-Scholes partial differential equation in the case of a riskless rate of interest $r \neq 0$.

From the Martingale Representation Theorem 5 above we know that the Black-Scholes option price *process*

$$C(S_t, T - t)_{0 \leq t \leq T} \tag{85}$$

is a martingale under the measure Q defined in (74). Hence, denoting by $(\widetilde{W}_t)_{0 \leq t \leq T}$ a standard Brownian motion under Q, using $dS_t = \sigma S_t d\widetilde{W}_t$, and working under the measure Q, we obtain from Itô's formula

$$dC_t = dC(S_t, T - t) = \frac{\partial C}{\partial S} \sigma S_t d\widetilde{W}_t + \left(\frac{\sigma^2}{2} S_t^2 \frac{\partial^2 C}{\partial S^2} + \frac{\partial C}{\partial t} \right) dt. \tag{86}$$

We first observe, using again $\sigma S_t d\widetilde{W}_t = dS_t$, that — similarly as in the context of Bachelier — the replicating trading strategy $H_t(\omega)$ is given by $\frac{\partial C}{\partial S}(S_t(\omega), T - t)$. In the lingo of finance this quantity is called the "Delta" of the option (which depends on S_t and t), and the trading strategy H is called "delta-hedging".

Next we pass to the drift term: as it must vanish, we arrive at the "Black-Scholes partial differential equation"

$$\frac{\partial C}{\partial t}(S, T - t) = -\frac{\sigma^2}{2} S^2 \frac{\partial^2 C}{\partial S^2}(S, T - t), \text{ for } S \geq 0, t \geq 0. \tag{87}$$

This is the multiplicative analogue of the heat equation (68) and may, in fact, easily be reduced to a heat equation (with drift) by passing to logarithmic coordinates $x = \ln(S)$.

Exactly as in Bachelier's case we may proceed by solving the partial differential equation (87) for the boundary condition $C(S, T - T) = C(S, 0) = (S - K)_+$ and $C(0, t) = 0$ to obtain the Black-Scholes formula.

In the lingo of finance, the quantity $-\frac{\partial C}{\partial t}$ is called the "Theta" and the quantity $\frac{\partial^2 C}{\partial S^2}$ the "Gamma" of the option. Hence the p.d.e. (87) allows for the following economic interpretation: the loss of value of the option, when time to maturity $T - t$ decreases (and S remains fixed), is equal to the "convexity" or "gamma" of the option price (as a function of S) at time t, normalized by

$\frac{\sigma^2}{2}S^2$ (in the case of the Bachelier model the normalisation was simply $\frac{\sigma^2}{2}$). This has a good economic interpretation and today's option traders think in these terms. They speak about "selling or buying convexity" or rather "going gamma-short or gamma-long" which amounts to the same thing. The interpretation of (87) is that, for the buyer of an option, the convexity of the function $C(S, T - t)$ in the variable S corresponds to a kind of insurance with respect to price movements of S. As there is no such thing as a free lunch, this insurance costs (proportional to the second derivative) and a positive $\frac{\sigma^2}{2}S^2\frac{\partial^2 C}{\partial S^2}$ is reflected by a negative partial derivative $\frac{\partial C}{\partial t}$ of $C(S, T - t)$ with respect to time t.

Let us illustrate this fact by reasoning once more heuristically with infinitesimal movements of Brownian motion: we want to explain the infinitesimal change of the option price when "time increases by an infinitesimal while the stock price S remains constant". To do so we apply the heuristic analogue of the Brownian bridge: consider the infinitesimal interval $[t, t + 2dt]$ and assume that the driving Q-Brownian motion $\widetilde{W}$ moves in the first half $[t, t + dt]$ from $\widetilde{W}_t$ to $\widetilde{W}_t + \epsilon_t\sqrt{dt}$, where ϵ_t is a random variable with $Q[\epsilon_t = 1] = Q[\epsilon_t = -1] = \frac{1}{2}$, while in the second half $[t + dt, t + 2dt]$ it moves back to $\widetilde{W}_t$. What happens during this time interval to a "hedger" who proceeds according to the Black-Scholes trading strategy H described above, which replicates the option? At time t she holds $\frac{\partial C}{\partial S}(S_t, T - t)$ units of the stock. Following first the scenario $\epsilon_t = +1$, the stock has a price of $S_t + \sigma S_t\sqrt{dt}$ at time $t + dt$. Appart from being happy about this up movement, the hedger now (i.e., at time $t + dt$) adjusts the portfolio to hold $\frac{\partial C}{\partial S}(S_t + \sigma S_t\sqrt{dt}, T - (t + dt))$ units of stock, which results in a net buy of $\frac{\partial^2 C}{\partial S^2}(S_t, T - t)\sigma S_t\sqrt{dt}$ units of stock, where we neglect terms of smaller order than $\sqrt{dt}$. In the next half $[t + dt, t + 2dt]$ of the interval the stock price S drops again to the value $S_{t+2dt} = S_t$ and the hedger readjusts the portfolio by selling again the $\frac{\partial^2 C}{\partial S^2}(S_t, T - t)\sigma S_t\sqrt{dt}$ units of stock (neglecting again terms of smaller order than $\sqrt{dt}$). It seems at first glance that the gains made in the first half are precisely compensated by the losses in the second half, but a closer inspection shows that the hedger did "buy high" and "sell low": the quantitiy $\frac{\partial^2 C}{\partial S^2}(S_t, T - t)\sigma S_t\sqrt{dt}$ was bought at price $S_t + \sigma S_t\sqrt{dt}$ at time $t + dt$, and sold at price S_t at time $t + 2dt$, resulting in a total loss of

$$\left(\frac{\partial^2 C}{\partial S^2}(S_t, T - t)\sigma S_t\sqrt{dt}\right)\left(\sigma S_t\sqrt{dt}\right) = \sigma^2 S_t^2\frac{\partial^2 C}{\partial S^2}(S_t, T - t)dt. \qquad (88)$$

Going through the scenario $\epsilon_t = -1$, one finds that the hedger did first "sell low" and then "buy high" resulting in the same loss (where again we neglect infinitesimals resulting in effects (with respect to the final result) of smaller order than dt).

Keeping in mind that this was achieved during an interval of total length $2dt$ we have found a heuristic explanation for the Black-Scholes equation (87) (we also note that the same argument, applied to Bachelier's model, yields a heuristic explanation of the heat equation (68)). The general phenomenon behind this fact is that, in the case of convexity, the "wobbling" of Brownian motion, which is of order $\sqrt{dt}$ in an interval of length dt, causes the hedger to have systematic losses, which are proportional to $\frac{\partial^2 C}{\partial S^2}$ as well as to the increment $d\langle S \rangle_t$ of the quadratic variation process $\langle S \rangle_t = \int_0^t \sigma^2 S_u^2 du$ of the stock price process S.

5.) When deriving the Black-Scholes formula (76) we did not go through the (elementary but tedious) trouble of explicitly calculating (75). We shall now furnish an explicit derivation of the formula which has the merit of yielding an interpretation of the two probabilities appearing in (76). It also allows for a better understanding of the formula (for example, for the remarkable fact, that the parameter μ has disappeared) and which also dispenses us of some troubles in the calculation.

As observed in (75) above, the contingent claim $f(\omega) = (\widetilde{S}_T(\omega) - Ke^{-rT})_+$ (expressed in terms of the numeraire B_t) splits into

$$
\begin{aligned}
(\widetilde{S}_T - Ke^{-rT})_+ &= \widetilde{S}_T \chi_{\{\widetilde{S}_T \geq Ke^{-rT}\}} - Ke^{-rT} \chi_{\{\widetilde{S}_T \geq Ke^{-rT}\}} \qquad (89) \\
&= \widetilde{S}_T \chi_{\{S_T \geq K\}} - Ke^{-rT} \chi_{\{S_T \geq K\}} \\
&= f_1 - f_2.
\end{aligned}
$$

We have to calculate $\mathbf{E}_Q[f_1]$ and $\mathbf{E}_Q[f_2]$ under the risk-neutral measure Q defined in (74). This is easy for f_2 and we do not have to use the explicit form of the density (74) provided by Girsanov's theorem. It suffices to observe that $\widetilde{S}_t = S_0 \exp(\sigma \widetilde{W}_t - \frac{\sigma^2}{2}t)$ where $\widetilde{W}$ is a Brownian motion under Q. So

$$
X := \frac{\ln\left(\frac{\widetilde{S}_T}{S_0}\right) + \frac{\sigma^2}{2}T}{\sigma\sqrt{T}} \sim N(0,1) \qquad \text{under } Q, \qquad (90)
$$

whence

$$
\begin{aligned}
\mathbf{E}_Q[f_2] &= e^{-rT} K\, Q[\widetilde{S}_T \geq e^{-rT}K] \qquad (91) \\
&= e^{-rT} K\, Q\left\{ \frac{\ln\left(\frac{\widetilde{S}_T}{S_0}\right) + \frac{\sigma^2}{2}T}{\sigma\sqrt{T}} \geq \frac{\ln\left(\frac{e^{-rT}K}{S_0}\right) + \frac{\sigma^2}{2}T}{\sigma\sqrt{T}} \right\} \\
&= e^{-rT} K\, Q\left\{ X \geq \frac{\ln\left(\frac{e^{-rT}K}{S_0}\right) + \frac{\sigma^2}{2}T}{\sigma\sqrt{T}} \right\} \\
&= e^{-rT} K\, \Phi\left(\frac{\ln\left(\frac{S_0}{K}\right) + \left(r - \frac{\sigma^2}{2}\right)T}{\sigma\sqrt{T}} \right),
\end{aligned}
$$

which yields the second term of the Black-Scholes formula.

Why was the calculation of $\mathbf{E}_Q[f_2]$ so easy? Because the amount Ke^{-rT} is just a constant (expressed in terms of the present numéraire); hence the calculation of the expectation reduced to the calculation of the probability of an event, namely the probability that the option will be exercised, with respect to Q.

To proceed similarly with the calculation of $\mathbf{E}_Q[f_1]$ we make a change of numéraire, now choosing the risky asset S in the Black-Scholes model (69) as numéraire. Under this numéraire the model reads

$$\frac{B_t}{S_t} = S_0^{-1}e^{-\sigma W_t+(r-\mu+\frac{\sigma^2}{2})t} \tag{92}$$

$$\frac{S_t}{S_t} \equiv 1$$

where W is a standard Brownian motion under $\mathbb{P}$. The reader certainly has noticed the symmetry with (72). But what ist the probability measure $\check{Q}$ under which the process $\frac{B_t}{S_t}$ becomes a martingale? Using Girsanov we can explicitly calculate the density $\frac{d\check{Q}}{d\mathbb{P}}$; but, in fact, we don't really need this full information. All we need is to observe that we may write

$$\frac{B_t}{S_t} = S_0^{-1}e^{-\sigma \check{W}_t-\frac{\sigma^2}{2}t}, \tag{93}$$

where $\check{W}$ is a standard Brownian motion under $\check{Q}$ (the reader worried by the minus sign in front of $\sigma \check{W}_t$ may note that $-\check{W}$ also is a standard Brownian motion under $\check{Q}$). We now apply the change of numéraire theorem (in the form of corollary 3) to calculate $\mathbf{E}_Q[f_1]$. In fact, we have only proved this theorem for the case of finite Ω, but we trust in the reader's faith that it also applies to the present case (for a thorough investigation for the validity of this theorem for general locally bounded semi-martingale models we refer to [DS 95]). Applying this theorem we obtain

$$\mathbf{E}_Q[f_1] = \mathbf{E}_Q\left[\widetilde{S}_T \chi_{\{\frac{S_T}{B_T}\geq e^{-rT}K\}}\right] \tag{94}$$

$$= S_0\,\mathbf{E}_{\check{Q}}\left[\frac{\widetilde{S}_T}{\widetilde{\check{S}}_T}\chi_{\{\frac{B_T}{S_T}\leq e^{rT}K^{-1}\}}\right]$$

$$= S_0\,\mathbf{E}_{\check{Q}}\left[\chi_{\{S_0^{-1}e^{-\sigma W_T-\frac{\sigma^2}{2}T}\leq e^{rT}K^{-1}\}}\right]$$

$$= S_0\,\check{Q}\left[S_0e^{\sigma \check{W}_T+\frac{\sigma^2}{2}T} \geq e^{-rT}K\right].$$

Noting that $\check{W}_T/\sqrt{T}$ is $N(0,1)$-distributed under $\check{Q}$, this expression is completely analogous to that appearing in (91), with the exception that now there is a plus in front of the term $\frac{\sigma^2}{2}T$. Hence we get

$$E_Q[f_1] = S_0 \Phi \left(\frac{\ln\left(\frac{S_0}{K}\right) + \left(r + \frac{\sigma^2}{2}\right) T}{\sigma\sqrt{T}} \right), \qquad (95)$$

which is the first term appearing in the Black-Scholes formula. We now may interpret $\Phi\left(\frac{\ln\left(\frac{S_0}{K}\right)+\left(r+\frac{\sigma^2}{2}\right)T}{\sigma\sqrt{T}}\right)$ as the probability, that the option will be exercised, with respect to $\check{Q}$.

4 The No-Arbitrage Theory for General Processes

We now again take up the theme of the no-arbitrage theory as developed in section 2: what can we deduce from applying the no-arbitrage principle with respect to pricing and hedging of derivative securities?

While we obtained satisfactory and mathematically rigorous answers to these questions in the case of a finite underlying probability space Ω in section 2, we saw in section 3, that the basic examples for this theory, the Bachelier and the Black-Scholes model, do not fit into this easy setting, as they involve Brownian motion.

In section 3 we coped with the difficulty either by using well-known results from stochastic analysis (e.g., the martingale representation theorem 5 for the Brownian filtration), or by appealing to the faith of the reader, that the results obtained in the finite case also carry over — mutatis mutandis — to more general situations, as we did when applying the change of numéraire theorem to the calculation of the Black-Scholes model.

In the present chapter we want to develop a "théorie génerale of no-arbitrage" applying to a general framework of stochastic processes. The development of Mathematical Finance since the work of Black, Merton and Scholes made it clear, that the relatively poor fit of the Black-Scholes model (as well as Bachelier's model) to empirical data (especially with respect to extremal behaviour, i.e., large changes in prices) makes it necessary for many applications, to pass to more general models; in some cases these models still have continuous paths, but also processes (in continuous time) with jumps are increasingly gaining importance.

We adopt the following general framework: let $S = (S_t)_{t\geq0}$ be an $\mathbb{R}^{d+1}$-valued stochastic process based on and adapted to the filtered probability space $(\Omega, \mathcal{F}, (\mathcal{F})_{t\geq0}, \mathbb{P})$. Again we assume that the zero coordinate S^0, called the bond, is normalised to $S_t^0 \equiv 1$.

We first will make a technical assumption, namely that the process S is *bounded*, i.e., that there exists a sequence $(\tau_n)_{n=1}^{\infty}$ of stopping times, increasing a.s. to $+\infty$, such that the stopped processes $S_t^{\tau_n} = S_{t\wedge\tau_n}$ are uniformly bounded, for each $n \in \mathbb{N}$. Note that continuous processes — or, more generally, càdlàg processes with uniformly bounded jumps — are locally bounded. This assumption will be very convenient for technical reasons, and only at the end of this section we shall indicate, how to extend to the general case of processes, which are not necessarily locally bounded.

We have chosen $[0, \infty[$ for the time index set in order to allow for maximal generality; of course this also covers the case of a compact interval $[0, T]$, which is relevant in most applications, by assuming that S_t is constant, for $t \geq T$. We shall always assume that the filtration $(\mathcal{F}_t)_{t\geq0}$ satisfies the usual assumptions of right continuity and saturatedness, and that S has a.s. càdlàg trajectories.

How to define the trading strategies H, which played a crucial role in the preceding sections? A very elementary approach, corresponding to the role of step functions in integration theory, is formalized by the subsequent concept.

Definition 9. *(compare, e.g., [P 90]) For a locally bounded stochastic process S we call an $\mathbb{R}^d$-valued process $H = (H_t)_{t \geq 0}$ a simple trading strategy (or, speaking more mathematically, a simple integrand), if H is of the form*

$$H = \sum_{i=1}^n h_i \chi_{]\tau_{i-1},\tau_i]}, \tag{96}$$

where $0 = \tau_0 \leq \tau_1 \leq \ldots \leq \tau_n$ are finite stopping times and h_i are $\mathcal{F}_{\tau_{i-1}}$-measureable, $\mathbb{R}^d$-valued functions.

We then may define, similarly as in definition 4, the stochastic integral $(H \cdot S)$ as the stochastic process

$$(H \cdot S)_t = \sum_{i=1}^n \left(h_i, S_{\tau_i \wedge t} - S_{\tau_{i-1} \wedge t} \right) \tag{97}$$

$$= \sum_{i=1}^n \sum_{j=1}^d h_i^j \left(S_{\tau_i \wedge t}^j - S_{\tau_{i-1} \wedge t}^j \right), \quad 0 \leq t < \infty,$$

and its terminal value as the random variable

$$(H \cdot S)_\infty = \sum_{i=1}^n (h_i, S_{\tau_i} - S_{\tau_{i-1}}). \tag{98}$$

We call H admissible if, in addition, the stopped process S^{τ_n} and the functions $h_1, \ldots, h_n$ are uniformly bounded.

This definition is a well known building block for developing a stochastic integration theory (see, e.g., [P 90]). It has a clear economic interpretation in the present context: at time τ_{i-1} an investor decides to adjust her portfolio in the assets $S^1, \ldots, S^j, \ldots, S^d$ by fixing her investment in asset S^j to be $h_i^j(\omega)$ units; we allow h_i^j to have arbitrary sign (holding a negative quantity means borrowing or "going short"), and to depend on the random element ω in an $\mathcal{F}_{\tau_{i-1}}$-measurable way, i.e., using the information available at time τ_{i-1}. The funds for adjusting the portfolio in this way simply are financed by taking the appropriate amount from (or putting into) the "cash box", modeled by the numéraire $S^0 \equiv 1$. The investor holds this portfolio fixed up to time τ_i. During this period the value of the risky stocks S^j, $j = 1, \ldots, d$, changed from $S_{\tau_{i-1}}^j(\omega)$ to $S_{\tau_i}^j(\omega)$ resulting in a total gain (or loss) given by the random variable $(h_i, S_{\tau_i} - S_{\tau_{i-1}})$. At time τ_i, for $i < n$, the investor readjusts the portfolio again and at time τ_n she liquidates the portfolio, i.e., converts all her positions into the numéraire. Hence the random variable $(H \cdot S)_{\tau_n} = (H \cdot S)_\infty$ models the total gain (in units of the numéraire S_0) which she finally, i.e., at time τ_n, obtained by adhering to the strategy H; the process $(H \cdot S)_t$ models the gains accumulated up to time t.

The concept of a simple trading strategy is designed in a purely algebraic way, avoiding limiting procedures, in order to be on safe grounds.

The next crucial ingredient in developing the theory is the proper generalisation of the notion of an equivalent martingale measure.

Definition 10. *A probability measure Q on $\mathcal{F}$ which is equivalent (resp. absolutely continuous with respect) to $\mathbb{P}$ is called an* equivalent *(resp. absolutely continuous) local martingale measure, if S is a local martingale under Q.*

We denote by $\mathcal{M}^e(S)$ (resp. $\mathcal{M}^a(S)$) the family of all such measures, and say that S satisfies the condition of the existence of an equivalent local martingale measure (EMM), if $\mathcal{M}^e(S) \neq \emptyset$.

Note that, by our assumption of local boundedness of S, we have that S is a local Q-martingale, iff S^τ is a Q-martingale for each stopping time τ such that S^τ is uniformly bounded.

Why did we use the notion of a local martingale instead of the more familiar notion of a martingale? The reason is, that it is the natural degree of generality. The subsequent easy lemma (whose proof is an obvious consequence of the chosen concepts and left to the reader) shows that this notion serves just as well as the notion of a martingale for the present purpose of a no-arbitrage theory. On the other hand, the restriction to the notion of martingale measures would make it impossible to formulate the general version of the fundamental theorem of asset pricing (theorem 1 below), as may bee seen from easy examples (see, e.g., [DS 94a]).

Lemma 2. *A locally bounded semi-martingale S is a local martingale under Q iff*

$$\mathbf{E}_Q\left[(H \cdot S)_\infty\right] = 0, \tag{99}$$

for each admissible simple trading strategy H.

For later use we note that the "=" in (99) may equivalently be replaced by "$\leq$" (or "$\geq$").

We define the subspace K^{simple} of $L^\infty(\Omega, \mathcal{F}, \mathbb{P})$ of contingent claims available at price zero via an admissible simple trading strategy by

$$K^{\text{simple}} = \{(H \cdot S)_\infty : H \text{ simple, admissible}\} \tag{100}$$

and by C^{simple} the convex cone in $L^\infty(\Omega, \mathcal{F}, \mathbb{P})$ of contingent claims dominated by some $f \in K$

$$C^{\text{simple}} = K^{\text{simple}} - L^\infty_+ = \left\{f - k : f \in K^{\text{simple}}, k \geq 0\right\}. \tag{101}$$

Definition 11. *S satisfies the* no-arbitrage condition (NA) *with respect to simple integrands, if $K^{\text{simple}} \cap L^\infty_+(\Omega, \mathcal{F}, \mathbb{P}) = \{0\}$ (or, equivalently, $C^{\text{simple}} \cap L^\infty_+(\Omega, \mathcal{F}, \mathbb{P}) = \{0\}$).*

We want to prove a fundamental theorem of asset pricing analogous to theorem 1 above. But now things are more delicate and the notion of *(NA)* defined above is not sufficiently strong to imply this result:

Proposition 2. *The condition (EMM) of the existence of an equivalent local martingale measure implies the condition (NA) of no-arbitrage with respect to simple integrands, but not vice versa.*

Proof *(EMM)* $\Rightarrow$ *(NA)*: this is an immediate consequence of lemma 2, noting that for $Q \sim P$, and a non-negative function $f \geq 0$, which does not vanish almost surely, we have $\mathbf{E}_Q[f] > 0$.

(NA) $\not\Rightarrow$ *(EMM)*: we give an easy counterexample which is just an infinite random walk.

Let $t_n = 1 - \frac{1}{n+1}$ and define the $\mathbb{R}$-valued process S to start at $S_0 = 1$, and to be constant except for jumps at the points t_n which are defined as

$$\Delta S_{t_n} = 2^{-n} \epsilon_n \tag{102}$$

such that $(\epsilon_n)_{n=1}^\infty$ are independent random variables taking the values $+1$ or -1 with probabilities

$$P[\epsilon_n = 1] = \frac{1 + \alpha_n}{2}, \quad P[\epsilon_n = -1] = \frac{1 - \alpha_n}{2}, \tag{103}$$

where $(\alpha_n)_{n=1}^\infty$ is a sequence in $]-1, +1[$ to be specified below.

Clearly this well-defines a bounded process S, for which there is a unique measure Q on $(\Omega, \mathcal{F}) = (\{-1, 1\}^{\mathbb{N}}, \text{Borel } (\{-1, 1\}^{\mathbb{N}}))$, under which S is a martingale; this measure is given by

$$Q[\epsilon_n = 1] = Q[\epsilon_n = -1] = \frac{1}{2}, \tag{104}$$

and $(\epsilon_n)_{n=1}^\infty$ are independent under Q.

By a result of Kakutani (see, e.g. [W 91]) we know that Q is either equivalent to P, or P and Q are mutually singular, depending on whether $\sum_{n=1}^\infty \alpha_n^2 < \infty$ or not.

Taking, for example, $\alpha_n = \frac{1}{2}$, for all $n \in \mathbb{N}$, we have constructed a process S on $(\Omega, \mathcal{F}, \mathbb{P})$, for which there is no equivalent (local) martingale measure Q. On the other hand, it is an easy and instructive exercise to show that, *for simple trading strategies*, there are no-arbitrage opportunities for the process S. $\square$

The example in the above proof shows, why the no-arbitrage condition defined in 11 is too narrow: it is intuitively rather obvious that by a sequence of properly scaled bets on a (sufficiently) biased coin one can "produce something like an arbitrage", while a finite number of bets (as formalized by definition 9) does not suffice to do so.

But here we are starting to move on thin ice, and it will be the crucial issue to find a mathematically precise framework, in which the above intuitive insight can be properly formalized.

A decisive step in this direction was done in the work of D. Kreps [K 81], who realized that the purely algebraic notion of no-arbitrage with respect to simple integrands has to be complemented with a topological notion:

Definition 12. *(compare [K 81]) S satisfies the condition of* no free lunch *(NFL), if the closure $\overline{C}$ of C^{simple}, taken with respect to the weak-star topology of $L^\infty(\Omega, \mathcal{F}, \mathbb{P})$, satisfies*

$$\overline{C} \cap L^\infty(\Omega, \mathcal{F}, \mathbb{P}) = \{0\}. \tag{105}$$

This strengthening of the condition of no-arbitrage is taylor-made so that the subsequent version of the fundamental theorem of asset pricing holds true.

Theorem 6 (Kreps - Yan). *A locally bounded process S satisfies the condition of no free lunch (NFL), iff condition (EMM) of the existence of an equivalent local martingale measure is satisfied:*

$$(NFL) \iff (EMM). \tag{106}$$

Proof *(EMM)* $\Rightarrow$ *(NFL)*: This is still the easy part. By lemma 2 we have $\mathbf{E}_Q[f] \leq 0$, for each $Q \in \mathcal{M}^e(S)$ and $f \in C^{\mathrm{simple}}$, and this inequality also extends to the weak-star closure $\overline{C}$. On the other hand, if *(EMM)* would hold true and *(NFL)* were violated, there would exist a $Q \in \mathcal{M}^e(S)$ and $f \in \overline{C}$, $f \geq 0$ not vanishing almost surely, whence $\mathbf{E}_Q[f] > 0$, a contradiction.

(NFL) $\Rightarrow$ *(EMM)*: We follow the strategy of the proof for the case of finite Ω, but have to refine the argument:

Step 1 (Hahn-Banach argument): We claim that, for fixed $f \in L^\infty_+$, $f \not\equiv 0$, there is $g \in L^1_+$ which, viewed as a linear functional on L^∞, is less than or equal to zero on $\overline{C}$, and such that $\langle f, g \rangle > 0$. To see this, apply the separation theorem (e.g., [Sch 66, th. II, 9.2]) to the σ^*-closed convex set $\overline{C}$ and the compact set $\{f\}$ to find $g \in L^1$ and $\alpha < \beta$ such that $g|_{\overline{C}} \leq \alpha$ and $\langle f, g \rangle > \beta$. Since $0 \in C$ we have $\alpha \geq 0$. As $\overline{C}$ is a cone, we have that g is zero or negative on $\overline{C}$ and, in particular, nonnegative on L^∞_+, i.e. $g \in L^1_+$. Noting that $\beta > 0$ we have proved step 1.

Step 2 (Exhaustion Argument): Denote by $\mathcal{G}$ the set of all $g \in L^1_+$, $g \leq 0$ on C. Since $0 \in \mathcal{G}$ (or by step 1), $\mathcal{G}$ is nonempty.

Let S be the family of (equivalence classes of) subsets of Ω formed by the supports of the elements $g \in \mathcal{G}$. Note that S is closed under countable unions, as for a sequence $(g_n)_{n=1}^\infty \in \mathcal{G}$, we may find strictly positive scalars $(\alpha_n)_{n=1}^\infty$, such that $\sum_{n=1}^\infty \alpha_n g_n \in \mathcal{G}$. Hence there is $g_0 \in \mathcal{G}$ such that, for $S_0 = \{g_0 > 0\}$, we have

$$P(S_0) = \sup\{P(S) : S \in \mathcal{G}\}. \tag{107}$$

We now claim that $P(S_0) = 1$, which readily shows, that g_0 is strictly positive almost surely. Indeed, if $P(S_0) < 1$, then we could apply step 1 to $f = \chi_{(\Omega \setminus S_0)}$ to find $g_1 \in \mathcal{G}$ with

$$\langle f, g_1 \rangle = \int_{\Omega \setminus S_0} g_1(\omega) dP(\omega) > 0 \tag{108}$$

Hence, $g_0 + g_1$ would be an element of $\mathcal{G}$ whose support has P-measure strictly bigger than $P(S_0)$, a contradiction.

Normalize g_0 so that $\|g_0\|_1 = 1$ and let Q be the measure on $\mathcal{F}$ with Radon-Nikodym derivative $dQ/dP = g_0$. We conclude from lemma 2 that Q is a local martingale measure for S, so that $\mathcal{M}^e(S) \neq 0$. $\square$

Some comments on the Kreps-Yan theorem seem in order: this theorem was obtained by D. Kreps [K 81] in a more general setting and under a — rather mild — additional separability assumption; the reason for the need of this assumption is that D. Kreps did not use the above exhaustion argument, but rather some sequential procedure relying on the separability of $L^1(\Omega, \mathcal{F}, \mathbb{P})$. Independently, and at about the same time, Ji-An Yan [Y 80] proved in a different context, namely the characterisation of semi-martingales as good integrators, and without a direct relation to finance, a general theorem. C. Stricker [S 90] observed, that Yan's theorem may be applied, to quickly yield the above theorem without any separability assumption. We therefore took the liberty to name it after these two authors.

The message of the theorem is, that the assertion of the "fundamental theorem of asset pricing" 1 is valid for general processes, if one is willing to interpret the notion of "no-arbitrage" in a somewhat liberal way, crystallized in the notion of "no free lunch" above.

What is the economic interpretation of a "free lunch"? By definition S violates the assumption (NFL), if there is a function $g_0 \in L_+^\infty(\Omega, \mathcal{F}, \mathbb{P})$, $g_0 \neq 0$, and nets $(g_\alpha)_{\alpha \in I}, (f_\alpha)_{\alpha \in I}$ in $L^\infty(\Omega, \mathcal{F}, \mathbb{P})$, such that $f_\alpha = (H^\alpha \cdot S)_\infty$ for some admissible, simple integrand H^α, $g_\alpha \leq f_\alpha$, and $\lim_{\alpha \in I} g_\alpha = g_0$, the limit converging with respect to the weak-star topology of $L^\infty(\Omega, \mathcal{F}, \mathbb{P})$. Speaking economically: an arbitrage opportunity would be the existence of a trading strategy H such that $(H \cdot S)_\infty \geq 0$, almost surely, and $P[(H \cdot S)_\infty > 0] > 0$. Of course, this is the dream of each arbitrageur, but we have seen, that — for the purpose of the fundamental theorem to hold true — this is asking for too much (at least, if we only allow for simple admissible trading strategies). Instead, a free lunch is the existence of a contingent claim $g_0 \geq 0$, $g_0 \neq 0$, which may, in general, not be written as (or dominated by) a stochastic integral $(H \cdot S)_\infty$ with respect to a simple admissible integrand; but there are contingent claims g_α "close to g_0", which can be obtained via the trading strategy H^α, and subsequently "throwing away" the amount of money $f_\alpha - g_\alpha$.

This triggers the question whether we can do somewhat better than the above — admittedly complicated — procedure. Can we find a requirement sharpening the notion of "no free lunch", i.e., being closer to the original notion of "no-arbitrage" and such that a — properly formulated — version of the "fundamental theorem" still holds true?

Here are some mathematically precise questions related to our attempt to make the process of taking the weak-star closure more understandable:

(i) is it possible, in general, to replace the net $(g_\alpha)_{\alpha\in I}$ above by a sequence $(g_n)_{n=0}^\infty$?

(ii) can we choose the net $(g_\alpha)_{\alpha\in I}$ (or, hopefully, the sequence $(g_n)_{n=0}^\infty$) such that $(g_\alpha)_{\alpha\in I}$ remains bounded in $L^\infty(P)$ (or at least such that the negative parts $((g_\alpha)_-)_{\alpha\in I}$ remain bounded)? This latter issue is crucial from an economic point of view, as it pertains to the question whether the approximation of f by $(g_\alpha)_{\alpha\in I}$ can be done *respecting a finite credit line.*

(iii) is it really necessary to allow for the "throwing away of money"?

It turns out that questions (i) and (ii) are intimately related and, in general, the answer to these questions is no. In fact, the study of the pathologies of the operation of taking the weak-star closure is an old theme of functional analysis. On the very last pages of S. Banach's original book ([B 32]) the following example is given: there is a separable Banach space X such that, for every given fixed number $n \geq 1$ (say $n = 35$), there is a convex cone C in the dual space X^*, such that $C \subsetneqq C^{(1)} \subsetneqq C^{(2)} \subsetneqq \ldots \subsetneqq C^{(n)} = C^{(n+1)} = \overline{C}$, where $C^{(k)}$ denotes the sequential weak-star closure of $C^{(k-1)}$, i.e., the limits of weak-star convergent *sequences* $(x_i)_{i=0}^\infty$, with $x_i \in C^{(k-1)}$, and $\overline{C}$ denotes the weak-star closure of C. In other words, by taking the limits of weak-star convergent *sequences* in C we do not obtain the weak-star closure of C immediately, but we have to repeat this operation precisely n times, when finally this process stabilizes to arrive at the weak-star closure $\overline{C}$.

In Banach's book this construction is done for $X = c_0$ and $X^* = l^1$ while our present context is $X = L^1(P)$ and $X^* = L^\infty(P)$. Adapting the ideas from Banach's book, it is possible to construct a semi-martingale S such that the corresponding convex cone C^{simple} has the following property: taking the weak-star sequential closure $(C^{\text{simple}})^{(1)}$, the resulting set intersects $L^\infty_+(P)$ only in $\{0\}$; but doing the operation twice, we obtain the weak-star closure $C^{(2)} = \overline{C}$, and $\overline{C}$ intersects $L^\infty_+(P)$ in a non-trivial way (see [DS 94, example 7.8]). Hence we cannot reduce to sequences $(g_n)_{n=0}^\infty$ in the definition of *(NFL)*. The construction of this example uses a process with jumps; for continuous processes the situation is, in fact, nicer, and in this case it is possible to give positive answers to questions (i) and (ii) above (see [S 90], [D 92] and [DS 94]).

As regards question (iii), the dividing line again is the continuity of the process S (see [S 90] and [D 92] for positive results for continuous processes, and [S 94] for a counterexample S, where S is a process with jumps).

Summing up the above discussion: the theorem of Kreps and Yan is a beautiful and mathematically precise extension of the fundamental theorem of asset pricing 1 to a general framework of stochastic processes in continuous time. However, in general, the concept of passing to the weak-star closure does not allow for a clear-cut economic interpretation. It is therefore desirable to prove versions of the above theorem, where the closure with respect to the weak-star topology is replaced by the closure with respect to some finer topology (ideally the topology of uniform concergence, which allows for an obvious and convincing economic interpretation).

To do so, let us contemplate once more, where the above encountered difficulties related to the weak-star topology originated from: they are essentially caused by our restriction to consider only *simple, admissible trading strategies.* These nice and simple objects can be defined without any limiting procedure, but we should not forget, that they are only auxiliary gimmicks, playing the same role as step functions in integration theory. The concrete examples of trading strategies encountered in section 3 for the case of the Bachelier and the Black-Scholes model led us already out of this class: of course, they are not simple trading strategies.

Hence we have to pass to a suitable class of more general trading strategies than just the simple, admissible ones. Among other pleasant and important features, this will have the following effect on the corresponding sets C and K: these sets will turn out to be "closer to their closures" (ideally they will already be closed in the relevant topology), than the above considered sets C^{simple} and K^{simple}; the reason is that the passage from simple to more general intergrands *involves already a limiting procedure.*

Let us do in a more systematic way our search for an appropriate class of trading strategies:

First of all, one has to restrict the choice of the integrands H to make sure that the process $H \cdot S$ exists. Besides the qualitative restrictions coming from the theory of stochastic integration, one has to avoid problems coming from so-called doubling strategies. This was already noted in the paper by Harrison and Pliska (1979). To explain this remark, let us consider the classical doubling strategy. We toss a coin, and when heads comes up, the player is paid 2 times his bet. If tails comes up, the player loses his bet. The strategy is well known: the player doubles his bet until the first time he wins. If he starts with 1 €, his final gain (= last pay out - total sum of the preceding bets) is 1 € almost surely. He has an almost sure win. The probability that heads will eventually show up, is indeed one (even if the coin is not fair). However, his accumulated losses are not bounded from below. Everybody, especially the casino boss, knows that this is a very risky way of winning 1 €. This type of strategy has to be ruled out: there should be a lower bound on the player's loss. The described doubling strategy is known for centuries and in French it is still referred to as "la martingale".

Here is the definition of the class of intergrands which turns out to be appropriate for our purposes.

Definition 13. *Fix an $\mathbb{R}^{d+1}$-valued stochastic process $S = (S_t)_{t\geq 0}$ as defined in the beginning of this section, which we now also assume to be a semi-martingale. An $\mathbb{R}^d$-valued predictable process $H = (H_t)_{t\geq 0}$ is called an admissible integrand for the semi-martingale S, if*

(i) H is S-integrable, i.e., the stochastic integral $H \cdot S = ((H \cdot S)_t)_{t \geq 0}$ is well-defined in the sense of stochastic integration theory for semi-martingales,
(ii) there is a constant M such that

$$(H \cdot S)_t \geq -M, \qquad a.s., \text{ for all } t \geq 0. \tag{109}$$

Let us comment on this definition: we place ourselves into the "théorie générale" of integration with respect to semi-martingales: here we are on safe grounds as the theory developed, in particular by P.-A. Meyer and his school, tells us precisely what it means that a predictable process H is S-integrable (see, e.g., [P 90]). But in order to do so we have to make sure that S is a semi-martingale: this is precisely the class of processes allowing for a satisfactory integration theory, as we know from the theorem of Bichteler and Dellacherie.

How natural is the assumption, that S is a semi-martingale, from an economic point of view? In fact, it fits very naturally into the present no-arbitrage framework: it is shown in ([DS 94, theorem 7.2]) that, for a locally bounded, càdlàg process S, the assumption, that the closure of C^{simple} *with respect to the norm topology of* $L^\infty(P)$ intersects $L^\infty(P)_+$ only in $\{0\}$, implies already that S is a semi-martingale. This assumption therefore is implied by a very mild strengthening of the no-arbitrage condition for simple, admissible integrands. Loosely speaking, the message of this theorem is that a no-arbitrage theory only makes sense, if we start with a semi-martingale model for the financial market S.

As regards condition (ii) in the above definition, this is a strong and economically convincing requirement to rule out the above discussed doubling strategy, as well as similar schemes, which try to make a final gain at the cost of possibly going very deep into the red. Condition (ii) goes back to the original work of Harrison and Pliska [HP 81]: there is a finite credit line M obliging the investor to finance her trading in such a way, that this credit line is respected at all times $t \geq 0$.

Definition 14. *Let*

$$K = \{(H \cdot S)_\infty : H \text{ admissible and } (H \cdot S)_\infty = \lim_{t \to \infty} (H \cdot S)_t \text{ exists a.s.}\}, \tag{110}$$

which forms a convex cone of functions in $L^0(\Omega, \mathcal{F}, \mathbb{P})$, and

$$C = \{g \in L^\infty(P) : g \leq f \text{ for some } f \in K\}. \tag{111}$$

We say that S satisfies the condition of no free lunch with vanishing risk *(NFLVR), if*

$$\overline{C} \cap L^\infty_+(P) = \{0\}, \tag{112}$$

where $\overline{C}$ now denotes the closure of C with respect to the norm topology of $L^\infty(P)$.

Comparing the present definition to the notion of "no free lunch" *(NFL)*, the weak-star topology has been replaced by the topology of uniform convergence. Taking up again the discussion following the Kreps-Yan theorem 6, we now find a better economic interpretation: S allows for a *free lunch with vanishing risk*, if there is $f \in L_+^\infty(P)\backslash\{0\}$ and sequences $(f_n)_{n=0}^\infty = ((H^n \cdot S)_\infty)_{n=0}^\infty \in K$, for a sequence $(H^n)_{n=0}^\infty$ of admissible integrands, and $(g_n)_{n=0}^\infty$ satisfying $g_n \le f_n$, such that

$$\lim_{n\to\infty} \|f - g_n\|_\infty = 0. \tag{113}$$

In particular the negative parts $((g_n)_-)_{n=0}^\infty$ tend to zero uniformly, which explains the term *"vanishing risk"*.

We now have all the ingredients to formulate a *general version of the fundamental theorem of asset pricing*.

Theorem 7. ([DS 94, corr.1.2]) *The following assertions are equivalent for an $\mathbb{R}^{d+1}$-valued locally bounded semi-martingale model $S = (S_t)_{t\ge 0}$ of a financial market:*

(i) *(EMM), i.e., there is a probability measure Q, equivalent to P, such that S is a local martingale under Q.*

(ii) *(NFLVR), i.e., S satisfies the condition of no free lunch with vanishing risk.*

The present theorem is a sharpening of the Kreps-Yan theorem, as it replaces the weak-star convergence in the definition of "no free lunch" by the economically more convincing notion of uniform convergence. The price to be paid for this improvement is, that now we have to place ourselves into the context of *general admissible*, instead of *simple admissible* integrands.

The proof of theorem 7 as given in [DS 94] is surprisingly long and technical; despite of several attempts, no essential simplification of this proof has been achieved so far. We are not able to go in detail through this proof, but we shall try to give a "guided tour" through it, which should motivate and help the interested reader to find her way through the arguments in [DS 94].

We start by observing that the implication (i) $\Rightarrow$ (ii) still is the easy one: supposing that S is a local martingale under Q and H is an admissible trading strategy, we may deduce from a result of Ansel-Stricker ([AS 94], see also [E 80]) and the fact that $H \cdot S$ is bounded from below, that $H \cdot S$ is a local martingale under Q, too. Using the boundedness from below of $H \cdot S$, we also conclude that $H \cdot S$ is a Q-super-martingale, so that

$$\mathbf{E}_Q[(H \cdot S)_\infty] \le 0. \tag{114}$$

Hence $\mathbf{E}_Q[g] \le 0$, for all $g \in C$, and this equality extends to the norm closure $\overline{C}$ of C (in fact, it also extends to the weak-star closure of C, but we don't need this stronger result for the proof of the present theorem).

Summing up, we have proved that *(EMM)* implies *(NFLVR)*.

Before passing to the reverse implication let us still have a closer look at the crucial inequality (114): its message is that the notion of equivalent local martingale measures Q and admissible integrands H has been designed in such a way, that the basic intuition behind the notion of a martingale holds true: *you cannot win in average by betting on a martingale*. Note, however, that the notion of admissible integrands does not rule out the possibility *to lose in average by betting on S*. An example, already noted in [HP 81], is the so-called "suicide stategy H". Consider a simplified roulette, where red and black both have probability $\frac{1}{2}$, and as usual, when winning, your bet is doubled. The strategy consists in placing one € on red and then walking to the bar and regarding the roulette from a distance: if it happens that consecutively only red turns up in the next couple of games, you may watch a huddle of chips piling up with exponential growth. But, inevitably, i.e., with probability one, black will eventually turn up, which will cause the huddle — including your original € — to disappear. Translating this story into the language of stochastic integration, we have a martingale S (in fact, a random walk) and an admissible trading strategy H such that we have a strict inequality in (114). Of course, the present process $H \cdot S$ corresponding to the "suicide strategy", is just the process corresponding to the "doubling strategy" with opposite sign.

We now discuss the hard implication *(NFLVR)* $\Rightarrow$ *(EMM)* of theorem 7. It is reduced to the subsequent theorem wich may be viewed as the "abstract" version of theorem 7:

Theorem 8. ([DS 94, theorem 4.2]) *In the setting of theorem 7 assume that (ii) holds true, i.e., that S satisfies (NFLVR).*
Then the cone $C \subseteq L^\infty(P)$ is weak-star closed.

The fact that theorem 8 implies theorem 7 now follows immediately from the Kreps-Yan theorem: theorem 8 tells us that we don't have to bother about passing to the weak-star closure of C any more, as assumption (ii) of theorem 7 implies that C *already is weak-star closed*. In other words, our program of choosing the *"right"* class of admissible integrands was successful: the "passage to the limit" which was necessary in the context of the Kreps-Yan theorem, i.e., the passage from C^{simple} to its weak-star closure, is already taken care of by the "passages to the limit" in the stochastic integration theory of general admissible integrands for the semi-martingale S.

In fact, theorem 8 tells us that — under the assumption of *(NFLVR)* — C equals precisely the weak-star closure of C^{simple} (the fact that C^{simple} is weak-star dense in C follows from the general theory of stochastic integration, which is based on the idea of approximating a general integrand by simple integrands).

By rephrasing theorem 7 in the form of theorem 8, we did not come closer to a proof yet. But we see more clearly, what the heart of the matter is: for a net $(H^\alpha)_{\alpha \in I}$ of admissible integrands, $f_\alpha = (H^\alpha \cdot S)_\infty$ and $g_\alpha \leq f_\alpha$ such that $(g_\alpha)_{\alpha \in I}$ weak-star converges to $f \in L^\infty(P)$, we have to show that we can find an admissible integrand H such that $f \leq (H \cdot S)_\infty$. This will prove theorem 8 and therefore 7. Loosely speaking, we have to be able to pass from a net $(H^\alpha)_{\alpha \in I}$ of admissible trading strategies to a limiting admissible trading stategy H.

The first good news on our way to prove this result is that in the present context we may reduce from the case of a general net $(H^\alpha)_{\alpha \in I}$ to the case of a sequence $(H^n)_{n=0}^\infty$. This follows from a good old friend from functional analysis, the theorem of Krein-Smulian (see, e.g., [Sch 66]): this theorem implies that a convex set C in a dual Banach space X^* is weak-star closed, iff it is relatively weak-star closed in each bounded subset of X^*. Using some easy additional facts from general functional analysis (see [DS 94, theorem 2.1]) we may conclude that the convex cone C in $L^\infty(\Omega, \mathcal{F}, \mathbb{P})$ is weak-star closed iff it is weak-star sequentially closed. The reader should note the subtle difference to the example from Banach's book discussed after the Kreps-Yan theorem 6 above: to pass from a convex set $C \subseteq L^\infty(\Omega, \mathcal{F}, \mathbb{P})$ to its weak-star closure, *it does*, in general, *not suffice* to add all the weak-star sequential limits. But to check, whether a convex set C is already weak-star closed, *it does suffice* to check, whether the weak-star sequential limits remain within C.

Once we have reduced to the case of sequences $(H^n)_{n=0}^\infty$ we may exploit another good friend from functional analysis, the theorem of Banach-Steinhaus (also called principle of uniform boundedness): if *a sequence* $(g_n)_{n=0}^\infty$ *in* X^* is weak-star convergent, the norms $(\|g_n\|)_{n=0}^\infty$ remain bounded. This result implies that we may reduce to the case that the sequence $(H^n)_{n=0}^\infty$ admits a uniform bound M such that $H^n \cdot S \geq -M$, for all $n \in \mathbb{N}$.

Putting together these reductions from general functional analysis, it will suffice for the proof of theorem 8 to prove the following result:

Proposition 3. *Under the hypotheses of theorem 8, let* $(H^n)_{n=0}^\infty$ *be a sequence of admissible integrands such that*

$$(H^n \cdot S)_t \geq -1, \qquad a.s., \text{ for } t \geq 0 \text{ and } n \in \mathbb{N}.. \tag{115}$$

Also assume that $f_n = (H^n \cdot S)_\infty$ *converges almost surely to* f. *Then there is an admissible integrand* H *such that*

$$(H \cdot S)_\infty \geq f. \tag{116}$$

To convince ourselves that proposition 3 indeed implies theorem 8, we still have to justify one more reduction step which is contained in the statement of proposition 3: we may reduce to the case, when $(f_n)_{n=0}^\infty$ converges almost surely. This is done by an elementary lemma in the spirit of Komlos' theorem ([DS 94, lemma A 1.1]). In its simplest form it states the follwing: Let $(g_n)_{n=0}^\infty$ be an arbitrary sequence of random variables uniformly bounded from below. Then we may find convex combinations $h_n \in \text{conv}(f_n, f_{n+1,...})$ converging almost surely to an $\mathbb{R} \cup \{+\infty\}$-valued random variable f. For more refined variations on this theme see [DS 99].

Note that the passage to convex combinations does not cost anything in the present context, where our aim is to find a limit to a given sequence in a locally convex vector space; hence the above lemma allows us to reduce to the case where we may assume, in addition to (115), that $(f_n)_{n=0}^{\infty} = ((H^n \cdot S)_{\infty})_{n=0}^{\infty}$ converges almost surely to a function $f : \Omega \mapsto \mathbb{R}\cup\{+\infty\}$. Using the assumtion *(NFLVR)* we can show in the present context that f is a.s. finitely valued.

Summing up, proposition 3 is a statement about the possibility of passing to a (kind of) limit H, for a given sequence $(H^n)_{n=0}^{\infty}$ of admissible integrands. The crucial hypothesis is the uniform one-sided boundedness (115); apart from this strong assumption, we only have an information on the a.s. convergence of *the terminal values* $((H^n \cdot S)_{\infty})_{n=0}^{\infty}$, but we do not have any a priori information on the convergence of *the processes* $((H^n \cdot S)_{t\geq0})_{n=0}^{\infty}$.

Let us compare proposition 3 with the literature. An important theorem of J. Memin [M 80] states the following: if a sequence of stochastic integrals $((H^n \cdot S)_{t\geq0})_{n=0}^{\infty}$ on a given semi-martingale S converges with respect to the semi-martingale topology, then the limit exists (as a semi-martingale) and is of the form $H \cdot S$ for some S-integrable predictable process H.

This theorem finally will play an important role in proving proposition 3; but we still have a long way to go, before we can apply it, as the assumptions of proposition 3 a priori do not tell us anything about the convergence of the sequence of processes $((H^n \cdot S)_{t\geq0})_{n=0}^{\infty}$.

Another line of results in the spirit of proposition 3 assumes that the process S is a (local) martingale. The arch-example is the theorem of Kunita-Watanabe (see, e.g. [P 90] or [Y 78]): suppose that S is a locally L^2-bounded martingale, that each $(H^n \cdot S)_{t\geq0}$ is an L^2-bounded martingal, and that the sequence $((H^n \cdot S)_{t\geq0})_{n=0}^{\infty}$ is Cauchy in the Hilbert space of square-integrable martingales (equivalently: that the sequence of terminal values $((H^n \cdot S)_{\infty})_{n=0}^{\infty}$ is Cauchy in the Hilbert space $L^2(\Omega, \mathcal{F}, \mathbb{P})$). Then the limit exists (as a square-integrable martingale) and it is of the form $(H \cdot S)_{t\geq0}$.

As the proof of this theorem is very simple and allows for some insight into the present theme, we sketch it (assuming, for simplicity, that S is $\mathbb{R}$-valued): denote by $\langle S \rangle_t$ the predictable, quadratic variation process of the L^2-bounded martingale S, which defines a finite measure $d\langle S \rangle_t$ on the sigma-algebra $\mathcal{P}$ of predictable subsets of $\Omega \times \mathbb{R}_+$. Denoting by $L^2(\Omega \times \mathbb{R}_+, \mathcal{P}, d\langle S \rangle_t)$ the corresponding Hilbert space, the stochastic integration theory is designed in such a way that we have the isometric identity

$$\|H\|_{L^2(\Omega \times \mathbb{R}_+, \mathcal{P}, d\langle S \rangle_t)} = \|(H \cdot S)_{\infty}\|_{L^2(\Omega, \mathcal{F}, \mathbb{P})}, \qquad (117)$$

for each predictable process H, for which the left hand side of (117) is finite.

Hence the assumption that $((H^n \cdot S)_{t\geq0})_{n=0}^{\infty}$ is Cauchy in the Hilbert space of square-integrable martingales is tantamount to the assumption that $(H^n)_{n=0}^{\infty}$ is Cauchy in $L^2(\Omega \times \mathbb{R}_+, \mathcal{P}, d\langle S \rangle_t)$. Now, once more, the stochastic integration theory is designed in that way that $L^2(\Omega \times \mathbb{R}_+, \mathcal{P}, d\langle S \rangle_t)$ consists precisely of the S-integrable, predictable processes H such that $H \cdot S$ is an L^2-bounded martingale. Hence by the completeness of the Hilbert space $L^2(\Omega \times \mathbb{R}_+, \mathcal{P}, d\langle S \rangle_t)$ we can pass from the Cauchy-sequence $(H^n)_{n=0}^\infty$ to its limit $H \in L^2(\Omega \times \mathbb{R}_+, \mathcal{P}, d\langle S \rangle_t)$, thus finishing the sketch of the proof of the Kunita-Watanabe theorem.

The above argument shows in a nice and transparent way how to deduce from a completeness property of the *space of predictable integrands H* a completeness property of the corresponding *space of stochastic integrals $H \cdot S$*. In the context of the theorem of Kunita-Watanabe, the functional analytic background for this argument is reduced to the — almost trivial — isometric identification of the two corresponding Hilbert spaces in (117).

Using substantially more refined arguments, M. Yor [Y 78] was able to extend this result to the case of Cauchy sequences $(H^n \cdot S)_{n=0}^\infty$ of martingales bounded in L^p, for arbitrary $1 \le p \le \infty$, the most delicate and interesting case being $p = 1$.

After this review of some of the previous literature on the topic of completeness of the space of stochastic integrals, let us turn back to propostion 3.

Unfortunately the theorems of Kunita-Watanabe and Yor do not apply to its proof, as we don't assume that S is a local martingale. It is precisely the point, that we finally want to *prove* that S is a local martingale with respect to some measure Q equivalent to P.

But in our attempt to build up some motivation for the proof of proposition 3, let us cheat for a moment and suppose that we know already that S is a local martingale under some equivalent measure Q and let $(H^n)_{n=0}^\infty$ be a sequence of S-integrable predictable processes satisfying (115). Using again the theorem of Ansel-Stricker [AS 94] we conclude that $(H^n \cdot S)_{n=0}^\infty$ is a sequence of local martingales; inequality (115) quickly implies that this sequence is bounded in $L^1(Q)$-norm:

$$\|H^n \cdot S\|_{L^1(Q)} := \sup\{\mathbf{E}[|(H^n \cdot S)_\tau|], \tau \text{ finite stopping time}\} \le 2, \text{ for } n \ge 0. \tag{118}$$

Let us cheat once more and assume that each $H^n \cdot S$ is in fact a uniformly integrable Q-martingale (instead of only being a local Q-martingale) and that $((H^n \cdot S)_\infty)_{n=0}^\infty$ is Cauchy with respect to the $L^1(Q)$-norm defined above (instead of only being bounded with respect to this norm).

Admitting the above "cheating steps" we are in a position to apply Yor's theorem to find a limiting process H to the sequence $(H^n)_{n=0}^\infty$ for which (116) holds true, where we even may replace the inequality by an equality. But, of course, this is only motivation, why proposition 3 should hold true, and we now have to find a proof without cheating.

We have taken some time for the above heuristic considerations to develop an intuition for the statement of proposition 3 and to motivate the general philosophy underlying its proof: *we want to prove results which are — at least more or less — known for (local) martingales S, but replacing the martingale assumption on S by the assumption that S satisfies (NFLVR)*.

As a starter we give the proof of a result which shows that, under the assumption of *(NFLVR)*, the technical condition imposed on the admissible integrand H in (110) is, in fact, automatically satisfied.

Lemma 3. ([DS 94, theorem 3.3]) *Let S satisfy (NFLVR) and H be an admissible integrand.*
Then

$$(H \cdot S)_\infty := \lim_{t \to \infty} (H \cdot S)_t \tag{119}$$

exists and is finite, almost surely.

This result is a good illustration for our philosophy: suppose *we know already* that the assumption of 3 implies that S is a local martingale under some Q equivalent to P. Then the conclusion follows immediately from known results: from Ansel-Stricker [AS 94] we know that $H \cdot S$ is a super-martingale. As $H \cdot S$ is bounded from below, Doob's theorem (see, e.g., [W 91]) implies the almost sure convergence of $(H \cdot S)_t$ as $t \to \infty$ to an a.s. finite random variable.

Our goal is to replace these martingale arguments by some arguments relying only on *(NFLVR)*. The nice feature is that these arguments also allow for an economic interpretation.

Proof of Lemma 3 As in the usual proof of Doob's super-martingale convergence theorem we consider the number of up-crossings: to show almost sure convergence of $(H \cdot S)_t$, for $t \to \infty$, it will suffice to show that, for any $\beta < \gamma$, the P-measure of the set $\{\omega : (H \cdot S)_t(\omega) \text{ upcrosses }]\beta, \gamma[\text{ infinitely often}\}$ equals zero.

So suppose to the contrary that there is $\beta < \gamma$ such that the set

$$A = \{\omega : (H \cdot S)_t \text{ upcrosses }]\beta, \gamma[\text{ infinitely often}\} \tag{120}$$

satisfies $P[A] > 0$. The economic interpretation of this situation is the following: an investor knows at time zero that, when following the trading strategy H, with probability $P[A] > 0$ her wealth will infinitely often be less than or equal to β as well as more than or equal to γ. A smart investor will realize that this offers a free lunch with vanishing risk, as she can modify H to obtain a very rewarding trading strategy K.

Indeed, define inductively the sequence of stopping times $(\sigma_n)_{n=0}^\infty$ and $(\tau_n)_{n=0}^\infty$ by $\sigma_0 = \tau_0 = 0$ and, for $n \geq 1$,

$$\sigma_n = \inf\{t \geq \tau_{n-1} : (H \cdot S)_t \leq \beta\}, \tag{121}$$
$$\tau_n = \inf\{t \geq \sigma_n : (H \cdot S)_t \geq \gamma\}.$$

The set A then equals the set where, σ_n and τ_n are finite, for each $n \in \mathbb{N}$ (as usual, the inf over the empty set is taken to be $+\infty$).

What every investor wants to do is to "buy low and sell high"; the above stopping times allow her to do that in a systematic way: define $K = H\mathbf{1}_{\{\cup_{n=1}^{\infty}]\sigma_n, \tau_n]\}}$, which clearly is a predictable S-integrable process. A more verbal description of K goes as follows: the investor starts by doing nothing (i.e., making a zero-investment into the risky assets $S^1, \ldots, S^d$) until the time σ_1 when the process $(H \cdot S)_t$ has dropped below β (If $\beta \geq 0$, we have $\sigma_1 = 0$)). At this time she starts to invest according to the rule prescribed by the trading strategy H; she continues to do so until time τ_1 when $(H \cdot S)_t$ first has passed beyond γ. Note that, if $\tau_1(\omega)$ is finite, our investor following the strategy K has at least gained the amount $\gamma - \beta$. At time τ_1 (if it happens to be finite) the investor clears all her positions and does not invest into the risky assets until time σ_2, when she repeats the above scheme.

One easily verifies (arguing either "mathematically" or "economically") that the process $K \cdot S$ is uniformly bounded from below and satisfies

$$(K \cdot S)_t \geq -M \qquad\qquad \text{a.s., for all } t, \qquad\qquad (122)$$

where M is the uniform lower bound for $(H \cdot S)$, and

$$\lim_{t \to \infty} (K \cdot S)_t = \infty. \qquad\qquad \text{a.s. on } A. \qquad\qquad (123)$$

Hence K describes a trading scheme, where the investor can lose at most a fixed amount of money, while, with strictly positive probability, she ultimately becomes infinitely rich. Intuitively speaking, this is "something like an arbitrage", and it is an easy task to formally deduce from these properties of K a "free lunch with vanishing risk": for example, it suffices to define $K^n = \frac{1}{n} K \mathbf{1}_{]0, \tau_n \wedge T_n]}$, for a sequence of (deterministic) times $(T_n)_{n=0}^{\infty}$, to let $f_n = (K^n \cdot S)_{\infty} = (K^n \cdot S)_{\tau_n \wedge T_n}$ and to define $g_n = f_n \wedge (\gamma - \beta)\mathbf{1}_B$ where $B = \bigcap_{n=0}^{\infty}\{\tau_n \leq T_n\}$. If $(T_n)_{n=1}^{\infty}$ tends to infinity sufficiently fast, we have $P[B] > 0$, and one readily verifies that $(g_n)_{n=1}^{\infty}$ converges uniformly to $(\gamma - \beta)\mathbf{1}_B$.

Summing up, we have shown that *(NFLVR)* implies that, for $\beta < \gamma$, the process $H \cdot S$ almost surely upcrosses the interval $]\beta, \gamma[$ only finitely many times. Whence $(H \cdot S)_t$ converges almost surely to a random variable $(H \cdot S)_{\infty}$ with values in $\mathbb{R} \cup \{\infty\}$. The fact that $(H \cdot S)_{\infty}$ is a.s. finitely valued follows from another application (similar but simpler than above) of the assumption of *(NFLVR)*, which we leave to the reader. □

After all these preparations we finally start to sketch the main arguments underlying the proof of proposition 3. The strategy is to obtain from assumption (115) and from suitable modifications of the original sequence $(H^n)_{n=0}^{\infty}$, still denoted by $(H^n)_{n=0}^{\infty}$, more information on the convergence of the sequence of *processes* $(H^n \cdot S)_{n=0}^{\infty}$. Eventually we shall be able to reduce to the case where $(H^n \cdot S)_{n=0}^{\infty}$ converges in the semi-martingale topology; at this stage Memin's theorem will give us the desired limiting trading strategy H.

So, what can we deduce from assumption (115) and the a.s. convergence of $(f_n)_{n=0}^{\infty} = ((H^n \cdot S)_{\infty})_{n=0}^{\infty}$ for the convergence of the sequence of processes $(H^n \cdot S)_{n=0}^{\infty}$? The unpleasant answer is: a priori, we cannot deduce anything. To see this, recall the "suicide" strategy H which we have discussed in the context of inequality (114) above: it designs an admissible way to lose one €. Speaking mathematically, the corresponding stochastic integral $H \cdot S$ starts at $(H \cdot S)_0 = 0$, satisfies $(H \cdot S)_t \geq -1$ a.s., for all $t \geq 0$, and $(H \cdot S)_{\infty} = -1$. But clearly this is not the only admissible way to lose one € and there are many other trading strategies K on the process S having the same properties. A trivial example is, to first wait without playing for a fixed number of games of the roulette, and to start the suicide strategy only after this waiting period; of course, this is a (slightly) different way of losing one €.

Speaking mathematically, this means that — even when S is a martingale, as it is the case in the example of the suicide strategy — the condition $(H \cdot S)_t \geq -1$. a.s., for all $t > 0$, and the final outcome $(H \cdot S)_{\infty}$ do not determine the process $H \cdot S$. In particular there is no hope to derive from (115) and the a.s. convergence of the sequence of random variables $((H^n \cdot S)_{\infty})_{n=0}^{\infty}$ a convergence property of the sequence of processes $(H^n \cdot S)$.

The idea to remedy the situation is to remark the following fact: the suicide strategy is a silly investment and obviously there are better trading strategies, e.g., not to gamble at all. By discarding such "silly investments", we hopefully will be able to improve the situation.

Here is the way to formalize the idea of discarding "silly investments": Denote by D the set of all random variables h such that there is a random variable $f \geq h$ and a sequence $(H^n)_{n=0}^{\infty}$ of admissible trading strategies satisfying (115), and such that $(H^n \cdot S)_{\infty}$ converges a.s. to f. We call f_0 a maximal element of D if the conditions $h \geq f_0$ and $h \in D$ imply that $h = f_0$.

For example, in the context of the "suicide strategy", $f \equiv -1$ is an element of D, but not a maximal element. A maximal element dominating f is, for example, $f_0 \equiv 0$.

More generally, it is not hard to prove under the assumptions of proposition 3 that, for a given $f = (H \cdot S)_{\infty} \geq -1$, where H is an admissible integrand, there is a maximal element $f_0 \in D$ dominating f (see [DS 94, lemma 4.3]).

The point of the above concept is that, in the proof of proposition 4.12, we may assume without loss of generality that f is a maximal element of D. Under this additional assumption it is indeed possible to derive from the a.s. convergence of the sequence of random variables $((H^n \cdot S)_{\infty})_{n=0}^{\infty}$ some information on the convergence of the sequence of processes $((H^n \cdot S)_{t \geq 0})_{n=0}^{\infty}$.

As the proof of this result is another nice illustration of our general approach of replacing "martingale arguments" by "economically motivated arguments" relying on the assumption *(NFLVR)*, we sketch the argument.

Lemma 4. ([DS 94, lemma 4.5]) *Under the assumptions of proposition 3 suppose, in addition, that f is a maximal element of D.*
Then the sequence of random variables

$$F_{n,m} = \sup_{t \geq 0} |(H^n \cdot S)_t - (H^m \cdot S)_t| \tag{124}$$

tends to zero in probability, as $n, m \to \infty$.

Proof Suppose to the contrary that there is $\alpha > 0$, and sequences $(n_k, m_k)_{k \geq 1}$ tending to ∞ s.t., for each k, we have $\mathbb{P}[\sup_{t \geq 0}((H^{n_k} \cdot S)_t - (H^{m_k} \cdot S)_t) > \alpha] \geq \alpha$.

Define the stopping times τ_k as

$$\tau_k = \inf\{t : (H^{n_k} \cdot S)_t - (H^{m_k} \cdot S)_t \geq \alpha\}, \tag{125}$$

so that we have $\mathbb{P}[\tau_k < \infty] \geq \alpha$.

Define L^k as $L^k = H^{n_k} \mathbf{1}_{[0,\tau_k]} + H^{m_k} \mathbf{1}_{]\tau_k,\infty[}$. Clearly the process L^k is predictable and $L^k \cdot S \geq -1$.

Translating the formal definition into prose: the trading strategy L^k consists in following the trading strategy H^{n_k} up to time τ_k, and then switching to H^{m_k}. The idea is that L^k produces a sensibly better final result $(L^k \cdot S)_\infty$ than either $(H^{n_k} \cdot S)_\infty$ or $(H^{m_k} \cdot S)_\infty$, which will finally lead to a contradiction to the maximality assumption on f.

Why is L^k "sensibly better" than H^{n_k} or H^{m_k}? For large k, the random variables $(H^{n_k} \cdot S)_\infty$ as well as $(H^{m_k} \cdot S)_\infty$ will both be close to f in probability; for the sake of the argument, assume that both are in fact equal to f (keeping in mind that the difference is "small with respect to convergence in probability"). A moment's reflection reveals that this implies that the random variables $(L^k \cdot S)_\infty$ equal f plus the random variable $((H^{n_k} \cdot S)_{\tau_k} - (H^{m_k} \cdot S)_{\tau_k}) \mathbf{1}_{\{\tau_k < \infty\}}$. The latter random variable is non-negative and with probability α greater than or equal to α; this means that this difference between f and $(L^k \cdot S)_\infty$ is not "small with respect to convergence in probability"; this is, what we had in mind when saying that L^k is a "sensible" impovement as compared to H^{n_k} or H^{m_k}.

Modulo some technicalities, which are worked out in [DS 94, lemma 4.5], this gives the desired contradiction to the maximality assumption on f, thus finishing the (sketch of the) proof of lemma 4.   □

Lemma 4 is our first step towards a proof of proposition 3: it gives some information on the convergence of the sequence of processes $(H^n \cdot S)_{n=0}^\infty$ in terms of the maximal functions defined in (124). But the assertion that these maximal functions tend to zero in probability is still much weaker than the convergence of $(H^n \cdot S)_{n=0}^\infty$ with respect to the semi-martingale topology, which we finally need in order to be able to apply Memin's theorem. There is still a long way to go!

But it is time to finish this "guided tour" and to advise the interested reader to find the remaining part of the proof on pages 482–494 of [DS 94]. We hope that we have succeeded to give some motivation for the proof and for the "economically motivated" arguments underlying it.

To finish this section we return to the issue, that we always have assumed that the process S is *locally bounded*. What happens if we drop this — technically very convenient — assumption?

Before starting to answer this question, we remark that it is not only of "academic" interest. It is also important from the point of view of applications: once one leaves the framework of continuous processes S — and there are good empirical reasons to do so — it is also natural to allow for the jumps to be unbounded. As a concrete example we mention the family of ARCH (Auto Regressive Conditional Heteroskedastic) processes and their relatives (GARCH, EGARCH etc.), which are very popular in the econometric literature. These are processes in discrete time where the conditional distribution of the jumps is Gaussian. In particular, these processes are not locally bounded. There are many other examples of processes which fail to be locally bounded, used in the modelling of financial markets.

The answer to the above question is as we expect it to be: *mutatis mutandis* the fundamental theorem of asset pricing 7 and the related theorems obtained in its proof carry over to the case of not necessarily locally bounded $\mathbb{R}^{d+1}$-valued semi-martingales S. Not coming as a surprise, the techniques of the proofs have to be refined: in particular, we cannot entirely reduce the situation to the study of the space $L^{\infty}(\Omega, \mathcal{F}, \mathbb{P})$, and the weak-star and norm topology of this space: there is no possibility any more to reduce to the case of (one-sided) bounded stochastic integrals and we therefore have to use larger spaces than $L^{\infty}(\Omega, \mathcal{F}, \mathbb{P})$. Yet it turns out — and this is slightly surprising — that the duality between $L^{\infty}(P)$ and $L^1(P)$ still remains the central issue of the proof.

Here is the statement of the extension of the fundamental theorem of asset pricing as obtained in [DS 98].

Theorem 9. ([DS 98, corr.1.2]) *The following assertions are equivalent for an $\mathbb{R}^{d+1}$-valued semi-martingale model $S = (S_t)_{t \geq 0}$ of a financial market:*

(i) (ESMM), i.e., there is a probability measure Q equivalent to P such that S is a sigma-martingale under Q.

(ii) (NFLVR), i.e., S satisfies the condition of no free lunch with vanishing risk.

There is a slight change in the statement of the theorem as compared to the statement of theorem 7: the term "local martingale" in the definition of *(EMM)* was replaced by the term "sigma-martingale" thus replacing the acronym *(EMM)* by *(ESMM)*. On the other hand, condition (ii) remained completely unchanged.

The notion of a sigma-martingale is a generalisation of the notion of a local martingale:

Definition 15. *[DS 98] An $\mathbb{R}^n$-valued semi-martingale $S = (S_t)_{t \geq 0}$ is called a sigma-martingale if there is a predictable process $g = (g_t)_{t \geq 0}$, taking its values in $]0, 1]$, such that the stochastic integral $g \cdot S$ is a martingale.*

It is easy to verify that a local martingale satisfies the above condition. More delicate is the fact that there are examples of sigma-martingales which fail to be local martingales: this was shown in a famous and ingenious example by M. Emery [E 80].

It is shown in [DS 98] that the notion of sigma-martingales makes good sense economically in the present context. Indeed, the "only if" implication of lemma 2 above extends to not necessarily locally bounded semi-martingales, if we replace the term local martingale by the term sigma-martingale. For this as well as for the (rather technical) proof of theorem 9 we refer to [DS 98].

5 Some Applications of the Fundamental Theorem of Asset Pricing

The crucial message of theorem 7 and the results obtained in the course of the proof is not only, that the version of the FTAP, as obtained by Harrison-Pliska for the case of finite Ω (theorem 1 above) and subsequently extended by several authors (we refer to [DS 94] for references on the literature), carries over — mutatis mutandis — to the general semi-martingale setting. For the applications, the additional information provided by theorem 8 pertaining to the weak-star closedness of the set C turns out to be at least as relevant.

As a typical example we show that, once the weak-star closedness of C is established by theorem 8, it is straight forward to deduce the extension of theorem 2 on *Pricing by No-Arbitrage* from the setting of finite Ω to the present semi-martingale setting.

We start with the analogue of proposition 1: for the sake of coherence we again place us into the setting of locally bounded processes as in the previous section; but we remark that the subsequent results also extend to the non locally bounded case (see [DS 98]).

Proposition 4. *Suppose that the locally bounded $\mathbb{R}^{d+1}$-valued semi-martingale $S = (S_t)_{t\geq 0}$ satisfies (NFLVR). Then the polar of C, taken with respect to the duality between $L^\infty(\mathbb{P})$ and $L^1(\mathbb{P})$, and identifying a $\mathbb{P}$-absolutely continuous measure Q with its Radon-Nikodym derivative $\frac{dQ}{d\mathbb{P}}$, is equal to $\mathrm{cone}(\mathcal{M}^a(S))$, and $\mathcal{M}^e(S)$ is dense in $\mathcal{M}^a(S)$ with respect to the norm topology of $L^1(\mathbb{P})$. Hence the following assertions are equivalent for an element $g \in L^\infty(\Omega, \mathcal{F}, \mathbb{P})$:*

(i) $g \in C$,
(ii) $\mathbf{E}_Q[g] \leq 0$, for all $g \in \mathcal{M}^a(S)$,
(iii) $\mathbf{E}_Q[g] \leq 0$, for all $g \in \mathcal{M}^e(S)$,

Proof First note that, similarly as in lemma 1, a probability measure Q, absolutely continuous with respect to $\mathbb{P}$, is in $\mathcal{M}^a(S)$ iff $\mathbf{E}_Q[g] \leq 0$, for all $g \in C$: the necessity of this condition was shown in (114); for the sufficiency we use the local boundedness of S and lemma 2 to obtain that the condition $\mathbf{E}_Q[g] \leq 0$, for $g \in C$, implies in particular that S is a local martingale under Q. In other words, the polar of C equals $\mathrm{cone}(\mathcal{M}^a(S))$.

The bipolar theorem [Sch 66] therefore implies that an element g of $L^\infty(\Omega, \mathcal{F}, \mathbb{P})$ is in the *weak-star closure* of C iff condition (ii) is satisfied. By theorem 8 we know that C is already weak-star closed, hence (i) is equivalent to (ii).

The density of $\mathcal{M}^e(S)$ in $\mathcal{M}^a(S)$ and therefore the equivalence of (ii) and (iii) follows by the same argument as in proposition 1 above. □

We now carry the argument underlying theorem 2 over to the present setting.

To maintain in line with the formulation of theorem 2, it is convenient to introduce some notation. A given $f \in L^\infty(\Omega, \mathcal{F}, \mathbb{P})$ is called *super-hedgeable* (resp. *strictly super-hedgeable*) at price $a \in \mathbb{R}$, if there is an admissible trading strategy H s.t. $f \le a + (H \cdot S)_\infty$ (resp. s.t., in addition, we have $P[f < a + (H \cdot S)_\infty] > 0$). In other words, f is super-hedgeable (resp. strictly super-hedgeable) at price a, if $f - a$ is in C (resp. if, in addition, $f - a$ is not a maximal element of C). Accordingly, we say that f is *sub-hedgeable*, (resp. *strictly sub-hedgeable*) at price a if $-(f - a)$ is in C (resp. if, in addition, $-(f - a)$ is not maximal in C). A real number a is called an *arbitrage free price* for f, if f is neither strictly super- nor strictly sub-hedgeable at price a.

Denoting by S_+ the set of prices a at which f is super-hedgeable, it is rather obvious that S_+ is an interval, its upper bound being equal to ∞, and its lower bound being an element of the interval [ess inf (f), ess sup (f)]. It is less obvious that S_+ is closed, but this fact is a straightforward consequence of theorem 8: If $(f - a_n) \in C$, for each n, and $\lim_{n \to \infty} a_n = a$, then $f - a \in C$. Hence there is $\beta \in \mathbb{R}$ s.t. $S_+ = [\beta, \infty[$.

Denoting by S_- the set of prices at which S is sub-hedgeable we similarly obtain that $S_- =] - \infty, a]$, for some $\alpha \in \mathbb{R}$. As, for any $Q \in \mathcal{M}^e(S)$, we have $\mathbf{E}_Q[f] \le \beta$ and $\mathbf{E}_Q[f] \ge \alpha$, (apply (114)), we observe that $\alpha \le \beta$, as soon as $\mathcal{M}^e(S) \ne \emptyset$.

Using the notation (30) and (31) we also have $\beta = \overline{\pi}(f)$ and $\alpha = \underline{\pi}(f)$. Indeed, we just have remarked the inequalities $\overline{\pi}(f) \le \beta$ and $\underline{\pi}(f) \ge \alpha$. Conversely, we know from theorem 8 and proposition 4 that, for $a < \beta$, we may find $Q \in \mathcal{M}^e(S)$ such that $\mathbf{E}_Q[f - a] > 0$, as $f - a$ is not in C. Hence for $\overline{\pi}(f) := \sup\{\mathbf{E}_Q[f] : Q \in \mathcal{M}^e(S)\}$, we obtain the inequality $\overline{\pi}(f) \ge \beta$; the same argument implies that $\underline{\pi}(f) \le \alpha$.

Having established $\alpha = \underline{\pi}(f)$ and $\beta = \overline{\pi}(f)$, we need a little extra argument for the proper treatment of the boundary cases α and β.

Lemma 5. *Under the above assumptions suppose in addition that $\alpha < \beta$. Then f is strictly super-hedgeable at price β and strictly sub-hedgeable at price α. Hence, for $Q \in \mathcal{M}^e(S)$, we have $\mathbf{E}_Q[f] \in]\alpha, \beta[$.*

Proof We know that f is super-hedgeable at price β, i.e., there is an admissible trading strategy H such that $f \le \beta + (H \cdot S)_\infty$. To show that f is, in fact, strictly super-hedgeable at price β, define the stopping time τ by

$$\tau = \inf\{t : (H \cdot S)_t \ge 1 + \text{ess sup}(f)\}. \tag{126}$$

Clearly $\hat{H} := H 1_{[0,\tau]}$ also is a super-hedging strategy for f.

Now we distinguish two cases: either $P[\tau < \infty] > 0$. Then the trading strategy $\hat{H}$ strictly super-hedges f. Or $P[\tau < \infty] = 0$; in this case we have that $H = \hat{H}$ and that $H \cdot S$ is a bounded process; therefore $H \cdot S$ is a uniformly integrable martingale under each $Q \in \mathcal{M}^a(S)$. This implies that the original strategy H defines a strict super-hedge for f, i.e., $P[f < \beta + (H \cdot S)_\infty] > 0$.

Indeed, otherwise we would have that $\mathbf{E}_Q[f] = \mathbf{E}_Q[\beta + (H \cdot S)_\infty] = \beta$, for each $Q \in \mathcal{M}^e(S)$, in contradiction to the assumption $\alpha < \beta$.

Summing up, we have shown that f is strictly super-hedgeable at price β; applying the same argument to $-f$ we see that f is strictly sub-hedgeable at price α.

The final statement of the lemma is now obvious. $\square$

Taking up again the discussion preceding lemma 5, we distingnish two cases: either $\alpha < \beta$, in which case lemma 5, tells us that the arbitrage-free prices for f consist of the open interval $]\alpha, \beta[$. We then also have that $]\alpha, \beta[=]\underline{\pi}(f), \overline{\pi}(f)[$. In the case $\alpha = \beta$ we have that there is an admissible trading strategy H such that $f \leq \alpha + (H \cdot S)_\infty$. Fixing an arbitrary $Q \in \mathcal{M}^e(S)$, we must have $\mathbf{E}_Q[f] = \alpha$, so that $H \cdot S$ must be a uniformly integrable martingale under Q (it is a Q-super-martingale verifying $\mathbf{E}_Q[(H \cdot S)_\infty] = \mathbf{E}_Q[(H \cdot S)_0] = 0$). Hence $(H \cdot S)_t = \mathbf{E}_Q[f - a | \mathcal{F}_t]$, which shows in particular that the process $H \cdot S$ is bounded. Therefore H as well as $-H$ are admissible trading strategies.

Summing up: we have proved the subsequent extension of theorem 2 (compare [DS 95, theorem 5.7]) to the present semi-martingale setting, which carries over almost verbatim from the setting of finite Ω.

Theorem 10 (Pricing by No-Arbitrage). *Assume that the locally bounded semi-martingale $S = (S_t)_{t \geq 0}$ satisfies (NFLVR) and let*

$$\overline{\pi}(f) = \sup \{\mathbf{E}_Q[f] : Q \in \mathcal{M}^e(S)\}, \tag{127}$$

$$\underline{\pi}(f) = \inf \{\mathbf{E}_Q[f] : Q \in \mathcal{M}^e(S)\}, \tag{128}$$

Either $\underline{\pi}(f) = \overline{\pi}(f)$, in which case $f = \pi(f) + (H \cdot S)_\infty$, where $\pi(f) = \overline{\pi}(f) = \underline{\pi}(f)$ and H is a predictable process such that the process $H \cdot S$ is bounded.

Or $\underline{\pi}(f) < \overline{\pi}(f)$, in which case $\{\mathbf{E}_Q[f] : Q \in \mathcal{M}^e(S)\}$ equals the open interval $]\underline{\pi}(f), \overline{\pi}(f)[$, which in turn equals the set of arbitrage-free prices for the contingent claim f.

In the formulation of the above theorem we have restricted ourselves to the case of bounded random variables $f \in L^\infty(\Omega, \mathcal{F}, \mathbb{P})$. One may also extend it — mutatis mutandis — to the case of functions f which are uniformly bounded from below or, more generally, bounded from below by some fixed random variable w having appropriate integrability conditions (see, e.g., [J 92], [AS 94] and [DS 98]).

Let us briefly review some other applications of theorem 8. A rather subtle consequence, requiring quite a bit of additional work, is the subsequent extension of the optional decomposition theorem 4 to a general semi-martingale setting as given by D. Kramkov([K 96]):

Theorem 11 (Optional Decomposition). *Let $S = (S_t)_{t \geq 0}$ be a locally bounded $\mathbb{R}^{d+1}$-valued semi-martingale satisfying (NFLVR), and let $V =$*

$(V_t)_{t\geq 0}$ be a non-negative, adapted, càdlàg process, defined on the filtered stochastic base $(\Omega, \mathcal{F}, (\mathcal{F}_t)_{t\geq 0}, \mathbb{P})$.

The following assertions are equivalent:

(i) V is a super-martingale, for each $Q \in \mathcal{M}^e(S)$.

(i') V is a super-martingale, for each $Q \in \mathcal{M}^a(S)$.

(ii) V may be decomposed into $V = V_0 + H \cdot S - C$, where H is an admissible trading strategy and $C = (C_t)_{t\geq 0}$ is an increasing, càdlàg, adapted process starting at $C_0 = 0$.

The above theorem extends the "baby version" for finite Ω presented in theorem 4 above. A first non-trivial version of this theorem was given by N. El Karoui and M.-C. Quenez [KQ 95] in the context of a filtration generated by an n-dimensional Brownian motion, using techniques from stochastic control. The version stated above was proved by D. Kramkov [K 96]. Subsequently H. Föllmer and Y. Kabanov [FK 98] extended the result to the case of non locally bounded semi-martingales; their method uses a Lagrange multiplier technique and does not rely on theorem 8. Finally, F. Delbaen and the present author [DS 99] also removed the assumption of non-negativity of V; their proof is similar in spirit to Kramkov's original one and heavily relies on theorem 10. We shall now present the basic idea of this proof.

Sketch of proof of theorem 11 As in theorem 4 above we only have to show the implication (i) $\Rightarrow$ (ii). Fix an increasing sequence of finite meshes $M^n = \{0, t_1^n, \ldots, t_{N_n}^n\}$, such that $\cup_{n=1}^{\infty} M^n$ is dense in $\mathbb{R}_+$. For example, we may take $N_n = n2^n$ and $t_i^n = \frac{i}{2^n}$, for $i = 1, \ldots, N_n$.

For fixed $n \in \mathbb{N}$ and $i = 1, \ldots, N_n$, we proceed similarly as in the proof of theorem 4 above: we consider the process $(S_t)_{t_{i-1}^n \leq t \leq t_i^n}$ and apply theorem 10: the condition

$$\mathbb{E}_Q[V_{t_i^n} | \mathcal{F}_{t_{i-1}^n}] \leq V_{t_{i-1}^n}, \qquad \text{for } Q \in \mathcal{M}^e(S), \tag{129}$$

implies that there is an admissible predictable process $(H_t^{n,i})_{t_{i-1}^n < t \leq t_i^n}$, supported by $]t_{i-1}^n, t_i^n]$, such that

$$V_{t_i^n} \leq V_{t_{i-1}^n} + (H^{n,i} \cdot S)_{t_i^n}. \tag{130}$$

In fact, we have to apply theorem 10 conditionally with respect to the sigma-algebra $\mathcal{F}_{t_{i-1}^n}$; but this conditional extension of theorem 10 does not present any difficulty.

Fixing $n \in \mathbb{N}$, letting $H^n := \Sigma_{i=1}^{N_n} H^{n,i}$ and, defining $\Delta C_i^n := V_{t_{i-1}^n} + (H^{n,i} \cdot S)_{t_i^n} - V_{t_i^n}$ for $i = 1, \ldots, N_n$, we obtain the following objects: an admissible trading strategy $H^n = (H_t^n)_{t\geq 0}$, indexed by $\mathbb{R}_+$, and an adapted increasing process $C^n = (C_t^n)_{t\in M^n}$, indexed by the finite time index set M^n, such that

$$V_t = (H^n \cdot S)_t - C_t^n, \qquad \text{for } t \in M_n. \tag{131}$$

This is not yet quite what we want to have, as we want to find a predictable process $H = (H_t)_{t \geq 0}$ and an adapted càdlàg process $C = (C_t)_{t \geq 0}$, indexed by $t \in \mathbb{R}_+$, such that (131) holds true for all $t \in \mathbb{R}_+$.

But it is clear what we have to do to achieve this goal: we have to *pass to the limit* of the sequence $(H^n)_{n=0}^{\infty}$. Hence, again, we face our usual problem: how to pass from a sequence $(H^n)_{n=0}^{\infty}$ of admissible integrands to a limit H? Similarly as in the context of the proof of the fundamental theorem of asset pricing, the only essential information on the sequence of admissible trading strategies $(H^n)_{n=1}^{\infty}$ is that they have a uniform lower bound: indeed, one easily deduces from the assumption $V \geq 0$ that $H^n \cdot S \geq -V_0$, for all $n \in \mathbb{N}$.

Hence the basic problem of the proof of the present theorem is very similar in spirit to the theme of the proof of theorem 8. It turns out that, refining some of these arguments, it is indeed possible to find a limiting strategy H above. For the details we refer to [K 96] or [DS 99]. □

As a final application of theorem 8 we mention the topic of utility optimization in financial markets. Roughly speaking, one fixes a utility function U on $\mathbb{R}$, i.e. an increasing, strictly concave function $U : \mathbb{R} \to \mathbb{R} \cup \{-\infty\}$, and an initial endowment $x \in \mathbb{R}$. A typical problem consists in finding, for a fixed horizon T, a trading strategy H maximizing

$$\mathbb{E}[U(x + (H \cdot S)_T)]. \tag{132}$$

We cannot go in detail into this rather extensive theory here and refer, e.g., to the survey paper [S 01]. We only mention that the modern way to deal with the problem of maximizing (132) is to use the duality theory of convex optimization in infinite-dimensional spaces. The crucial property in order to make this theory work, again, is the polar relation between the sets C and $\mathcal{M}^a(S)$ as stated in theorem 5. The heart of the matter therefore again is the weak-star closedness of C as stated in theorem 8.

References

[AS 94] J.P. Ansel, C. Stricker, (1994), *Couverture des actifs contingents et prix maximum.* Ann. Inst. Henri Poincaré, Vol. 30, pp. 303–315.

[B 00] L. Bachelier, (1900), *Théorie de la Speculation.* Ann. Sci. Ecole Norm. Sup., Vol. 17, pp. 21–86, [English translation in: The Random Character of stock prices (P. Cotner, editor), MIT Press, 1964].

[B 32] S. Banach, (1932), *Théorie des operations linéaires.* Monogr. Mat., Warszawa 1. Reprint by Chelsea Scientific Books.

[BS 73] F. Black, M. Scholes, (1973), *The pricing of options and corporate liabilities.* Journal of Political Economy, Vol. 81, pp. 637–659.

[CK 00] J.-M. Courtault, Y. Kabanov, B. Bru, P. Crépel, I. Lebon, A. Le Marchand, (2000), *Louis Bachelier: On the Centenary of "Theorie de la Spéculation".* Mathematical Finance, Vol. 10, No. 3, pp. 341–353.

[CRR 79] J. Cox, S. Ross, M. Rubinstein, (1979), *Option pricing: a simplified approch*. Journal of Financial Economics, Vol. 7, pp. 229–263.

[D 92] F. Delbaen, (1992), *Representing martingale measures when asset prices are continuous and bounded*. Mathematical Finance, Vol. 2, pp. 107–130.

[DS 94] F. Delbaen, W. Schachermayer, (1994), *A General Version of the Fundamental Theorem of Asset Pricing*. Math. Annalen, Vol. 300, pp. 463–520.

[DS 94a] F. Delbaen, Schachermayer, W. (1994) *Arbitrage and free lunch with bounded risk for unbounded continuous processes*. Mathematical Finance, Vol. 4, pp. 343–348.

[DS 95] F. Delbaen, W. Schachermayer, (1995), *The No-Arbitrage Property under a change of numéraire*. Stochastics and Stochastic Reports, Vol. 53, pp. 213–226.

[DS 98] F. Delbaen, W. Schachermayer, (1998), *The Fundamental Theorem of Asset Pricing for Unbounded Stochastic Processes*. Mathematische Annalen, Vol. 312, pp. 215–250.

[DS 99] F. Delbaen, W. Schachermayer, (1999), *A Compactness Principle for Bounded Sequences of Martingales with Applications*. Proceedings of the Seminar of Stochastic Analysis, Random Fields and Applications, Progress in Probability, Vol. 45, pp. 137–173.

[E 80] M. Emery, (1980), *Compensation de processus à variation finie non localement intégrables*. Séminaire de Probabilités XIV, Springer Lecture Notes in Mathematics, Vol. 784, pp. 152–160.

[FK 98] H. Föllmer, Yu.M. Kabanov, (1998), *Optional decomposition and Lagrange multiplievs*. Finance and Stochastics, Vol. 2, pp. 69–81.

[H 99] J. Hull, (1999), *Options, Futures, and Other Derivatives*. 4th Edition , Prentice-Hall, Englewood Cliffs, New Jersey.

[HK 79] J.M. Harrison, D.M. Kreps, (1979), *Martingales and arbitrage in multiperiod securities markets*. J. Econ. Theory, Vol. 20, pp. 381–408.

[HP 81] J.M Harrison, S.R. Pliska, (1981), *Martingales and Stochastic intefrals in the theory of continuous trading*. Stoch. Proc. & Appl., Vol. 11, pp. 215–260.

[J 92] S.D. Jacka, (1992), *A martingale representation result and an application to incomplete financial markets*. Mathematical Finance, Vol. 2, pp. 239–250.

[K 81] D.M. Kreps, (1981), *Arbitrage and equilibrium in economies with infinitely many commodities*. J. Math. Econ., Vol. 8, pp. 15–35.

[K 96] D. Kramkov, (1996), *Optional decomposition of supermartingales and hedging contingent claims in incomplete security markets*. Probability Theory and Related Fields, Vol. 105, pp. 459–479.

[KQ 95] N. El Karoui, M.-C. Quenez, (1995), *Dynamic programming and pricing of contingent claims in an incomplete market*. SIAM J. Control Optim., Vol. 33, pp. 29–66.

[LL 96] D. Lamberton, B. Lapeyre, (1996), *Introduction to Stochastic Calculus Applied to Finance*. Chapman & Hall, London.

[M 73] R.C. Merton, (1973), *Theory of rational option pricing*. Bell J. Econom. Manag. Sci., Vol. 4, pp. 141–183.

[M 80] J. Memin, (1980), *Espaces de semi-martingales et changement de probebilité*. Zeitschrift für Wahrscheinlichkeitstheorie und verwandte Gebiete, Vol. 52, pp. 9–39.

[P 90] Protter, P. (1990) *Stochastic Integration and Differential Equations*. Applications of Mathematics, Vol. 21, Springer, Berlin, Heidelberg, New York.

[RY 91] D. Revuz, M. Yor, (1991), *Continuous Martingales and Brownian Motion.*
Springer, Berlin, Heidelberg, New York.

[S 65]   P.A. Samuelson, (1965), *Proof that properly anticipated prices fluctuate randomly.* Industrial Management Review, Vol. 6, pp. 41–50.

[S 90]   C. Stricker, (1990), *Arbitrage et lois de martingale.* Ann. Inst. H.Poincaré
Probab. Statist., Vol. 26, pp. 451–460.

[S 94]   W. Schachermayer, (1994), *Martingale Measures for discrete time processes
with infinite horizon.* Math. Finance, Vol. 4, No. 1, pp. 25–55.

[S 01]   W. Schachermayer, (2001), *Optimal Investment in Incomplete Financial
Markets.* Mathematical Finance: Bachelier Congress 2000 (H. Geman, D.
Madan, St.R. Pliska, T. Vorst, editors), Springer, pp. 427–462.

[Sch 66] H.H. Schäfer, (1966), *Topological Vector Spaces.* Graduate Texts in Mathematics.

[Sh 99] A.N. Shiryaev, (1999), *Essentials of Stochastic Finance. Facts, Models, Theory.* World Scientific.

[T 00]   M.S. Taqqu, (2000), *Bachelier and his Times: A Conversation with Bernard
Bru.* Finance & Stochastics, Vol. 5, Issue 1.

[W 91]   D. Williams, (1991), *Probability with Martingales.* Cambridge University
Press.

[Y 80]   J.A. Yan, (1980), *Caracterisation d' une classe d'ensembles convexes de L^1
ou H^1.* Lect. Notes Mathematics, Vol. 784, pp. 220–222.

[Y 78]   M. Yor, (1978), *Sous-espaces denses dans L^1 ou H^1.* Lect. Notes Mathematics, Vol. 649, pp. 265–309.

Part III

Michel Talagrand: Mean field models for spin glasses:
a first course

Table of Contents

1 Introduction

Consider a large collection of random variables $(H_s)_{s \leq M}$. What is the value of the largest of them? More generally, what is the structure of the "few largest" values? Certainly the question is too general. Even if the variables are identically distributed, the answer depends upon both the distribution and the correlation structure of the variables. When the variables are independent, then of course, everything can be computed. To move beyond that case, one should ask what correlation structures are of interest. A collection of random variables is, in other words, a stochastic process. The examples that first come to mind are indexed by the real line. The correlation structure one will consider will naturally take advantage of the structure of the real line, a "one-dimensional" object.

The direction purposed here will be completely different and probably completely new to the reader. The correlation structure considered will be high dimensional, in a sense that will soon be clear. This type of structures appear naturally in problems of combinatorial optimization in a random setting. One can dream that there exists an entire theory in the above direction; the present monograph presents only a few steps of the study of a typical example.

The motivation behind the present research is the work of physicists on mean field models for spin glasses, and in particular the book "Spin glass theory and beyond" by Mézard, Parisi, Virasoro. The physicists have built a truly remarkable theory, which they believe applies not only to the models they have introduced, but also to a number of problems of fundamental interest in probability. Due to space (and energy) limitation, these models will not be discussed here, and we refer the reader to [T6] for a non-technical discussion.

The predictions of the physicists are most likely true. A reader of this volume is however probably well aware that studying a problem by the methods of theoretical physics (in the present case heuristic arguments backed by numerical simulation) and providing mathematical arguments are two distinct endeavors. We make a definite choice for the later approach.

The author has devoted a major effort since 1994 to the task of providing proofs to some of the physicists' statements. This has resulted in a number of very long and very technical papers. The author's concern is that they will remain forever hardly accessible to others. The main goal of the present notes is to address this concern. Our overwhelming objective is readability. No knowledge of physics, or statistical mechanics is required, or even probably useful to read these notes. There are no prerequisite of any kind, beside a familiarity with the basic concepts of probability theory. In fact, the most advanced tool that will not be proved is Hölder's inequality. With the exception of a few results that are probably still far from their final form, every proof is given in complete detail. The usual dreadful "it is easy to see..." have not been permitted to creep in. In order to improve readability we will discuss only the SK model and its cousin, the p-spin interaction model.

A regrettable consequence of this (necessary) choice is that it will not be possible to introduce the reader to what is possibly the greatest charm of the topic: the existence of several models, which yield to the same overall approach, but seem to require different technical tools. But of course the reader having mastered the present notes should be well armed to explore this aspect by reading the relevant research papers.

Let us now briefly describe the content of each chapter.

Chapter 2 is devoted to a toy model, Derrida's Random Energy Model, that helps understanding what this is all about. This model is not treated in the simplest possible manner, but is also an opportunity to introduce some tools.

Chapter 3 studies the high temperature case of the SK model without external field. Some special symmetries make this case particularly simple and well understood.

The main motivation is that the methods developed there also apply to the p-spin interaction model, considered in Chapter 4. In this chapter, we prove the basic a priori estimates that will be the starting point of our study of this model below the critical temperature.

In Chapter 5, we return to the SK model, this time with external field. This is a much more interesting situation. We give a simple and complete proof of the mysterious "replica-symmetric" formula and we describe in great detail the structure behind them. This chapter is essentially self contained.

In Chapter 6, we investigate in much greater detail the structure behind the replica-symmetric formula. We replace crude moment inequalities by sharp exponential inequalities. Most importantly, the proofs there, rather than taking advantage of specific features of the SK model, rely on tools of much wider applicability.

In Chapter 7 we compute the inter-spin correlation. We explain the occurrence of the AT line, the conjectured boundary between the high and low temperature region. We also argue that the validity of the replica-symmetric solution on the entire high-temperature region is not self-evident (as the physicists think) but rather is a difficult problem.

In Chapter 8 we return to the p-spin interaction model. We study it *below* the critical temperature. We prove that the configuration space spontaneously decomposes in small pieces that we call the lumps. We prove that these are as much separated from each other as they can be.

The structure of this model below (but not too far below) the critical temperature is now (January 2000) rigorously understood in much greater detail than is presented here [T7]. Having decided at the onset that this monograph would contain only proofs that were being written a second time, we have not included these recent results here. For the same reason, the reader is referred to [T8] for another major progress (April 2000) on the region of validity of the replica-symmetric solution. The reader having penetrated the present monograph should however be well prepared to read [T7, T8].

Acknowledgment. The author expresses his deepest gratitude to Professor Anne Boutet de Monvel who made the publication of this monograph possible by accepting the unrewarding labor of typing it. He also expresses his thanks to the many participants of the school that helped to correct countless typos and inaccuracies.

2 What this is all about: the REM

Consider the space $\Sigma_N = \{-1, 1\}^N$. A point $\sigma = (\sigma_1, \ldots, \sigma_N)$ in Σ_N will be called a configuration. It physically represents the position (= configuration) of N "spins" $\sigma_i = \pm 1$. We will have a collection of r.v. $H_N(\sigma)$, and we will be interested in the smallest, or the few smallest, of the values $H_N(\sigma)$ as σ ranges over all possible configurations. The physical idea is that $H_N(\sigma)$ represents the "energy" of the configuration σ. The function H_N is called the Hamiltonian. The energies $H_N(\sigma)$ are random due to some kind of "disorder" that we try to understand. To study the small values of $H_N(\sigma)$ we will weight them using the Boltzmann factors $\exp(-\beta H_N(\sigma))$, where $\beta \geq 0$ is a parameter. We will then consider Gibbs' (probability) measure G_N on Σ_N, that will be given by

$$(2.1) \qquad G_N(\sigma) = \frac{1}{Z_N} \exp(-\beta H_N(\sigma)),$$

where Z_N is the normalization factor, called the *partition function*,

$$(2.2) \qquad Z_N = Z_N(\beta) = \sum_\sigma \exp(-\beta H_N(\sigma)),$$

where the sum is over all configurations σ. The physical idea is that Gibbs' measure quantifies the thermal fluctuations of the N-spin system with energy levels $H_N(\sigma)$ when it is in thermal equilibrium with a heat bath at temperature $T = 1/\beta$. That is, $G_N(\sigma)$ is the probability to find the system in configuration σ after it has been left undisturbed a long time. The $-$ sign is to respect the conventions of physics, where energies are minimized, not maximized.

Gibbs' measure is a basic concern of statistical mechanics. The reader however will know very soon all the statistical mechanics he will need for these notes. The reason is that disorder (= randomness) is the overwhelming feature here, and that the vast body of knowledge of standard "non disordered" statistical mechanics seems to be largely irrelevant.

Gibbs' measure is a probability measure. It is itself random. Thus, we have two levels of randomness, and this topic is in some sense a probabilist's paradise.

Given a (typical) occurrence of the disorder, we will try to understand the structure of G_N. The disorder does not evolve with the thermal fluctuations. It is "frozen" or "quenched" as the physicists say. The word "glass" conveys that idea of "frozen disorder".

The subtilty of definition (2.1) is not apparent at first sight. The normalization will be chosen so that $\max_\sigma(-H_N(\sigma))$ is of order N so the sum (2.2) is the sum of 2^N terms, some of which small, others of exponential order, and it is not clear a priori from which of these comes the main contribution. Computation of Z_N is a major objective, and this computation is essentially equivalent to understanding G_N, because the derivatives of $\log Z_N$ with respect to the various parameters we will consider are expressed as integrals with respect to G_N. For example

$$(2.3) \qquad \frac{\mathrm{d}}{\mathrm{d}\beta} \log Z_N(\beta) = \frac{1}{Z_N(\beta)} \sum_\sigma -H_N(\sigma) \exp\left(-\beta H_N(\sigma)\right)$$
$$= \langle -H_N(\sigma) \rangle.$$

In this formula, as well as in the rest of the notes, $\langle \cdot \rangle$ denotes average with respect to Gibbs' measure, i.e.

$$(2.4) \qquad \langle f(\sigma) \rangle = \frac{1}{Z_N} \sum_\sigma f(\sigma) \exp(-\beta H_N(\sigma)).$$

Throughout the notes, we will assume that

(2.5)   The family $(H_N(\sigma))$ is a jointly Gaussian family of r.v.

This family will not always be centered. We will also assume

$$(2.6) \qquad \forall \sigma, \quad \mathsf{E}(H_N^2(\sigma)) - (\mathsf{E}\,H_N(\sigma))^2 \le \frac{N}{2}.$$

It should be apparent that E will denote expectation in the variables $H_N(\sigma)$. The corresponding probability will be denoted by P.

It is of course impossible to even *think* the word Gaussian without immediately mentioning the most important property of Gaussian processes, that is concentration of measure. We will use the following convenient formulation ([I-S-T]).

Proposition 2.1. *Consider a function F on $\mathbb{R}^M$, and assume that its Lipschitz constant is at most A, i.e.*

$$(2.7) \qquad x, y \in \mathbb{R}^M \implies |F(x) - F(y)| \leq A\|x - y\|$$

where $\|x\|$ is the Euclidean norm of x. Then if $g = (g_1, \ldots, g_M)$ where the r.v. g_i are i.i.d. $\mathcal{N}(0,1)$ we have, for each $u > 0$,

$$(2.8) \qquad P(|F(g) - \mathsf{E}\, F(g)| \geq u) \leq \exp\left(-\frac{u^2}{2A^2}\right).$$

A statement of similar strength (with the loss of a factor 2 in the exponent, loss that is irrelevant here) will be proved at the end of Chapter 6.

Throughout the paper, we will write

$$(2.9) \qquad p_N = p_N(\beta) = \frac{1}{N} \mathsf{E} \log Z_N(\beta).$$

Thus p_N is the "expected density of the logarithm of the partition function". It would be interesting to find a nice name for p_N. We find it safer to avoid taking unnecessary risks here. We will call this quantity "p_N". The quantity p_N is closely related to the "free energy" considered by physicists. The free energy is $-\frac{1}{\beta} p_N(\beta)$. The factor $\frac{1}{\beta}$ is a nuisance for mathematics. This had led to the name "free energy" for p_N in previous papers but this confusing terminology is abandoned here.

It will turn out that $p_N(\beta)$ is of order 1. The following proposition shows that the single number $p_N(\beta)$ is a good way to capture information about the r.v. $\frac{1}{N} \log Z_N(\beta)$.

Proposition 2.2. *Under (2.5), (2.6) we have for $u \geq 0$,*

$$(2.10) \qquad P\left(\left|\frac{1}{N} \log Z_N(\beta) - p_N(\beta)\right| \geq u\right) \leq \exp\left(-\frac{Nu^2}{\beta^2}\right).$$

Proof. It is elementary that for $M = 2^N$ we have a representation

$$(2.11) \qquad H_N(\sigma) = \mathsf{E}\, H_N(\sigma) + a(\sigma) \cdot g$$

where $a(\sigma) \in \mathbb{R}^M$, g is as in Proposition 2.1, $\cdot$ denotes the dot product and $\|a(\sigma)\|^2 \leq N/2$. The function

$$F(x) = \frac{1}{N} \log\left(\sum_\sigma \exp\left(-\beta(\mathsf{E}\, H_N(\sigma) + a(\sigma) \cdot x)\right)\right)$$

satisfies (2.7) with $A = \frac{\beta}{\sqrt{2N}}$, because

$$|a(\sigma) \cdot x - a(\sigma) \cdot y| \leq \sqrt{\frac{N}{2}} \|x - y\|.$$

The result follows from (2.8). $\qquad\qquad\square$

Comment. The reader who does not find (2.11) obvious should be reassured. We do not provide details because in all cases where (2.10) will actually be used, the representation (2.11) will be obvious from the very definition of H_N.

Here is another simple observation:

Proposition 2.3. *The (random) functions $\beta \mapsto \frac{1}{N} \log Z_N(\beta)$ are convex, and so is p_N.*

Proof. Let f be a positive function on a measure space. Then, the map $\beta \mapsto \log(\int f^\beta \, d\mu)$ is convex by Hölder's inequality. □

Thus, the random function $\frac{1}{N} \log Z_N(\beta)$ is convex and has small fluctuations. The quantity of information that can be extracted from this simple fact is amazing [**G-G**].

Here is another simple fact.

Proposition 2.4. *For each $\beta \geq 0$ we have*

$$(2.12) \qquad p_N(\beta) \leq \frac{\beta^2}{4} + \frac{1}{N} \log \sum_\sigma \exp(-\beta \, \mathsf{E} \, H_N(\sigma)).$$

Proof. We use Jensen's inequality

$$(2.13) \qquad \mathsf{E} \log Z_N \leq \log \mathsf{E} \, Z_N$$

and we use (2.6) and the fact that for a (centered) Gaussian r.v. g we have

$$(2.14) \qquad \mathsf{E} \, e^g = e^{\mathsf{E} \, g^2 / 2}. \qquad \qquad □$$

There is a crucial point that must be explained here.

Proposition 2.2 shows that $p_N(\beta)$ is the correct way to obtain information about $\frac{1}{N} \log Z_N(\beta)$. However, in general $\mathsf{E} \, Z_N$ is *not* the correct information about Z_N. This is because a large part of $\mathsf{E} \, Z_N$ comes from very rare events. In other words, the mean and the median of Z_N are very different. This is true even when measured on a logarithmic scale. In that scale, the median and the average of Z_N are fairly represented respectively by

$$(2.15) \qquad p_N = \frac{1}{N} \, \mathsf{E} \log Z_N \quad \text{and} \quad \frac{1}{N} \log \mathsf{E} \, Z_N.$$

These two numbers will be of order 1, but different. The interesting (and difficult) problem is to understand the median of Z_N. The fundamental difference between these numbers (2.15) could be called the *lottery phenomenon*. This name reflects the situation of a person holding one of (say) 10^6 tickets of a lottery that offers a single (huge) prize. His expected gain might be sizable; but his typical gain is zero, significantly lower. The notion of "expected gain" implies that the experiment is repeated many times, and is not the way to look at a single lottery drawing.

For simplicity, we now assume

(2.16) $E\, H_N(\sigma) = 0$ for each σ,

so that the variables $H_N(\sigma)$ are centered and $E\, H_N^2(\sigma) \leq N/2$.

Proposition 2.5. *If $\beta \geq 0$, we have*

(2.17) $$p_N(\beta) \leq \frac{\beta^2}{4} + \log 2.$$

and if $\beta \geq 2\sqrt{\log 2}$ we have

(2.18) $$p_N(\beta) \leq \beta\sqrt{\log 2}.$$

Proof. We first observe that (2.17) is a consequence of (2.12) under (2.16). Next we observe that

$$Z_N \geq \exp\big(\beta \max_\sigma(-H_N(\sigma))\big)$$

and thus

$$p_N(\beta) \geq \frac{\beta}{N}\, E \max_\sigma(-H_N(\sigma)).$$

Combining with (2.17) for $\beta = 2\sqrt{\log 2}$, we see that

(2.19) $$\frac{1}{N}\, E \max_\sigma(-H_N(\sigma)) \leq \sqrt{\log 2}.$$

Next, we use (2.3) to get

(2.20) $$\frac{d}{d\beta}p_N(\beta) = \frac{1}{N}\, E\langle -H_N(\sigma)\rangle$$

$$\leq \frac{1}{N}\, E\big(\max_\sigma(-H_N(\sigma))\big)$$

$$\leq \sqrt{\log 2},$$

using (2.18). Combining (2.17) and (2.20) proves (2.18). $\square$

The upper bound of Proposition 2.5 is a kind of worst case. For example, if $H_N(\sigma)$ does not depend upon σ, we have $p_N(\beta) = \log 2$ for each N, β. The interesting (and difficult) case will be when there is some (but not too much) correlation between the variables $H_N(\sigma)$. In the rest of the section, we will investigate the case where

(2.21)   The r.v. $(H_N(\sigma))$ are i.i.d. $\mathcal{N}(0, N/2)$.

This case is known as *Derrida's random energy model* (REM): the energy levels are random independent. It is not surprising that in this case we can get a complete picture of the situation. This is nonetheless a very instructive exercise.

Our first goal is to show that in this case, the upper bounds of Proposition 2.5 are the correct ones.

Proposition 2.6. *For the REM we have*

$$(2.22) \qquad \lim_{N \to \infty} p_N(\beta) = \frac{\beta^2}{4} + \log 2 \quad \textit{if } \beta \le 2\sqrt{\log 2}$$

$$(2.23) \qquad \lim_{N \to \infty} p_N(\beta) = \beta\sqrt{\log 2} \quad \textit{if } \beta \ge 2\sqrt{\log 2}.$$

Proof. Let us first prove (2.23). We simply use that

$$p_N(\beta) = \frac{1}{N} \, \mathsf{E} \log Z_N(\beta)$$

$$\ge \frac{1}{N} \, \mathsf{E} \log \exp(-\beta \min_{\sigma} H_N(\sigma))$$

$$= \frac{\beta}{N} \, \mathsf{E} \max_{\sigma}(-H_N(\sigma)).$$

That is, we take into account only the largest term of the sum (2.2). Thus to prove (2.23) it is enough to show that the estimate (2.19) is essentially correct when the r.v. $H_N(\sigma)$ are independent. (We do not detail this easy point, as we will prove more later.) Another route is that (2.23) follows from (2.22), the upper bound of (2.20), and the convexity of p_N.

Let us now turn to the proof of (2.22). To understand what happens, it is useful to note that in the left-hand side of (2.14), most of the contribution occurs for $g \simeq \mathsf{E} g^2$. More precisely we have the following

Lemma 2.7. *Consider an $\mathcal{N}(0, \tau^2)$ r.v. g. Then, if $c < \tau^2$,*

$$\mathsf{E} \, e^g 1_{\{g \le c\}} \le \frac{1}{2} \exp\left(c - \frac{c^2}{2\tau^2}\right),$$

while, if $c > \tau^2$,

$$\mathsf{E} \, e^g 1_{\{g \ge c\}} \le \frac{1}{2} \exp\left(c - \frac{c^2}{2\tau^2}\right).$$

Thus, if $a > 0$,

$$(2.24) \qquad \mathsf{E} \, e^g 1_{\{|g - \tau^2| \ge a\}} \le e^{\tau^2/2} \exp\left(-\frac{a^2}{2\tau^2}\right).$$

Proof. For example, if $c > \tau^2$

$$\mathsf{E}\, e^g 1_{\{g \geq c\}} = \frac{1}{\tau\sqrt{2\pi}} \int_c^\infty e^t e^{-t^2/(2\tau^2)} \, dt$$

$$= \frac{1}{\tau\sqrt{2\pi}} e^{\tau^2/2} \int_c^\infty \exp\left(-\frac{1}{2\tau^2}(t - \tau^2)^2\right) dt$$

$$= e^{\tau^2/2} \int_{(c-\tau^2)/\tau}^\infty \frac{1}{\sqrt{2\pi}} \exp\left(-\frac{t^2}{2}\right) dt$$

$$\leq \frac{1}{2} e^{\tau^2/2} \exp\left(-\frac{(c - \tau^2)^2}{2\tau^2}\right)$$

$$= \frac{1}{2} \exp\left(c - \frac{c^2}{2\tau^2}\right). \qquad \square$$

The meaning of (2.22) is that (on a certain logarithmic scale), most of the time

$$(2.25) \qquad \sum_\sigma \exp\left(-\beta H_N(\sigma)\right) \simeq 2^N \exp \frac{N\beta^2}{4}$$

and Lemma 2.7 shows that about the only way this can happen is that there are enough values of σ for which $-\beta H_N(\sigma) \geq \frac{N\beta^2}{2}$. This is because if $a > 0$ the expected contribution of the terms for which $-\beta H_N(\sigma) \leq \frac{N\beta^2}{2} - aN$ is hopelessly smaller than the right hand side of (2.25). This suggests use of the bound

$$(2.26) \qquad Z_N(\beta) \geq \exp \frac{N\beta^2}{2} \,\mathrm{card}\left\{\sigma \,\Big|\, -H_N(\sigma) \geq \frac{N\beta}{2}\right\}$$

so that

$$(2.27) \qquad \frac{1}{N} \log Z_N(\beta) \geq \frac{\beta^2}{2} + \frac{1}{N} \log \mathrm{card}\left\{\sigma \,\Big|\, -H_N(\sigma) \geq \frac{N\beta}{2}\right\}.$$

We now turn to the task of bounding below the card involved in (2.26). Essentially the only known method to do that is called (now) the *second moment method*, and relies upon the following principle.

Proposition 2.8. *Consider a r.v. $Y \geq 0$. Then*

$$(2.28) \qquad \mathsf{P}\left(Y \geq \frac{1}{2}\mathsf{E}Y\right) \geq \frac{1}{4}\frac{(\mathsf{E}Y)^2}{\mathsf{E}Y^2}.$$

This was apparently used first in the work of Paley and Zygmund on trigonometric series.

Proof. If $A = \{Y \geq \frac{1}{2}\mathsf{E}Y\}$ then

$$\mathsf{E}Y \leq \mathsf{E}(Y 1_{A^c}) + \mathsf{E}(Y 1_A) \leq \frac{1}{2}\mathsf{E}Y + \mathsf{E}(Y 1_A).$$

Thus

$$\frac{EY}{2} \le E(Y1_A) \le (EY^2)^{1/2} P(A)^{1/2}$$

using Cauchy-Schwarz. The result follows. $\square$

Proof of (2.22). We will use (2.28) for

(2.29) $$Y = \text{card}\left\{\sigma \mid -H_N(\sigma) \ge \frac{N\beta}{2}\right\} = \sum_\sigma 1_{\{-H_N(\sigma) \ge \frac{N\beta}{2}\}}$$

and thus

(2.30) $$EY = 2^N \Phi(\beta\sqrt{N/2}),$$

where, if g is $\mathcal{N}(0,1)$,

(2.31) $$\Phi(t) = P(g \ge t) = \frac{1}{\sqrt{2\pi}} \int_t^\infty \exp\left(-\frac{u^2}{2}\right) du.$$

Now, from (2.29)

$$Y^2 = \sum_{\sigma,\sigma'} 1_{\{-H_N(\sigma) \ge \frac{N\beta}{2}, -H_N(\sigma') \ge \frac{N\beta}{2}\}}$$

so that

(2.32) $$EY^2 = \sum_{\sigma,\sigma'} A(\sigma,\sigma'),$$

where

$$A(\sigma,\sigma') = P\left(\left\{-H_N(\sigma) \ge \frac{N\beta}{2}, -H_N(\sigma') \ge \frac{N\beta}{2}\right\}\right).$$

If $\sigma' = \sigma$, we have

$$A(\sigma,\sigma) = \Phi(\beta\sqrt{N/2}).$$

If $\sigma \ne \sigma'$, by independence, we have

$$A(\sigma,\sigma') = \Phi(\beta\sqrt{N/2})^2.$$

Combining these estimates we get from (2.32) that

(2.33) $$EY^2 \le 2^N \Phi(\beta\sqrt{N/2}) + 2^{2N}\Phi(\beta\sqrt{N/2})^2 = EY + (EY)^2.$$

We now use the well known fact that for $t > 0$,

(2.34) $$\frac{1}{L(1+t)} \exp\left(-\frac{t^2}{2}\right) \le \Phi(t) \le \frac{1}{2}\exp\left(-\frac{t^2}{2}\right).$$

Here, as well as in the rest of the paper, L denotes a positive number, not necessarily (and even rarely) the same at each occurrence. Thus, given $\beta < 2\sqrt{\log 2}$, for N large we have

(2.35) $$EY \ge \frac{2^N}{L\sqrt{N}} e^{-\beta^2 N/4} \ge 1$$

and (2.28), (2.33) show that

$$P\left(Y \geq \frac{2^N}{2L\sqrt{N}}\, e^{-\beta^2 N/4}\right) \geq \frac{1}{8}.$$

Combining with (2.27) we get

$$P\left(\frac{1}{N}\log Z_N(\beta) \geq \frac{\beta^2}{4} + \log 2 - \frac{1}{N}\log(LN)\right) \geq \frac{1}{8}.$$

To prove (2.21), we then observe that by (2.10), we have

$$P\left(\frac{1}{N}\log Z_N(\beta) \leq p_N(\beta) + \frac{2\beta}{\sqrt{N}}\right) \geq 1 - e^{-4} > \frac{7}{8},$$

and thus

$$p_N(\beta) \geq \frac{\beta^2}{4} + \log 2 - \frac{2\beta}{\sqrt{N}} - \frac{1}{N}\log(LN). \qquad \square$$

Comments. 1. We have not tried to give the simplest proof or the sharpest result, but we have tried to prepare for future results. Rather than using Proposition 2.8, one can use that $E(Y - EY)^2 \leq EY \ll (EY)^2$ to see that $P\left(Y \leq \frac{EY}{2}\right)$ is in fact very small.

2. The reader might wonder why we did not use (2.28) for $Y = Z_N$. This is because it is untrue that $E Z_N^2$ and $(E Z_N)^2$ are of the same order of magnitude if $\beta > \sqrt{2\log 2}$. To make (2.28) work directly, one has to use a truncation argument, that our approach was designed to bypass.

3. This scheme of proof would work even if we only knew that $EY^2 \leq KN^2(EY)^2$. This remark will be used in Chapter 4 (see (4.13)).

Having computed p_N for Derrida's REM, we turn to the study of Gibbs' measure, and we first consider the case $\beta < 2\sqrt{\log 2}$. Two obvious ways of looking at Gibbs' measure are respectively the "local way" involving only finitely many coordinates and the "global way" involving all coordinates. These yield very different results. We first consider the "local" point of view. Given a subset I of $\{1, \ldots, N\}$, we denote

by $G_I \ (= G_{N,I}(\beta))$ the projection of Gibbs' measure on $\{-1, 1\}^I$;

by μ_I the uniform probability on $\{-1, 1\}^I$;

by $|G_I - \mu_I|$ the total variation distance between G_I and μ_I.

Proposition 2.9. If $\beta < 2\sqrt{\log 2}$ and $\delta < 1 - \beta^2/(4\log 2)$, there is a number K depending upon β, δ only such that

$$(2.36) \qquad \text{card}\, I \leq \delta N \implies E\,|G_I - \mu_I| \leq \exp\left(-\frac{N}{K}\right).$$

Throughout these notes K will denote a quantity independent of N, but that might depend upon other parameters. This quantity need not be the same at each occurrence. (We recall that L, on the other hand, denotes a universal constant).

Before proving the proposition, we show that the condition

$$\delta < 1 - \beta^2/(4\log 2)$$

is not absurd. In fact, (2.22) and (2.24) imply that if $\gamma < \beta/2$ then (with probability close to one)

$$G_N(\{\boldsymbol{\sigma} \mid -H_N(\boldsymbol{\sigma}) \le N\gamma\}) \le \exp\left(-\frac{N}{K}\right),$$

that is, G_N is essentially supported by

(2.37) $$A = \{\boldsymbol{\sigma} \mid -H_N(\boldsymbol{\sigma}) \ge N\gamma\}$$

which has the property that

(2.38) $$\mathsf{E}\,\mathrm{card}\,A \le 2^N \exp(-N\gamma^2).$$

Thus, if $\mathrm{card}\,I > \delta'N$, where $\delta' > 1 - \beta^2/(4\log 2)$, we can find $\gamma < \beta/2$ such that

$$2^{\mathrm{card}\,I} \gg \mathsf{E}\,\mathrm{card}\,A$$

and there are not enough points in A to control all the atoms of $\{-1,1\}^I$, so (2.36) must fail.

Proof of Proposition 2.9. Of course this can't be hard, but there is a tiny difficulty due to the fact that second moment computations require truncation (see comment 2 above). We set $M = \mathrm{card}\,I$. The inverse images of the points of $\{-1,1\}^I$ under the projection from $\{-1,1\}^N$ to $\{-1,1\}^I$ form a family $\mathcal{B}$ of subsets of $\{-1,1\}^N$. It has 2^M elements, each of which contains 2^{N-M} points. We have to show that

(2.39) $$\mathsf{E}\sum_{B\in\mathcal{B}}\left|2^{-M} - \frac{S(B)}{Z_N}\right| \le \exp\left(-\frac{N}{K}\right)$$

where $S(B) = \sum_{\boldsymbol{\sigma}\in B}\exp(-\beta H_N(\boldsymbol{\sigma}))$.

Consider a number γ with $\gamma < \beta^2/4$, that will be determined later. It follows from (2.22), (2.10) that

$$\mathsf{P}\big(Z_N \le 2^N e^{N\gamma}\big) \le \exp\left(-\frac{N}{K}\right),$$

so that to prove (2.39) it is enough to prove that

(2.40) $$\mathsf{E}\sum_{B\in\mathcal{B}}|2^{-M}Z_N - S(B)| \le 2^N \exp N\left(\gamma - \frac{1}{K}\right).$$

Since $Z_N = \sum_{B\in\mathcal{B}} S(B)$, it suffices to show that for $B, B' \in \mathcal{B}$,

(2.41) $$\mathsf{E}|S(B) - S(B')| \le 2^{N-M} \exp N\left(\gamma - \frac{1}{K}\right).$$

This is an exercise. It requires truncation. Consider $\gamma' > \beta^2/2$, set

$$S_1(B) = \sum_{\sigma \in B} \exp(-\beta H_N(\sigma)) 1_{\{-\beta H_N(\sigma) \geq N\gamma'\}}$$

and $S_2(B) = S(B) - S_1(B)$. It follows from Lemma 2.7 that

$$\mathsf{E}\, S_1(B) \leq 2^{N-M} \exp N\left(\gamma' - \frac{\gamma'^2}{\beta^2}\right).$$

Thus, if we assume

(2.42) $\gamma' - \dfrac{\gamma'^2}{\beta^2} < \gamma,$

to prove (2.41) it suffices to prove that

(2.43) $\mathsf{E}|S_2(B) - S_2(B')| \leq 2^{N-M} \exp N\left(\gamma - \dfrac{1}{K}\right),$

or even

(2.44) $\mathsf{E}\big(S_2(B)\big)^2 - \big(\mathsf{E}\, S_2(B)\big)^2 \leq \left(2^{N-M} \exp N\left(\gamma - \dfrac{1}{K}\right)\right)^2.$

Using independence, for any σ,

$$\mathsf{E}\big(S_2(B)\big)^2 - (\mathsf{E}\, S_2(B))^2 \leq 2^{N-M}\, \mathsf{E} \exp(-2\beta H_N(\sigma)) 1_{\{-\beta H_N(\sigma) \leq N\gamma'\}}$$

$$\leq 2^{N-M} \exp N\left(2\gamma' - \frac{\gamma'^2}{\beta^2}\right),$$

using Lemma 2.7 with $\tau^2 = 2N\beta^2$, $c = 2N\gamma'$. Thus, to obtain (2.44), it suffices to have that (since $M/N = \delta$)

$$2\gamma' - \frac{\gamma'^2}{\beta^2} < (1 - \delta) \log 2 + 2\gamma.$$

If $(1 - \delta) \log 2 > \beta^2/4$, we can find $\gamma < \beta^2/4$ and $\gamma' > \beta^2/2$ that satisfy this and (2.42). $\square$

A more tricky question would be to find the largest value of δ such that

$$\mathsf{E} \sup_{\operatorname{card} I \leq \delta} |G_I - \mu_I| \ll 1.$$

After having looked at G_N "locally", what can we say "globally"? It follows from (2.37) and (2.38) that G_N is very far from the uniform measure on Σ_N. The reader has certainly already understood that the whole point of studying H_N is that it has few large values. These have a dominant influence on Gibbs' measure. The "peaks" of H_N reflect upon the structure of G_N at any temperature. We need however to look at the global structure of G_N to see this for small β (i.e. $\beta < 2\sqrt{\log 2}$). This point of view seems to have no motivations from the physical content of the theory, but possibly will raise interesting mathematical problems. It will be discussed again in Chapter 3.

We now turn to the study of Gibbs' measure at $\beta > 2\sqrt{\log 2}$. In that case, the most striking feature is that "Gibbs' measure is supported by a sequence of configurations" in the sense that given $\varepsilon > 0$, one can find k independent of N such that

(2.45) $$P\left(\max\{G(B) \mid \operatorname{card} B \leq k\} \leq 1 - \varepsilon\right) \leq \varepsilon.$$

We start with the observation that

(2.46) $$P\left(-H_N(\sigma) \geq u\right) = \frac{1}{\sqrt{\pi N}} \int_u^\infty \exp\left(-\frac{t^2}{N}\right) dt,$$

so that if we define a_N by $Na_N^2 = \log(2^N/\sqrt{N})$, we have

$$
\begin{aligned}
P\left(-H_N(\sigma) \geq Na_N + u\right) &= \frac{1}{\sqrt{\pi N}} \int_u^\infty \exp\left(-\frac{(t + Na_N)^2}{N}\right) dt \\
&\leq \frac{2^{-N}}{\sqrt{\pi}} \int_u^\infty \exp\left(-2ta_N\right) dt \\
&\leq 2^{-N} \exp\left(-2ua_N\right).
\end{aligned}
$$

Thus, given k, we have

$$
\begin{aligned}
\mathsf{E}\Big(\sum \exp(-\beta H_N(\sigma)) &\mathbf{1}_{\{-k+Na_N \geq -H_N(\sigma) \geq -k-1+Na_N\}}\Big) \\
&\leq 4 \exp\left(\beta Na_N - k(\beta - 2a_N)\right),
\end{aligned}
$$

and, by summation, since $\beta > 2a_N$,

$$\mathsf{E}\Big(\sum \exp\left(-\beta H_N(\sigma)\right) \mathbf{1}_{\{-H_N(\sigma) \leq Na_N - k\}}\Big) \leq K \exp\left(-\frac{k}{K}\right) \exp \beta Na_N.$$

The details of the rest of the proof of (2.45) will be left to the reader. If

$$U = \max(-H_N(\sigma))$$

one shows that $P(U \leq Na_N - u)$ goes to zero uniformly in N as $u \to \infty$. One then observes that

$$
\begin{aligned}
\sum\{G_N(\sigma) \mid -H_N(\sigma) &\leq Na_N - k\} \\
&\leq \exp(-\beta U) \sum \exp(-\beta H_N(\sigma)) \mathbf{1}_{\{-H_N(\sigma) \leq Na_N - k\}}.
\end{aligned}
$$

$\square$

The "tightness" result (2.45) can be completed by a limit theorem. Given a bounded Borel subset A of $\mathbb{R}$, the number $X_N(A)$ of values of σ for which $-H_N(\sigma) - Na_N$ belongs to A converges in law to a Poisson r.v. of expectation

$$\frac{1}{\sqrt{\pi}} \int_A \exp\left(-2t\sqrt{\log 2}\right) dt,$$

and, if B is another Borel subset, disjoint of A, $X_N(B)$ and $X_N(A)$ are asymptotically independent. In other words, the values taken by $-H_N(\sigma) -$

Na_N, for $\sigma \in \Sigma_N$ converge (in each interval $[u, +\infty)$) to a Poisson point process of intensity measure $d\mu = \exp(-2t\sqrt{\log 2})\, dt$. The values taken by the weights $\exp(-\beta H_N(\sigma) - \beta N a_N)$ converge (on each interval $[u, +\infty)$ where $u > 0$) to a Poisson point process having as intensity measure the image of μ under the map $x \mapsto \exp(-\beta x)$, which is the measure ν_m such that

$$(2.47) \qquad d\nu_m(x) = \frac{1}{\beta\sqrt{\pi}} x^{-m-1}\, dx$$

for $m = 2\sqrt{\log 2}/\beta < 1$. The important feature is that, since $m < 1$, we have

$$\int x\, d\nu_m(x) < \infty,$$

so that if we denote by (u_α) the family of numbers created by a realization of the Poisson point process (2.47), we have $\sum u_\beta < \infty$ a.s. We can then define the random weights

$$(2.48) \qquad v_\alpha = \frac{u_\alpha}{\sum u_\beta}.$$

If we agree to label them in decreasing order, we thus define a probability measure on the space of sequences $(v_\alpha)_{\alpha \geq 1}$, $v_\alpha \geq 0$ that sum to one. Going back to Gibbs' measure, the distribution of the numbers $G_N(\sigma)$, when ordered in decreasing order, converges to the distribution of the weights v_α (in a sense that the reader will make precise). The distribution of the weights v_α is very natural and interesting (see [P-Y]). It plays a fundamental role at low temperature [T7].

3 The Sherrington-Kirkpatrick model at high temperature

The Hamiltonian of the Sherrington-Kirkpatrick model (SK) is given by

$$(3.1) \qquad H_N(\sigma) = -\frac{1}{\sqrt{N}} \sum_{i<j} g_{ij}\sigma_i\sigma_j,$$

where $(g_{ij})_{i<j}$ are i.i.d. $\mathcal{N}(0,1)$. Thus, given another configuration σ' we have

$$(3.2) \qquad \mathsf{E}\big(H_N(\sigma)H_N(\sigma')\big) = \frac{1}{N} \sum_{i<j} \sigma_i\sigma_j\sigma_i'\sigma_j'$$

$$= \frac{N}{2}\Big(\frac{1}{N}\sum_{i\leq N}\sigma_i\sigma_i'\Big)^2 - \frac{1}{2},$$

and in particular

$$(3.3) \qquad \mathsf{E}\big(H_N^2(\sigma)\big) = \frac{N-1}{2}.$$

In (3.2) we see the first occurrence of

$$(3.4) \qquad R(\sigma,\sigma') = \frac{\sigma\cdot\sigma'}{N} = \frac{1}{N}\sum_{i\leq N}\sigma_i\sigma_i',$$

which is called the *overlap* of σ and σ'. The name is motivated by the relation

$$R(\sigma,\sigma') = \frac{2}{N}\operatorname{card}\{i \leq N \mid \sigma_i = \sigma_i'\} - 1.$$

The overlaps will play a fundamental role in the sequel.

In this chapter we study the SK model for $\beta < 1$. This is basically simple, and comes as a natural second step after Derrida's model. The reader should however keep in mind that many special features occur in this model, and that it definitely gives a false idea of our topic, as will start to be obvious in Chapter 5.

The SK model at inverse temperature $\beta < 1$ is well understood, in particular through the work of Comets and Neveu [C-N]. This pretty paper uses stochastic calculus. Unfortunately stochastic calculus has yet to demonstrate its use outside a few very special situations, so we will not talk about it (a good excuse for putting off one more time the task of learning something about it). We actually know how to prove some of the sharp results of [C-N] (central limit theorems) by other methods. These will be developed in Chapters 5 to 7, in situations that do not seem to be within the reach of stochastic calculus.

The really special feature of the present model is that

$$(3.5) \qquad \mathsf{E}(Z_N^2) \leq K(\beta)(\mathsf{E}\,Z_N)^2,$$

and proving this is our first goal. We first observe that, from (3.3)

$$(3.6) \qquad \mathsf{E}\,Z_N = 2^N \exp \frac{\beta^2}{4}(N-1).$$

Lemma 3.1. *If $\delta < 1$ we have*

$$(3.7) \qquad 2^{-N} \sum \exp \frac{\delta}{2N} \Big(\sum_{i \leq N} \sigma_i\Big)^2 \leq \frac{1}{\sqrt{1-\delta}}$$

In (3.7), the sum is of course over $\sigma \in \Sigma_N$. This will not be mentioned anymore.

Proof. For all t,

$$(3.8) \qquad 2^{-N} \sum \exp t \sum_{i \leq N} \sigma_i = (\mathrm{ch}\,t)^N \leq \exp \frac{Nt^2}{2},$$

using the elementary inequality $\mathrm{ch}\,t \leq \exp \frac{t^2}{2}$. Using (3.8) for $t = g\sqrt{\delta/N}$ where g is $\mathcal{N}(0,1)$ and taking expectation yield (3.7). □

Comment. I learned this pretty argument in [C-N]. Of course, one can also deduce tail estimates from (3.8), from which weaker bounds than (3.7) follow that do not require this trick.

Lemma 3.2. *If $\gamma + \beta^2 < 1$ we have*

$$(3.9) \qquad \mathsf{E}\Big(\sum \exp(-\beta H_N(\sigma) - \beta H_N(\sigma')) \exp \frac{\gamma}{2N}(\sigma \cdot \sigma')^2\Big)$$

$$\leq \frac{1}{\sqrt{1 - \beta^2 - \gamma}}(\mathsf{E}\, Z_N)^2.$$

Of course here it is understood that the summation is over all values of σ, σ'.

Proof. Using (3.2), (3.3), we see that

$$\mathsf{E}\big(H_N(\sigma) + H_N(\sigma')\big)^2 = N - 2 + \frac{1}{N}(\sigma \cdot \sigma')^2,$$

so that the left-hand side of (3.9) is at most equal to

$$\exp \frac{\beta^2(N-2)}{2} \sum \exp \frac{\delta}{2N}(\sigma \cdot \sigma')^2,$$

where $\delta = \beta^2 + \gamma$. Now, using (3.7), we have

$$\sum \exp \frac{\delta}{2N}(\sigma \cdot \sigma')^2 = 2^N \sum \exp \frac{\delta}{2N}\Big(\sum_{i \leq N} \sigma_i\Big)^2 \leq \frac{2^{2N}}{\sqrt{1 - \delta}},$$

which implies (3.9). □

If $\gamma = 0$, (3.9) implies

$$(3.10) \qquad\qquad \mathsf{E}(Z_N^2) \leq (1 - \beta^2)^{-1/2}(\mathsf{E}\, Z_N)^2.$$

Theorem 3.3 ([A-L-R]). *If $\beta < 1$, we have*

$$(3.11) \qquad\qquad \lim_{N \to \infty} p_N(\beta) = \frac{\beta^2}{4} + \log 2 .$$

Proof. Using Proposition 2.8 and (3.10) we see that

$$(3.12) \qquad\qquad \mathsf{P}\Big(Z_N \geq \frac{1}{2}\mathsf{E}\, Z_N\Big) \geq \frac{1}{4}\sqrt{1 - \beta^2},$$

so that, using (3.6)

$$\mathsf{P}\Big(\frac{1}{N} \log Z_N \geq \big(\frac{\beta^2}{4} + \log 2\big)\frac{N-1}{N}\Big) \geq \frac{1}{4}\sqrt{1 - \beta^2}.$$

The end of the proof is then as in the proof of (2.22): by Proposition 2.2, we have

$$\mathsf{P}\Big(\frac{1}{N} \log Z_N \leq p_N(\beta) + \frac{u\beta}{\sqrt{N}}\Big) \geq 1 - \exp(-u^2),$$

so that if $\exp(-u^2) < \sqrt{1 - \beta^2}/4$ we have

$$p_N(\beta) \geq \frac{\beta^2}{4} + \log 2 - \frac{u\beta}{\sqrt{N}} - \frac{K(\beta)}{N}. \qquad\qquad □$$

We now prove a more sophisticated result.

Theorem 3.4. *For each $\beta < 1$ there is a constant $K(\beta)$ such that for each N, each $u > 0$,*

$$(3.13) \qquad P\left(\log Z_N(\beta) < N\left(\frac{\beta^2}{4} + \log 2\right) - u\right) \leq K(\beta) \exp\left(-\frac{u^2}{K(\beta)}\right).$$

To understand this, we note that this really consists of two parts

$$(3.14) \qquad \left| p_N(\beta) - \frac{\beta^2}{4} - \log 2 \right| \leq \frac{K(\beta)}{N}$$

$$(3.15) \qquad P\left(\frac{1}{N}\log Z_N(\beta) \leq p_N(\beta) - u\right) \leq K(\beta) \exp\left(-\frac{N^2 u^2}{K(\beta)}\right).$$

Condition (3.15) gains a factor N in the exponent of the bound compared to Proposition 2.2. This is not a general phenomenon, but a very special feature of the present model. For this reason, Theorem 3.4 is not particularly important, but the proof is quite pretty. Before we start, we recall an essential notation, the bracket $\langle \cdot \rangle$, that denotes average with respect to Gibbs' measure

$$(3.16) \qquad \langle f(\sigma)\rangle = \int f(\sigma)\,dG_N(\sigma) = \frac{1}{Z_N}\sum f(\sigma)\exp(-\beta H_N(\sigma)).$$

A fundamental idea throughout the paper is that of *replicas*. This simply means that we will consider products of the probability space (Σ_N, G_N), where G_N corresponds to the *same* realization of the g_{ij}'s in each copy. Averages in a replica with respect to the corresponding power of G_N are also denoted by brackets, so that, for example,

$$\langle \exp\frac{\gamma}{2N}(\sigma \cdot \sigma')^2\rangle = \iint \exp\frac{\gamma}{2N}(\sigma \cdot \sigma')^2\,dG_N(\sigma)\,dG_N(\sigma')$$

$$(3.17) \qquad\qquad = \frac{1}{Z_N^2}\sum \exp\left(-\beta H_N(\sigma) - \beta H_N(\sigma') + \frac{\gamma}{2N}(\sigma \cdot \sigma')^2\right)$$

and (3.9) means that

$$(3.18) \qquad E\left(Z_N^2\langle \exp\frac{\gamma}{2N}(\sigma \cdot \sigma')^2\rangle\right) \leq \frac{1}{\sqrt{1 - \beta^2 - \gamma}}(E\,Z_N)^2.$$

Lemma 3.5. *If $\beta < 1$ we have*

$$P\left(Z_N \geq \frac{1}{2}E\,Z_N,\ \langle\frac{1}{N}(\sigma \cdot \sigma')^2\rangle \leq K(\beta)\right) \geq \frac{1}{K(\beta)}.$$

Comment. Here, as always, this means: "If $\beta < 1$, there exists a number $K(\beta)$ depending upon β only such that..."

Proof. Using (3.12), (3.18) for $\gamma = \frac{1-\beta^2}{2}$ and Chebyshev inequality we see that

$$P\left(Z_N \geq \frac{1}{2}\,\mathsf{E}\,Z_N, \; Z_N^2\langle \exp\frac{\gamma}{2N}(\sigma \cdot \sigma')^2\rangle \leq K(\beta)(\mathsf{E}\,Z_N)^2\right) \geq \frac{1}{K(\beta)}.$$

But since $e^x \geq x$ we have

$$\left.\begin{array}{l} Z_N \geq \dfrac{1}{2}\,\mathsf{E}\,Z_N \\[2mm] Z_N^2 \left\langle \exp\dfrac{\gamma}{2N}(\sigma \cdot \sigma')^2\right\rangle \leq K(\beta)(\mathsf{E}\,Z_N)^2 \end{array}\right\} \implies \left\langle \frac{1}{N}(\sigma \cdot \sigma')^2\right\rangle \leq K(\beta).$$

$\square$

The proof of Theorem 3.4 will use a simple algebraic fact that is not so intuitive. This (and similar facts) will play a fundamental role, so it is better brought to light now.

Lemma 3.6. *We have*

$$(3.19) \qquad\qquad \langle (\sigma \cdot \sigma')^2\rangle = \sum_{i,j}\langle \sigma_i\sigma_j\rangle^2$$

Proof. We write

$$\sigma \cdot \sigma' = \sum_{i\leq N} \sigma_i\sigma_i',$$

so

$$(\sigma \cdot \sigma')^2 = \sum_{i,j\leq N} \sigma_i\sigma_i'\sigma_j\sigma_j' = \sum_{i,j\leq N} \sigma_i\sigma_j\sigma_i'\sigma_j'.$$

Now

$$(3.20) \qquad\qquad \langle \sigma_i\sigma_j\sigma_i'\sigma_j'\rangle = \langle \sigma_i\sigma_j\rangle^2$$

because the bracket on the left deals with a product measure on G_N^2, with respect to which σ and σ' are independent r.v. $\square$

Comment. A very useful function of replica is that they transform products of integrals into integrals, such as in the formula

$$\langle f(\sigma)\rangle\langle f'(\sigma)\rangle = \langle f(\sigma)f'(\sigma')\rangle$$

which simply means that

$$\int f(\sigma)\,\mathrm{d}G_N(\sigma)\int f'(\sigma)\,\mathrm{d}G_N(\sigma) = \iint f(\sigma)f'(\sigma')\,\mathrm{d}G_N(\sigma)\,\mathrm{d}G_N(\sigma').$$

Proof of Theorem 3.4. It relies upon Proposition 2.1. To apply (2.8), it is convenient to reformulate Lemma 3.5 as follows. The subset B of $\mathbb{R}^M$ (for $M = \frac{N(N-1)}{2}$) such that

$$(3.21) \qquad (g_{ij}) \in B \Longleftrightarrow \begin{cases} Z_N \geq \frac{1}{2} E Z_N(\beta); \\ \langle (\sigma \cdot \sigma')^2 \rangle \leq K(\beta) N \end{cases}$$

satisfies

$$(3.22) \qquad \mathsf{P}\big((g_{ij}) \in B\big) \geq \frac{1}{K(\beta)}.$$

It is of course implicit here that the (g_{ij}) on the left are those used to define Z_N and Gibbs' measure on the right. We apply (2.8) to the function

$$F(x) = d(x, B) = \inf_{y \in B} d(x, y),$$

the Euclidean distance from x to B in $\mathbb{R}^M$. There, of course

$$(3.23) \qquad d^2(x, y) = \sum_{i < j} (x_{ij} - y_{ij})^2.$$

Thus, by (2.8), we have

$$(3.24) \qquad \mathsf{P}\big(|d(g, B) - \mathsf{E}\, d(g, B)| \geq u\big) \leq \exp\left(-\frac{u^2}{2}\right).$$

In particular, if $u < \mathsf{E}\, d(g, B)$, we have

$$\frac{1}{K(\beta)} \leq \mathsf{P}(g \in B) \leq \mathsf{P}\big(|d(g, B) - \mathsf{E}\, d(g, B)| \geq u\big) \leq \exp\left(-\frac{u^2}{2}\right),$$

so that $u \leq K(\beta)$ and thus $\mathsf{E}\, d(g, B) \leq K(\beta)$. It then follows from (3.24) that for $u \geq 0$

$$(3.25) \qquad \mathsf{P}\big(d(g, B) \geq u + K(\beta)\big) \leq \exp\left(-\frac{u^2}{2}\right).$$

To prove (3.13), it suffices to combine (3.25) with the following

$$(3.26) \qquad \log Z_N(g) \geq N\left(\frac{\beta^2}{4} + \log 2\right) - K(\beta)(d(g, B) + 1)$$

where $Z_N(g)$ means of course that Z_N corresponds to the choice $g = (g_{ij})$ of the r.v. Indeed, if (3.26) holds, and if

$$\log Z_N(g) \leq N\left(\frac{\beta^2}{4} + \log 2\right) - u,$$

then $d(g, B) \geq \frac{u - K(\beta)}{K(\beta)}$. Using (3.25) this is an event of probability at most

$$\exp\left(-\left(\min\left\{0, \frac{u - K(\beta)}{K(\beta)}\right\}\right)^2\right) \leq K(\beta) \exp\left(-\frac{u^2}{K(\beta)}\right).$$

To prove (3.26), since $\log Z_N(g') \geq (N-1)(\frac{\beta^2}{4} + \log 2)$ for g' in B, it suffices to prove that if $g \in \mathbb{R}^M$ we have

(3.27) $$\log Z_N(g) \geq \log Z_N(g') - K(\beta)d(g,g')$$

where d is given by (3.23). We start with the algebraic identity

(3.28) $$Z_N(g) = Z_N(g')\langle\exp-\frac{\beta}{\sqrt{N}}\sum_{i<j}(g_{ij}-g'_{ij})\sigma_i\sigma_j\rangle'$$

where $\langle \cdot \rangle'$ means that Gibbs' measure is computed for g'_{ij}. Using successively Jensen's inequality and Cauchy-Schwarz, we get

$$Z_N(g) \geq Z_N(g')\exp\left(-\frac{\beta}{\sqrt{N}}\sum_{i<j}(g_{ij}-g'_{ij})\langle\sigma_i\sigma_j\rangle'\right)$$

$$\geq Z_N(g')\exp\left(-\frac{\beta}{\sqrt{N}}d(g,g')\left(\sum_{i<j}\langle\sigma_i\sigma_j\rangle'^2\right)^{1/2}\right).$$

Using (3.19), and the fact that $\langle(\sigma\cdot\sigma')^2\rangle' \leq NK(\beta)$ by (3.22) we have proved (3.27). $\square$

Theorem 3.7. *If $\beta < 1$, there exists $K(\beta) < \infty$ such that*

(3.29) $$\mathsf{E}\langle\exp\frac{1-\beta^2}{8N}(\sigma\cdot\sigma')^2\rangle \leq K(\beta).$$

This implies a very good control of the overlap of two configurations σ, σ', when the overlap is viewed as a function on Σ_N^2, provided with $G_N^{\otimes 2}$, or, if one prefers, when σ, σ' are weighted according to their Gibbs weights. It will gradually become apparent that this type of information is of considerable importance. We will have to work very hard later to prove the extensions of (3.29) to other situations.

Proof of Theorem 3.7. Using Cauchy-Schwarz for $\langle \cdot \rangle$ we have

$$\langle\exp\frac{1-\beta^2}{8N}(\sigma\cdot\sigma')^2\rangle \leq \langle\exp\frac{1-\beta^2}{4N}(\sigma\cdot\sigma')^2\rangle^{1/2}.$$

Now

$$\mathsf{E}\langle\exp\frac{1-\beta^2}{4N}(\sigma\cdot\sigma')^2\rangle^{1/2} = \mathsf{E}\left(\frac{1}{Z_N}\left(Z_N^2\langle\exp\frac{1-\beta^2}{4N}(\sigma\cdot\sigma')^2\rangle\right)^{1/2}\right)$$

$$\leq K(\beta)\left(\mathsf{E}\frac{1}{Z_N^2}\right)^{1/2}\mathsf{E}Z_N,$$

using again Cauchy-Schwarz and (3.9) for $\gamma = \frac{1-\beta^2}{2}$. But (3.13) implies that for $u > 1$

$$\mathsf{P}\left(\frac{(\mathsf{E}Z_N)^2}{Z_N^2} > u\right) \leq K(\beta)\exp\left(-\frac{(\log u)^2}{K(\beta)}\right),$$

so that

$$E\left(\frac{E\,Z_N}{Z_N}\right)^2 \leq K(\beta).$$

$\square$

Theorem 3.7 has applications to a statement of the same nature as Proposition 2.9, but with card I about $\sqrt{N}$. We refer the reader to [T2] for this.

What about the global structure of G_N ? We will show that, in a strong sense, G_N is very far from the uniform measure μ on Σ_N. Let us denote by d the Hamming distance on Σ_N, given by

$$(3.30) \qquad d(\sigma, \sigma') = \text{card}\{i \leq N \mid \sigma_i \neq \sigma_i'\} = \frac{N - \sigma \cdot \sigma'}{2}.$$

Proposition 3.8. *If $\beta > 0$, with overwhelming probability one can find two sets A, $B \subset \Sigma_N$ with*

$$(3.31) \qquad \begin{cases} G_N(A) \geq \dfrac{2}{3}; \\ \mu(B) \geq \dfrac{2}{3}; \\ d(A, B) \geq \dfrac{N}{L} \min(1, \beta^2). \end{cases}$$

In this statement, as through these notes "with overwhelming probability" means that the probability of failure is $\leq \exp(-N/K)$. A consequence of (3.31) is that "the transportation cost of G_N to μ is of order N". (Don't worry if you don't know what this is.)

Proof of Proposition 3.8. We set $\beta' = \min(1, \beta)$. We define

$$A = \left\{\sigma \;\middle|\; -H_N(\sigma) \geq \frac{N\beta'}{3}\right\}$$
$$B = \left\{\sigma \;\middle|\; -H_N(\sigma) \leq \frac{N\beta'}{4}\right\},$$

and we prove that A, B satisfy (3.31). To prove that $G_N(A) \geq 2/3$ and in fact $G_N(A) \geq 1 - \exp(-N/K)$ with overwhelming probability, we use Lemma 2.7, Proposition 2.2, Theorem 3.3 (and the convexity of p_N if $\beta \geq 1$) to bound $G_N(A^c)$. To prove that $\mu(B) \geq 2/3$ (and in fact $\mu(B) \geq 1 - \exp(-N/K)$) with overwhelming probability, we will use Fubini Theorem. If we define

$$\Omega(\sigma) = \left\{H_N(\sigma) \geq \frac{N\beta'}{4}\right\},$$

then

$$E\left(\sum 1_{\Omega(\sigma)}\right) = \sum P(\Omega(\sigma)) \leq 2^N \exp\left(-\frac{N\beta'^2}{16}\right),$$

so that

$$P\left(2^{-N}\sum 1_{\Omega(\sigma)} \geq \exp\left(-\frac{N\beta'^2}{32}\right)\right) \leq \exp\left(-\frac{N\beta'^2}{32}\right).$$

To prove that $d(A, B) \geq \frac{N}{K}$ with overwhelming probability, we observe that if $\sigma \in A$, $\sigma' \in B$ then

(3.32) $$H_N(\sigma') - H_N(\sigma) \geq \frac{N\beta'}{12}.$$

Now, using (3.2), (3.3), we have

$$\begin{aligned}
\mathsf{E}(H_N(\sigma') - H_N(\sigma))^2 &= N\left(1 - \left(\frac{\sigma \cdot \sigma'}{N}\right)^2\right) \\
&= N\left(1 - \left(1 - \frac{2d(\sigma, \sigma')}{N}\right)^2\right) \\
&\leq 4d(\sigma, \sigma'),
\end{aligned}$$

where we have used (3.30) in the last line. Thus

$$P\left(H_N(\sigma') - H_N(\sigma) \geq \frac{N\beta'}{12}\right) \leq \exp\left(-\frac{N^2\beta'^2}{L\,d(\sigma, \sigma')}\right).$$

This shows that if $N\beta'^2 > LD$, with overwhelming probability there exist no pairs (σ, σ') with $d(\sigma, \sigma') < D$ that satisfy (3.32). $\qquad\square$

Definition 3.9. A probability measure G on Σ_N is ϑ-shattered if we can find two sets A, $B \subset \Sigma_N$ such that

(3.33) $$\begin{cases} G(A) \geq \dfrac{1}{3}, \\ G(B) \geq \dfrac{1}{3}, \\ d(A, B) \geq \vartheta N. \end{cases}$$

The idea there is that a ϑ-shattered probability utterly fails concentration of measure. It is better understood if one knows the following important result: nice probability measures on Σ_N (such as product measures) satisfy

$$\left.\begin{array}{l} \nu(A) \geq \dfrac{1}{3} \\ \nu(B) \geq \dfrac{1}{3} \end{array}\right\} \implies d(A, B) \leq L\sqrt{N}.$$

Conjecture 3.10. If $\beta > 0$, there is $\vartheta(\beta) > 0$ such that with overwhelming probability, G_N is ϑ-shattered.

Even though a physicist would probably consider this question as esoteric and far fetched, it seems that it possibly touches some central issues. If true, Conjecture 3.10 would mean that we can find two pieces A, B of the configuration space, that are well separated for d, and that both carry some mass. Moreover, such pieces would depend essentially upon the randomness. Certain central low temperature predictions of the physicists can be expressed with roughly the same words. It is conceivable that proving or disproving Conjecture 3.10 is really difficult.

Challenge problem 3.11. Construct a random probability measure ν on Σ_N that is ϑ-shattered for some $\vartheta > 0$ but satisfies

$$\mathsf{E} \iint \exp\Big(\frac{1}{NL}(\sigma \cdot \sigma')^2\Big)\, d\nu(\sigma)\, d\nu(\sigma') < L.$$

The difference between an open problem and a challenge problem is that I have no clue about the former but that I know (or at least, I thought at some point I would know if I really tried) how to do the later.

Challenge problem 3.12. Prove that if $\beta > 0$, there exists $\vartheta = \vartheta(\beta)$ such that the Gibbs' measure associated to Derrida's REM (Chapter 2) is ϑ-shattered with overwhelming probability.

What happens if $\beta > 1$? It is known that Theorem 3.3 fails, as is shown by the following

Theorem 3.13 ([C, K-T-J]). *If $\beta > 1$, then*

$$(3.34) \qquad \limsup_{N\to\infty} p_N(\beta) \leq \log 2 + \beta - \frac{3}{4} - \frac{1}{2}\log\beta.$$

Let us explain the basic idea. The starting point is that the joint law of a jointly Gaussian family of r.v. is determined by its covariance structure. Consider now independent $\mathcal{N}(0,1)$ r.v. $(g_i)_{i\leq N}$ (independent of the g_{ij}) and for $x \in \mathbb{R}^N$ consider

$$(3.35) \qquad H_N'(x) = -\frac{1}{\sqrt{N}}\sum_{i<j} g_{ij}x_i x_j + \frac{1}{\sqrt{2}}\sum_{i\leq N} g_i x_i^2.$$

An immediate computation shows that

$$\mathsf{E}\, H_N'(x)H_N'(y) = \frac{1}{2N}(x\cdot y)^2.$$

Thus, given any linear isometry R of $\mathbb{R}^N$ the joint law of the family $(H_N'(R(\sigma)))$ for σ in Σ_N is independent of R (since $R(x)\cdot R(y) = x\cdot y$) and thus

$$(3.36) \qquad \mathsf{E}\,\frac{1}{N}\log\sum_{\sigma}\exp\big(-\beta H_N'(R(\sigma))\big)$$

is independent of R. It is in particular equal to its value when $R = \mathrm{Id}$. In that case

$$H_N'(\sigma) = -\frac{1}{\sqrt{N}}\sum_{i<j} g_{ij}\sigma_i\sigma_j + \frac{1}{\sqrt{2}}\sum_i g_i.$$

The term $\sum_i g_i$ is independent of σ. We can take it through the log and it gives no contribution. Thus, for each R,

$$(3.37) \qquad p_N(\beta) = \mathsf{E} \frac{1}{N} \log \sum_\sigma \exp\bigl(-\beta H'_N(R(\sigma))\bigr).$$

Let us denote by $\mathrm{d}R$ the Haar ($=$ uniform) measure on the set of all rotations of $\mathbb{R}^N$. We integrate (3.36) with respect to $\mathrm{d}R$, and use Jensen's inequality to put the integral inside the log. We observe that

$$\int \sum_\sigma \exp\bigl(-\beta H'_N(R(\sigma))\bigr) \, \mathrm{d}R = 2^N \int_{S_N} \exp\bigl(-\beta H'_N(x)\bigr) \, \mathrm{d}\mu_N(x)$$

where μ_N is the uniform measure on $S_N = \{ x \in \mathbb{R}^N \mid x_1^2 + \cdots + x_N^2 = N \}$. Thus we have

$$(3.38) \qquad p_n(\beta) \le \log 2 + \mathsf{E} \frac{1}{N} \log \int_{S_N} \exp\bigl(-\beta H'_N(x)\bigr) \, \mathrm{d}\mu_N(x).$$

("Domination of the Ising model by the spherical model" in jargon.) The reason why the last term is easier to compute than $p_N(\beta)$ is that it has more symmetries. We can compute the integral in whatever basis we choose. The result depends only upon the eigenvalues of the matrix $(g'_{ij})_{i,j \le N}$ given by

$$\begin{cases} g'_{ij} = \dfrac{1}{2} g_{ij} & \text{if } i < j \\[2mm] g'_{ij} = \dfrac{1}{2} g_{ji} & \text{if } i > j \\[2mm] g'_{ii} = \dfrac{1}{\sqrt{2}} g_i. \end{cases}$$

These are well understood through a famous theorem of Wigner. They are (in a precise sense) essentially independent upon the realization of the (g'_{ij}), and follow "Wigner semi circle law". This allows to bound in the limit the right-hand side of (3.38) by the right-hand side of (3.34). We will refer the reader to [C] for this pretty computation, because the entire approach we are following here is somewhat unsatisfactory. Conditioning upon the eigenvalues of the matrix (g'_{ij}) is possibly not the right idea. Moreover, while Wigner's theorem is basically not difficult, it relies upon the very special structure of matrices (that can be multiplied), and it is unclear how to extend it, say to the situation of the next chapter.

What really happens for $\beta > 1$? The physicists have proposed an entire theory, of great complexity. It seems so much out of reach of the current rigorous methods that there is no point to even discuss it.

Challenge problem 3.14. Consider the Hamiltonian $H_N^{\text{REM}}(\sigma)$ of the REM and the Hamiltonian $H_N^{\text{SK}}(\sigma)$ of (3.1). Assume these are independent. Find the exact region of parameters β_1, β_2 for which

$$\lim_{N \to \infty} \frac{1}{N} \mathsf{E} \log \sum_\sigma \exp\left(-\beta_1 H_N^{\text{REM}}(\sigma) - \beta_2 H_N^{\text{SK}}(\sigma)\right) = \log 2 + \frac{1}{4}(\beta_1^2 + \beta_2^2).$$

4 The p-spin interaction model

The Hamiltonian of this model replaces the 2-spin interactions of the SK model by p-spin interactions. The Hamiltonian is given by

(4.1)
$$H_N(\sigma) = -\left(\frac{p!}{2N^{p-1}}\right)^{\frac{1}{2}} \sum g_{i_1,\dots,i_p} \sigma_{i_1} \dots \sigma_{i_p}.$$

The summation is over all possible choices of indices $i_1 < \dots < i_p$. The $g_{i_1,\dots,i_p}$ are i.i.d. $\mathcal{N}(0,1)$. We will soon explain why the case $p \geq 3$ is fundamentally different from the case $p = 2$. To understand the normalization factor in (4.1) we observe that for two configurations σ, σ', we have

(4.2)
$$E(H_N(\sigma)H_N(\sigma')) = \frac{p!}{2N^{p-1}} \sum_{i_1 < \dots < i_p} \sigma_{i_1} \dots \sigma_{i_p} \sigma'_{i_1} \dots \sigma'_{i_p}$$

$$= \frac{1}{2N^{p-1}} \sum_d \sigma_{i_1} \sigma'_{i_1} \sigma_{i_2} \sigma'_{i_2} \dots \sigma_{i_p} \sigma'_{i_p}$$

$$= \frac{N}{2}\left(\frac{\sigma \cdot \sigma'}{N}\right)^p - \frac{1}{2N^{p-1}} \sum_{nd} \sigma_{i_1} \sigma'_{i_1} \sigma_{i_2} \sigma'_{i_2} \dots \sigma_{i_p} \sigma'_{i_p}.$$

Here, $\sum_d$ denotes the sum over all possible choices of indexes $i_1, \dots, i_p$ that are all distinct, while $\sum_{nd}$ denotes the sum over all choices of indexes such that at least two of them coincide. In particular we have

(4.3)
$$\frac{N}{2} - K \leq E(H_N(\sigma)^2) \leq \frac{N}{2}.$$

(Following our convention, $K = K(p)$ depends upon p only).

If we view (4.2) as

$$E(H_N(\sigma)H_N(\sigma')) \simeq \frac{N}{2}\left(\frac{\sigma \cdot \sigma'}{N}\right)^p,$$

and we observe that since $|\sigma \cdot \sigma'| \leq N$, as p increases, there is less correlation between $H_N(\sigma)$ and $H_N(\sigma')$ if $\sigma \neq \pm\sigma'$. Thus, as $p \to \infty$ the model should resemble more and more the REM. This intuition is correct, but we must keep in mind that we cannot exchange the limits $p \to \infty$ and $N \to \infty$.

Let us note another consequence of (4.2).

Lemma 4.1.
$$E\big((H_N(\sigma) + H_N(\sigma'))^2\big) \leq N\big(1 + R(\sigma,\sigma')^p\big).$$

Let us define the number β_p by

$$(4.4) \qquad \beta_p = \sup\Big\{\beta \ \Big| \ \lim_{N \to \infty} p_N(\beta) = \frac{\beta^2}{4} + \log 2\Big\},$$

where of course $p_N(\beta)$ corresponds to the Hamiltonian (4.1). It follows from Theorems 3.3 and 3.13 that $\beta_2 = 1$. For $p \geq 3$, the value of β_p is not rigorously known, but we will approximate it. To state our result, we need the function

$$(4.5) \qquad \varphi(t) = \frac{1}{2}\big((1 + t)\log(1 + t) + (1 - t)\log(1 - t)\big).$$

This function is probably best understood by noting that $\varphi(0) = \varphi'(0) = 0$, $\varphi''(t) = 1/(1 - t^2)$. Thus $\varphi(t) \geq \frac{t^2}{2}$, and $\varphi(t) \simeq \frac{t^2}{2}$ for small t. This function arises through the following result, a special case of the Chernov bounds for the binomial law.

Lemma 4.2. *If $t \geq 0$, we have*

$$(4.6) \qquad \operatorname{card}\Big\{\sigma \ \Big| \ \sum_{i \leq N} \sigma_i \geq Nt\Big\} \leq 2^N \exp\big(-N\varphi(t)\big).$$

Proof. We observe that, for each λ

$$\sum \exp \lambda \sum_{i \leq N} \sigma_i = 2^N (\operatorname{ch}\lambda)^N = 2^N \exp N \log \operatorname{ch}\lambda,$$

so that the left-hand side of (4.6) is bounded by

$$2^N \inf_{\lambda \geq 0} \exp N(\log \operatorname{ch}\lambda - \lambda t) = 2^N \exp\big(-N\varphi(t)\big). \qquad \square$$

Theorem 4.3. *For $p \geq 2$ we have*

$$(4.7) \qquad \inf_{0 < t < 1} 2(1 + t^{-p})\varphi(t) \leq \beta_p^2 \leq 4\log 2.$$

The right-hand side inequality follows from (2.18).

Challenge problem 4.4. Show that in fact

$$\beta_p^2 < 4\log 2.$$

Since $\varphi(t) \geq \frac{t^2}{2}$, for $p = 2$, the left-hand side of (4.7) is 1, the correct value. Thus Theorem 4.3 recovers Theorem 3.3. This is not a surprise: the proofs are very similar.

Proof of Theorem 4.3. It will closely follow that of (2.22). Consider

$$(4.8) \qquad\qquad \beta^2 \leq \inf_{0<t<1} 2(1+t^{-p})\varphi(t).$$

We define Y as in (2.29) and using (4.3), (2.34), we have

$$EY \geq \frac{2^{N-1}}{K\sqrt{N}} \exp\left(-\frac{N^2\beta^2}{8(\frac{N}{2}-K)}\right).$$

Now,

$$\frac{N^2\beta^2}{8(\frac{N}{2}-K)} \leq \frac{N\beta^2}{4(1-\frac{2K}{N})} \leq \frac{N\beta^2}{4} + K,$$

so that

$$(4.9) \qquad\qquad EY \geq \frac{1}{K\sqrt{N}} 2^N \exp\left(-\frac{N\beta^2}{4}\right).$$

Now, we have

$$(4.10) \qquad\qquad EY^2 = \sum_{\sigma,\sigma'} A(\sigma,\sigma'),$$

where

$$A(\sigma,\sigma') = P\left(\left\{-H_N(\sigma) \geq \frac{N\beta}{2}, -H_N(\sigma') \geq \frac{N\beta}{2}\right\}\right)$$
$$\leq P\left(\left\{-(H_N(\sigma) + H_N(\sigma')) \geq N\beta\right\}\right)$$
$$\leq \exp\left(-\frac{N\beta^2}{2(1+R(\sigma,\sigma')^p)}\right)$$
$$\leq \exp\left(-\frac{N\beta^2}{2(1+|R(\sigma,\sigma')|^p)}\right),$$

using successively (2.34) and Lemma 4.1. We observe that $|\sigma \cdot \sigma'|$ is an integer. Thus if in (4.10) we regroup the pairs σ, σ' for which $|\sigma \cdot \sigma'| = tN$, we get

$$(4.11) \qquad EY^2 \leq \sum_{\substack{0 \leq t \leq 1 \\ Nt \text{ integer}}} \exp\left(-\frac{N\beta^2}{2(1+t^p)}\right) \text{card}\{\sigma,\sigma' \mid |\sigma \cdot \sigma'| = tN\}.$$

Now, by (4.3)

$$(4.12) \qquad \text{card}\{\sigma, \sigma' \mid |\sigma \cdot \sigma'| = tN\} = 2^N \, \text{card}\Big\{ \sigma \; \Big| \; \sum_{i \leq N} \sigma_i = Nt \Big\}$$

$$\leq 2^{2N+1} \exp(-N\varphi(t)).$$

Moreover, (4.8) implies that

$$\frac{\beta^2}{2(1+t^p)} + \varphi(t) \geq \frac{\beta^2}{2}$$

so that (4.9), (4.12) imply

$$(4.13) \qquad \mathbb{E} Y^2 \leq (N+1)2^{2N} \exp\Big(-\frac{N\beta^2}{2}\Big) \leq KN^2(\mathbb{E}Y)^2.$$

The proof is then completed as in Proposition 2.6.   □

Proposition 4.5. *If p is large enough we have*

$$(4.14) \qquad 2\sqrt{\log 2}\,(1 - 2^{-p}) \leq \beta_p \leq 2\sqrt{\log 2}.$$

This follows from Theorem 4.3 and (4.16) below.

Lemma 4.6. *If p is large enough, the following holds. If $v \geq 2^{-\frac{p}{2}}$ then*

$$(4.15) \qquad \inf_{0 \leq t \leq 1-v} 2(1 + t^{-p})\varphi(t) \geq 4\log 2 + \frac{p}{4}v.$$

Moreover

$$(4.16) \qquad \inf_{0 \leq t \leq 1} 2(1 + t^{-p})\varphi(t) \geq 4\log 2\Big(1 - \frac{2^{-p}}{\log 2}\Big) + O\big(p^2 2^{-2p}\big).$$

where $O(A)$ denotes a quantity such that $|O(A)| \leq LA$.

Proof. Since $\varphi(t) \geq \frac{t^2}{2}$, we have, by convexity,

$$2(1 + t^{-p})\varphi(t) \geq t^{2-p} \geq 1 + (p-2)(1-t) \geq 4\log 2 + \frac{p}{2}(1-t),$$

if p is large and $1 - t \geq \frac{4}{p}$. Thus we need only to consider values of $t \geq 1 - \frac{4}{p}$. Setting $u = 1 - t$, we observe that, by convexity,

$$1 + t^{-p} \geq 2 + pu$$
$$(1+t)\log(1+t) \geq 2\log 2 - (1 + \log 2)u,$$

and thus

$$2(1 + t^{-p})\varphi(t) \geq \psi(u) := (2 + pu)\big(2\log 2 - (1 + \log 2)u + u\log u\big).$$

Now,

$$(4.17) \qquad \psi(u) = 4\log 2 + 2u(p-1)\log 2 - 2u$$
$$+ 2u\log u - pu^2(1 + \log 2) + pu^2 \log u.$$

For $u \leq \frac{4}{p}$, we have

(4.18)

$$\psi(u) \geq 4\log 2 + u\left(2(p-1)\log 2 - 2 + 2\log u - 4(1+\log 2) - 4\log\frac{p}{4}\right)$$

$$\geq 4\log 2 + \frac{pu}{4},$$

provided $\log u \geq -\frac{p}{2}\log 2$ and p is large enough. This proves (4.15). To prove (4.16), we can assume $u \leq 2^{-\frac{p}{2}}$. In that case, rather than (4.18), we obtain

(4.19) $$\psi(u) \geq 4\log 2 + u\left[2(p-1)\log 2 - 2 + 2\log u\right.$$

$$\left. - p\,2^{-\frac{p}{2}}(1+\log 2) - p\,2^{-\frac{p}{2}}\log(2^{\frac{p}{2}})\right],$$

so that (for p large)

$$\psi(u) \geq 4\log 2 \quad \text{if} \quad u \geq 2^{-p+3}.$$

But, if $u \leq 2^{-p+3}$, then (4.17) yields

$$\psi(u) \geq 4\log 2 + 2u\left((p-1)\log 2 - 1\right) + 2u\log u + O(p^2 2^{-2p}),$$

and the right-hand side has its minimum at $u = 2^{-p+1}$, from which (4.16) follows. $\square$

Our purpose is not to study in detail the p-spin interaction model for $\beta \leq \beta_p$. Rather we will be interested in the case $\beta > \beta_p$.

Challenge problem 4.7. Find a convincing version of Theorem 3.13 for the present model.

Challenge problem 4.8. Find a convincing version of Proposition 2.9. What is the optimal size of I?

Challenge problem 4.9. More generally, which of properties proved in the setting of Chapter 3 for $p = 2$ can be generalized to the case $p \geq 3$?

Challenge problem 4.10. Prove that given $\beta > 0$, there is $p(\beta) > 0$, $\theta = \theta(\beta)$ such that if $p \geq p(\beta)$, the Gibbs' measure of the p-spin interaction model is θ-shattered (as in Definition 3.9) at inverse temperature β with overwhelming probability.

Open Problem 4.11. Does there exist $p > 0$ such that the Gibbs' measure of the p-spin interaction model is shattered for each $\beta > 0$?

We now turn to the main topic of this chapter, the search for estimates on the overlaps that will be valid not only for $\beta \leq \beta_p$ but also for β larger. To avoid having to distinguish cases we assume p even. In that case, Gibbs' measure is invariant by global symmetry around zero.

Our estimates on the overlap will rule out the possibility that the overlaps belong to certain intervals. This justifies the following definition.

Definition 4.12. We say that an interval I is negligible (at a given β) if

$$\mathbb{E}\,G_N^{\otimes 2}(\{\sigma,\sigma' \mid |R(\sigma,\sigma')| \in I\}) \leq K\exp\left(-\frac{N}{K}\right).$$

We will prove that certain intervals are negligible. This is the key to the low temperature results of Chapter 8, but will not be needed before that.

Our goal is the following (I am grateful to A. Bovier for having pointed out to me that my argument gives the result without any restriction on β)

Theorem 4.13. *There exists a number p_0 such that if $p \geq p_0$ then for all values of β, the interval $\left[2^{-\frac{p}{4}}, 1 - 2^{-\frac{p}{2}}\right]$ is negligible.*

We consider the function $\xi(\beta)$ given by

$$(4.20) \qquad \xi(\beta) = \begin{cases} \log 2 + \dfrac{\beta^2}{4} & \text{if } \beta \leq 2\sqrt{\log 2}, \\ \beta\sqrt{\log 2} & \text{if } \beta \geq 2\sqrt{\log 2}. \end{cases}$$

This function occurs in Propositions 2.5 and 2.6.

Lemma 4.14. *For all p sufficiently large, we have*

$$(4.21) \qquad \liminf_{N\to+\infty} p_N(\beta) \geq \xi(\beta) - 2^{-p}\beta.$$

Proof. The function $\beta \mapsto p_N(\beta)$ is convex, and by definition of β_p, for $\beta \leq \beta_p$ its limit is $\log 2 + \beta^2/4$. It is elementary to conclude that

$$\liminf_{N\to+\infty} p_N'(\beta_p) \geq \beta_p/2,$$

so that, using convexity of p_N again, for $\beta \geq \beta_p$ we have

$$\liminf_{N\to+\infty} p_N(\beta) \geq \log 2 + \frac{\beta_p^2}{4} + \frac{1}{2}\beta_p(\beta - \beta_p),$$

and thus, by an elementary computation

$$\liminf_{N\to+\infty} p_N(\beta) \geq \xi(\beta) - \frac{1}{2}(2\sqrt{\log 2} - \beta_p)\beta.$$

But it follows from (4.14) that $2\sqrt{\log 2} - \beta_p \leq 2^{-p+1}$.   □

We now define, for $|u| \leq 1$,

$$(4.22) \qquad A_N(u) = \sum \exp\left(-\beta(H_N(\sigma) + H_N(\sigma'))\right),$$

where the sum is over the pairs of configurations σ,σ' that satisfy $R(\sigma,\sigma') = u$.

Lemma 4.15. *We have*

(4.23)
$$\frac{1}{N} \operatorname{E} \log A_N(u) \le \xi_u(\beta)$$

where

$$\xi_u(\beta) = \frac{\beta^2}{2}(1 + u^p) + 2 \log 2 - \varphi(u)$$

if

(4.24)
$$\frac{\beta^2}{2}(1 + u^p) \le 2 \log 2 - \varphi(u)$$

and

$$\xi_u(\beta) = \beta \sqrt{2(1 + u^p)(2 \log 2 - \varphi(u))}$$

if (4.24) *fails.*

Proof. It is identical to that of Proposition 2.5 if one observes that when $R(\sigma, \sigma') = u$,
$$\operatorname{E}\big((H_N(\sigma) + H_N(\sigma'))^2\big) \le N(1 + u^p)$$
by Lemma 4.1, and that the number of pairs in the summation (4.22) is at most $2^{2N} \exp(-N\varphi(u))$. ☐

Using Proposition 2.2 (and its extension to the left-hand side of (4.23)), we see that to prove Theorem 4.13 it suffices to prove the following

Lemma 4.16. *If* $|u| \in [2^{-p/4}, 1 - 2^{-p/2}]$ *and if p is large enough, we have*

(4.25)
$$\xi_u(\beta) \le 2\xi(\beta) - \beta 2^{-\frac{p}{2}-4}.$$

Proof. We can assume $u \ge 0$ by symmetry. The basic fact is that if $|u| \in [2^{-p/4}, 1 - 2^{-p/2}]$ (and p is large enough) then

(4.26)
$$2u^p \log 2 - (1 + u^p)\varphi(u) \le -2^{-\frac{p}{2}-2}.$$

To prove this, we observe that if $u^{p-2} \le 1/8$, this is at most
$$\frac{u^2}{4} - \frac{u^2}{2} \le -\frac{u^2}{4} \le -2^{-\frac{p}{2}-2}.$$

If $u^{p-2} \ge 1/8$, we appeal to (4.15) with $v = 2^{-p/2}$ to see that
$$(1 + u^p)\varphi(u) \ge \left(2 \log 2 + \frac{p \, 2^{-p/2}}{4}\right)u^p$$
so that the left-hand side of (4.26) is at most
$$-\frac{p \, 2^{-p/2}}{4}\left(\frac{1}{8}\right)^{p/(p-2)} \le -2^{-\frac{p}{2}-2}$$
for p large enough. This proves (4.26).

To prove (4.25), we first assume that (4.24) fails. We observe that this implies

(4.27) $$\beta^2 \geq \log 2 \geq \frac{1}{2}.$$

We write

$$2(1 + u^p)(2 \log 2 - \varphi(u)) = 4 \log 2 + 2(2u^p \log 2 - (1 + u^p)\varphi(u))$$
$$\leq 4 \log 2 - 2^{-\frac{p}{2}-1}$$

by (4.26), and thus

$$\sqrt{2(1 + u^p)(2 \log 2 - \varphi(u))} \leq 2\sqrt{\log 2} - 2^{-\frac{p}{2}-3}.$$

Since $2\xi(\beta) \geq 2\beta\sqrt{\log 2}$, (4.25) follows by (4.27).

Finally, if (4.24) holds,

(4.28) $$\xi_u(\beta) = \frac{\beta^2}{2} + 2 \log 2 + \frac{\beta^2}{2} u^p - \varphi(u).$$

By (4.24),

(4.29) $$\frac{\beta^2}{2} \leq 2 \log 2 - \varphi(u)$$

so that, by (4.28),

$$\xi_u(\beta) \leq \frac{\beta^2}{2} + 2 \log 2 + 2u^p \log 2 - (1 + u^p)\varphi(u)$$
$$\leq \frac{\beta^2}{2} + 2 \log 2 - 2^{-\frac{p}{2}-2}$$

by (4.26). Now, by (4.24), $\beta^2 < 4 \log 2$, and $\beta^2/2 + 2 \log 2 = 2\xi(\beta)$.   □

Challenge problem 4.17. Show that if $\beta < \beta_p$ and $x > 0$, the interval $[x, 1]$ is negligible.

5 External field and the replica-symmetric solution

In this chapter we return to the SK model. One way to read Theorem 3.3 is that, if $\beta < 1$, we have

(5.1) $$\lim_{N \to \infty} \mathsf{E} \frac{1}{N} \log Z_N = \lim_{N \to \infty} \frac{1}{N} \log \mathsf{E} Z_N.$$

It is always true, by Jensen's inequality, that we have

$$\mathsf{E} \log Z_N \leq \log \mathsf{E} Z_N$$

and the content of (5.1) is that this trivial bound gives the correct order. On the other hand (5.1) does not hold for $\beta > 1$, as Theorem 3.13 shows. This is of course a more interesting situation.

As it turns out, (5.1) represents an accidental feature rather than a general one. For several models of considerable interest, it does not occur, even at high temperature. In this chapter we investigate a typical case, the SK model with external field. With the notation (3.1), the Hamiltonian is given by

(5.2) $$H_N(\sigma) = -\frac{1}{\sqrt{N}} \sum_{i<j} g_{ij} \sigma_i \sigma_j - h' \sum_{i \leq N} \sigma_i.$$

The last term represents the influence of an external field (that is, created by an apparatus independent of our sample of matter). The reason for the unexpected $'$ on the parameter h' (which measures the strength of the external field) is that we find it convenient to consider β and $h = \beta h'$ as independent parameters.

In this chapter we prove the following theorem, where g is standard normal.

Theorem 5.1. *There is a number β_0 such that if $\beta \leq \beta_0$, we have*

$$(5.3) \qquad \lim_{N \to \infty} p_N(\beta, h) = \log 2 + \frac{\beta^2}{4}(1 - q)^2 + \mathsf{E} \log \mathrm{ch}(\beta g \sqrt{q} + h)$$

where q is the unique solution of the equation

$$(5.4) \qquad\qquad q = \mathsf{E} \, \mathrm{th}^2(\beta g \sqrt{q} + h) .$$

Thus (to write the formulas of the physicists),

$$(5.5) \quad \lim_{N \to \infty} p_N(\beta, h) = \frac{\beta^2}{4}(1 - q)^2 + \frac{1}{\sqrt{2\pi}} \int_{-\infty}^{\infty} e^{-\frac{t^2}{2}} \log(2 \, \mathrm{ch}(\beta t \sqrt{q} + h)) \, dt.$$

On the other hand, it is very simple to show that

$$\lim_{N \to \infty} \frac{1}{N} \log \mathsf{E} \, Z_N = \log 2 + \frac{\beta^2}{4} + \log \mathrm{ch} \, h$$

and that (5.1) fails unless $h = 0$. (As we will show later, the value of q given by (5.4) minimizes the right-hand side of (5.3).) The remarkable character of a formula of the type (5.3) should be self apparent. This formula is not an accident, but reflects an underlying structure. It is the purpose of the present chapter to explain this structure and to prove (5.3). The formulae (5.3), (5.4) are called by physicists the "replica-symmetric" formulas.

The methods of this chapter are rather elementary. Considerably more powerful methods will be used in Chapter 6 and 7, and will provide shorter proofs of better results. Our motivation for presenting first an elementary (an leisurely) approach is that we feel that some of the important ideas are more apparent this way. These ideas have been very useful to the author.

If we think of the spins σ_i as functions on the probability space (Σ_N, G_N), the correlation of two spins is given by $\langle \sigma_i \sigma_j \rangle - \langle \sigma_i \rangle \langle \sigma_j \rangle$. The fundamental property behind (5.3) (and many other "replica-symmetric" formulas), is that in average over the disorder, these correlations are small, i.e.

$$(5.6) \qquad\qquad \lim_{N \to \infty} \mathsf{E} \big(\langle \sigma_i \sigma_j \rangle - \langle \sigma_i \rangle \langle \sigma_j \rangle \big)^2 = 0.$$

The purpose of the square is to prevent cancelation. By symmetry upon the sites, we can use any values for i, j.

Calculations are much simpler when one replaces products of integrals for Gibbs' measure by single integrals (this is what makes replicas so useful), and we first explain how to do this in the present setting.

One first idea is to use a procedure called "symmetrization" to replace centering. (This idea is very useful in many areas of probability.) If X, Y are r.v., and (X', Y') is an independent copy of the pair (X, Y), we have

$$(5.7) \qquad \mathsf{E}(XY) - \mathsf{E}(X)\,\mathsf{E}(Y) = \frac{1}{2}\mathsf{E}((X - X')(Y - Y')).$$

We consider two replicas σ^1, σ^2, they are independent under $G_N^{\otimes 2}$, so that we can use (5.7) to write

$$2(\langle \sigma_i \sigma_j \rangle - \langle \sigma_i \rangle \langle \sigma_j \rangle) = \langle (\sigma_i^1 - \sigma_i^2)(\sigma_j^1 - \sigma_j^2) \rangle.$$

The difference of two replicas σ^1, σ^2 will occur many times so we will simplify notation by setting

$$(5.8) \qquad \tilde{\sigma} = \sigma^1 - \sigma^2,$$

and we see that (5.6) is equivalent to

$$(5.9) \qquad \lim_{N \to \infty} \mathsf{E}\langle \tilde{\sigma}_i \tilde{\sigma}_j \rangle^2 = 0.$$

To remove this square, we bring in two new replicas σ^3, σ^4 and we set

$$(5.10) \qquad \sigma^* = \sigma^3 - \sigma^4$$

so that, since $\langle \sigma_i^* \sigma_j^* \rangle = \langle \tilde{\sigma}_i \tilde{\sigma}_j \rangle$,

$$(5.11) \qquad \langle \tilde{\sigma}_i \tilde{\sigma}_j \rangle^2 = \langle \tilde{\sigma}_i \tilde{\sigma}_j \rangle \langle \sigma_i^* \sigma_j^* \rangle = \langle \tilde{\sigma}_i \tilde{\sigma}_j \sigma_i^* \sigma_j^* \rangle.$$

It is often useful to use symmetry among coordinates. We define

$$(5.12) \qquad C_N = C_N(\beta, h) = \mathsf{E}\left\langle \left(\frac{\tilde{\sigma} \cdot \sigma^*}{N} \right)^2 \right\rangle$$

Writing

$$(5.13) \qquad \begin{aligned} \frac{\tilde{\sigma} \cdot \sigma^*}{N} &= \frac{1}{N} \sum_i \tilde{\sigma}_i \sigma_i^* \\ \left(\frac{\tilde{\sigma} \cdot \sigma^*}{N} \right)^2 &= \frac{1}{N^2} \sum_{i,j} \tilde{\sigma}_i \tilde{\sigma}_j \sigma_i^* \sigma_j^*, \end{aligned}$$

we see that by symmetry upon the coordinates,

$$(5.14) \qquad C_N = \frac{1}{N} \mathsf{E}\langle \tilde{\sigma}_1^2 \sigma_1^{*2} \rangle + \frac{N-1}{N} \mathsf{E}\langle \tilde{\sigma}_1 \tilde{\sigma}_2 \sigma_1^* \sigma_2^* \rangle.$$

Thus, (5.6) is equivalent to $\lim_{N \to \infty} C_N = 0$. One could say that when using

$$\mathsf{E}\left\langle \left(\frac{\tilde{\sigma} \cdot \sigma^*}{N} \right)^2 \right\rangle$$

one has a "global" point of view, while when using

$$\mathsf{E}(\langle \tilde{\sigma}_1 \tilde{\sigma}_2 \rangle^2) = \mathsf{E}\langle \tilde{\sigma}_1 \tilde{\sigma}_2 \sigma_1^* \sigma_2^* \rangle$$

one has a "local" point of view (two spins involved). It seems fruitful to use both.

We will study C_N using the *cavity method*. This is simply induction upon N. One removes a spin to reduce to an $(N-1)$-spin system (thereby creating a cavity). This method is of fundamental importance, both in the rigorous work of mathematicians, and in the heuristic arguments of physicists.

Let us consider a sequence $(g_i)_{i \leq N}$ of i.i.d. standard normal r.v., that is independent of all the g_{ij}. Consider $\sigma_{N+1} \in \{-1, +1\}$. We have

$$(5.15) \qquad -\beta H_N(\sigma) + \sigma_{N+1}\Big(\frac{\beta}{\sqrt{N}} \sum_{i \leq N} g_i \sigma_i + h\Big)$$

$$= \frac{\beta}{\sqrt{N}} \sum_{1 \leq i < j \leq N+1} g_{ij} \sigma_i \sigma_j + h \sum_{i \leq N+1} \sigma_i$$

where we set $g_{i,N+1} = g_i$. If we set

$$(5.16) \qquad \varrho = (\sigma, \sigma_{N+1}) = (\sigma_i)_{i \leq N+1} \in \Sigma_{N+1},$$

we see that the right-hand side of (5.15) is

$$(5.17) \qquad -\beta' \, H_{N+1}(\varrho),$$

where

$$(5.18) \qquad \beta' = \beta \, \frac{\sqrt{N+1}}{\sqrt{N}}.$$

That is, (5.15) is the Hamiltonian of an $(N+1)$-spin system at a slightly lower temperature. We denote by $G_{N+1} = G_{N+1}(\beta')$ the corresponding Gibbs' measure, and by $\langle \cdot \rangle'$ average with respect to this measure.

Consider now a function f on Σ_{N+1}. Using the notation (5.16), we can write either $f = f(\varrho)$, $\varrho \in \Sigma_{N+1}$ or $f = f(\sigma, \sigma_{N+1})$, $\sigma \in \Sigma_N$, $\sigma_{N+1} = \pm 1$. We then have the following algebraic identity

$$(5.19) \qquad \langle f(\varrho) \rangle' = \frac{1}{Z} \Big\langle \operatorname*{Av}_{\sigma_{N+1}=\pm 1} f(\sigma, \sigma_{N+1}) \mathcal{E} \Big\rangle,$$

where

$$(5.20) \qquad \mathcal{E} = \mathcal{E}(\sigma, \sigma_{N+1}) = \exp \sigma_{N+1}\Big(\frac{\beta}{\sqrt{N}} \sum_{i \leq N} g_i \sigma_i + h\Big),$$

and where

$$(5.21) \qquad Z = \Big\langle \operatorname*{Av}_{\sigma_{N+1}=\pm 1} \mathcal{E}(\sigma, \sigma_{N+1}) \Big\rangle.$$

(The reader should not confuse Z with the partition function Z_N.)

What these formulas mean is that we average first over $\sigma_{N+1} = \pm 1$; the resulting quantity depends on σ only. We then average over σ in G_N.

We will need a formula similar to (5.19), when f is a function on $(\Sigma_{N+1}^k, G_{N+1}^{\otimes k})$. In that case, we have

$$(5.22) \qquad \langle f(\varrho^1, \ldots, \varrho^k) \rangle' = \frac{1}{Z^k} \langle \mathrm{Av}\, f(\varrho^1, \ldots, \varrho^k) \mathcal{E}_k \rangle,$$

where now

$$(5.23) \qquad \mathcal{E}_k = \exp \sum_{l \leq k} \sigma_{N+1}^l \Big(\frac{\beta}{\sqrt{N}} \sum_{i \leq N} g_i \sigma_i^l + h \Big).$$

The notation (5.22) uses two conventions, that we will be valid throughout the paper.

First, the relation between ϱ^l and $(\sigma^l, \sigma_{N+1}^l)$ is as in (5.16),

$$(5.24) \qquad \varrho^l = (\sigma_1^l, \ldots, \sigma_N^l, \sigma_{N+1}^l).$$

Second, the average is over $\sigma_{N+1}^1, \ldots, \sigma_{N+1}^k = \pm 1$. The result of this averaging depends only upon $\sigma^1, \ldots, \sigma^k$, and is then integrated for $G_N^{\otimes k}$.

When one averages over $\sigma_{N+1}^1, \ldots, \sigma_{N+1}^k$ these are simply dumb variables, $\sigma_{N+1}^l = \pm 1$, so we will lighten notation by writing $\varepsilon_l = \sigma_{N+1}^l, \tilde{\varepsilon} = \sigma_{\tilde{N}+1}, \varepsilon^* = \sigma_{N+1}^*$ inside such averages. Then we write of course (5.23) as

$$\mathcal{E}_k = \exp \sum_{l \leq k} \varepsilon_l \Big(\frac{\beta}{\sqrt{N}} \sum_{i \leq N} g_i \sigma_i^l + h \Big).$$

Proposition 5.2. *There is a number $\beta_0 \geq 1/3$ such that if $\beta \leq \beta_0$, then for all $h \geq 0$, we have*

$$(5.25) \qquad\qquad NC_N(\beta, h) \leq 32.$$

Proof. From (5.14), we write

$$C_N(\beta, h) \leq \frac{16}{N} + \frac{N-1}{N} \mathsf{E} \langle \sigma_{\tilde{1}} \sigma_1^* \sigma_{\tilde{N}} \sigma_N^* \rangle,$$

and changing N in $N+1$,

$$C_{N+1}(\beta', h) \leq \frac{16}{N+1} + \frac{N}{N+1} \mathsf{E} \langle \sigma_{\tilde{1}} \sigma_1^* \sigma_{\tilde{N}+1} \sigma_{N+1}^* \rangle'.$$

We then use (5.22) on the last term to obtain

$$(5.26) \qquad \mathsf{E} \langle \sigma_{\tilde{1}} \sigma_1^* \sigma_{\tilde{N}+1} \sigma_{N+1}^* \rangle' = \mathsf{E} \frac{1}{Z^4} \langle \mathrm{Av}\, \sigma_{\tilde{1}} \sigma_1^* \tilde{\varepsilon} \varepsilon^* \mathcal{E}_4 \rangle,$$

where Av is over $\varepsilon_1, \ldots, \varepsilon_4$. We observe that $\mathrm{Av}\, \mathcal{E}(\sigma) \geq 1$, so that $Z \geq 1$. Also

$$\langle \mathrm{Av}\, \sigma_{\tilde{1}} \sigma_1^* \tilde{\varepsilon} \varepsilon^* \mathcal{E}_4 \rangle = \langle \mathrm{Av}\, \sigma_{\tilde{1}} \varepsilon \tilde{\mathcal{E}}_2 \rangle^2 \geq 0$$

and thus,

$$(5.27) \qquad \mathsf{E} \langle \sigma_{\tilde{1}} \sigma_1^* \sigma_{\tilde{N}+1} \sigma_{N+1}^* \rangle' \leq \mathsf{E} \langle \mathrm{Av}\, \sigma_{\tilde{1}} \sigma_1^* \tilde{\varepsilon} \varepsilon^* \mathcal{E}_4 \rangle.$$

The previous argument does not take advantage of the fact that Z can be much bigger than one. But removing this denominator makes the computation of the expectation much easier. Before we start this computation we observe an important fact. Since $\varepsilon_l \in \{-1, +1\}$, if $\varepsilon^\sim \neq 0$, $\varepsilon^* \neq 0$ we have $\varepsilon_2 = -\varepsilon_1$, $\varepsilon_4 = -\varepsilon_3$, so that

$$\varepsilon^\sim \varepsilon^* \mathcal{E}_4 = \varepsilon^\sim \varepsilon^* \mathcal{E}'$$

where

$$\mathcal{E}' = \exp \frac{\beta}{\sqrt{N}} \sum_{i \leq N} g_i(\varepsilon_1 \sigma_i^\sim + \varepsilon_3 \sigma_i^*).$$

Now

$$\operatorname*{Av}_{\varepsilon_2 = \pm 1} \varepsilon^\sim = \varepsilon_1, \qquad \operatorname*{Av}_{\varepsilon_4 = \pm 1} \varepsilon^* = \varepsilon_3$$

and thus

$$(5.28) \qquad \langle \operatorname{Av} \sigma_1^\sim \sigma_1^* \varepsilon^\sim \varepsilon^* \mathcal{E}_4 \rangle = \langle \operatorname{Av} \sigma_1^\sim \sigma_1^* \varepsilon_1 \varepsilon_3 \mathcal{E}' \rangle.$$

We denote by E_g the expectation in $(g_i)_{i \leq N}$ only. Thus

$$(5.29) \qquad \mathcal{E}'' := \mathsf{E}_g \mathcal{E}' = \exp \frac{\beta^2}{2N} \sum_{i \leq N} \left((\sigma_i^\sim)^2 + (\sigma_i^*)^2 + 2\varepsilon_1 \varepsilon_3 \sigma_i^\sim \sigma_i^*\right).$$

Since the disorder in $\langle \cdot \rangle$ is independent of the $(g_i)_{i \leq N}$, we have

$$\mathsf{E}_g \langle \operatorname{Av} \sigma_1^\sim \sigma_1^* \varepsilon_1 \varepsilon_3 \mathcal{E}' \rangle = \langle \operatorname{Av} \sigma_1^\sim \sigma_1^* \varepsilon_1 \varepsilon_3 \, \mathsf{E}_g \, \mathcal{E}' \rangle.$$

From (5.28) and (5.29), we then get that

$$(5.30) \qquad \mathsf{E} \langle \operatorname{Av} \sigma_1^\sim \sigma_1^* \varepsilon^\sim \varepsilon^* \mathcal{E}_4 \rangle = \mathsf{E} \langle \operatorname{Av} \sigma_1^\sim \sigma_1^* \varepsilon_1 \varepsilon_3 \mathcal{E}'' \rangle$$
$$= \mathsf{E} \langle \sigma_1^\sim \sigma_1^* \operatorname{Av} \varepsilon_1 \varepsilon_3 \mathcal{E}'' \rangle.$$

We then use symmetry among the sites to see that

$$(5.31) \qquad \mathsf{E} \langle \operatorname{Av} \sigma_1^\sim \sigma_1^* \varepsilon_1 \varepsilon_3 \mathcal{E}'' \rangle = \mathsf{E} \left\langle \left(\frac{1}{N} \sum_{i \leq N} \sigma_i^\sim \sigma_i^*\right) \operatorname{Av} \varepsilon_1 \varepsilon_3 \mathcal{E}'' \right\rangle.$$

In view of the notation $\sigma^1 \cdot \sigma^2 = \sum_{i \leq N} \sigma_i^1 \sigma_i^2$, it is natural to write

$$(5.32) \qquad \sum_{i \leq N} (\sigma_i^\sim)^2 = \|\sigma^\sim\|^2 ; \qquad \sum_{i \leq N} (\sigma_i^*)^2 = \|\sigma^*\|^2.$$

Thus the right-hand side of (5.31) is

$$(5.33) \quad \frac{1}{N} \mathsf{E} \left\langle \sigma^\sim \cdot \sigma^* \operatorname{Av} \varepsilon_1 \varepsilon_3 \exp \frac{\beta^2}{2N} \left(\|\sigma^\sim\|^2 + \|\sigma^*\|^2 + 2\varepsilon_1 \varepsilon_3 \sigma^\sim \cdot \sigma^*\right) \right\rangle$$
$$= \frac{1}{N} \mathsf{E} \left\langle \sigma^\sim \cdot \sigma^* \operatorname{sh}\left(\frac{\beta^2}{N} \sigma^\sim \cdot \sigma^*\right) \exp \frac{\beta^2}{2N} (\|\sigma^\sim\|^2 + \|\sigma^*\|^2) \right\rangle.$$

Now we use the fact that for $|x| \leq 4$, we have $x \, \text{sh} \, \beta^2 x \leq \frac{x^2}{4} \text{sh} \, 4\beta^2$, and that $\|\sigma^{\sim} \cdot \sigma^*\|^2$, $\|\sigma^{\sim}\|^2$, $\|\sigma^*\|^2 \leq 4N$. From (5.31), (5.33) we obtain

$$\mathsf{E}\langle \text{Av} \, \sigma_1^{\sim} \sigma_1^* \varepsilon_1 \varepsilon_3 \mathcal{E}'' \rangle \leq \frac{1}{4} \text{sh} \, 4\beta^2 \exp 8\beta^2 \left\langle \left(\frac{\sigma^{\sim} \cdot \sigma^*}{N}\right)^2 \right\rangle.$$

If we recall (5.25), we see that

(5.34) $C_{N+1}(\beta', h) \leq \dfrac{16}{N+1} + \dfrac{1}{4}\left(\dfrac{N}{N+1}\right)^2 \text{sh} \, 4\beta^2 \exp 8\beta^2 C_N(\beta, h).$

We consider now β_0 such that

$$\frac{1}{4} \text{sh} \, 4\beta_0^2 \exp 8\beta_0^2 \leq \frac{1}{2}$$

(e.g. $\beta_0 = \frac{1}{3}$). Thus, if $\beta' \leq \beta_0$ we have

$$\frac{1}{4} \text{sh} \, 4\beta'^2 \exp 8\beta'^2 \leq \frac{1}{2}$$

and

$$(N+1)C_{N+1}(\beta', h) \leq 16 + \frac{1}{2}NC_N(\beta, h).$$

If we set

$$a_N = \sup_{\beta \leq \beta_0} NC_N(\beta, h),$$

since $\beta' > \beta$ we see that

$$a_{N+1} \leq 16 + \frac{1}{2}a_N$$

and thus by induction over N that $a_N \leq 32$.   □

It should be pointed out that Proposition 5.2 is a very special case of a powerful result of [F-Z]; but this special proof we gave is particularly simple.

Let us now explore consequences of Proposition 5.2. We set

(5.35) $$D_N = D_N(\beta, h) = \mathsf{E}\left\langle \left(\frac{\sigma^{\sim} \cdot \sigma^3}{N}\right)^2 \right\rangle.$$

Lemma 5.3. *We have*

(5.36) $$C_N \leq 4D_N$$

(5.37) $$D_N \leq \sqrt{C_N}.$$

Proof. To prove (5.36), using that $(a+b)^2 \leq 2a^2 + 2b^2$, we have

$$C_N \leq 2\mathsf{E}\left\langle \left(\frac{\sigma^{\sim} \cdot \sigma^3}{N}\right)^2 \right\rangle + 2\mathsf{E}\left\langle \left(\frac{\sigma^{\sim} \cdot \sigma^4}{N}\right)^2 \right\rangle = 4D_N.$$

To prove (5.37), we write

$$\sigma^{\sim} \cdot \sigma^3 = \sum_{i \leq N} \sigma_i^{\sim} \sigma_i^3,$$

so that, using Cauchy-Schwarz twice,

$$D_N = \frac{1}{N^2} \mathsf{E} \sum_{i,j \le N} \langle \sigma_i^\sim \sigma_i^3 \sigma_j^\sim \sigma_j^3 \rangle$$

$$= \frac{1}{N^2} \sum_{i,j \le N} \mathsf{E}(\langle \sigma_i^\sim \sigma_j^\sim \rangle \langle \sigma_i^3 \sigma_j^3 \rangle)$$

$$\le \frac{1}{N^2} \sum_{i,j \le N} (\mathsf{E}\langle \sigma_i^\sim \sigma_j^\sim \rangle^2)^{\frac{1}{2}} (\mathsf{E}\langle \sigma_i^3 \sigma_j^3 \rangle^2)^{\frac{1}{2}}$$

$$\le \frac{1}{N^2} \sum_{i,j \le N} (\mathsf{E}\langle \sigma_i^\sim \sigma_j^\sim \rangle^2)^{\frac{1}{2}}$$

$$\le \left(\frac{1}{N^2} \sum_{i,j \le N} \mathsf{E}\langle \sigma_i^\sim \sigma_j^\sim \rangle^2 \right)^{\frac{1}{2}}$$

$$= \left(\frac{1}{N^2} \sum_{i,j \le N} \mathsf{E}\langle \sigma_i^\sim \sigma_j^\sim \sigma_i^* \sigma_j^* \rangle \right)^{\frac{1}{2}}$$

$$= \left(\mathsf{E}\langle (\sum_{i \le N} \frac{1}{N} \sigma_i^\sim \sigma_i^*)^2 \rangle \right)^{\frac{1}{2}}$$

$$= \sqrt{C_N}. \qquad \qquad \square$$

Thus, it follows from (5.25) and Proposition 5.2 that $D_N \le 8/\sqrt{N}$ for $\beta \le \beta_0$. It is in fact true, as we will show later, that $D_N \le L/N$. This however cannot be proved by general principles.

Challenge problem 5.4. Prove by an example that (5.37) is essentially sharp. That is, construct a random measure G'_N on $\{-1, 1\}^N$ such that if we define C_N and D_N by replacing G_N by G'_N, then (5.37) is essentially optimal.

As a consequence, we will not be able to deduce that $D_N \le L/N$ from (5.24) but we will have to make a specific analysis, along the lines of Proposition 5.2. It is unfortunately harder to relate D_N and D_{N-1} than C_N and C_{N-1}.

As simplicity is our main goal in the present chapter, we will proceed through (5.37). This does not give the correct rates of convergences, so we will not attempt to get any rate of convergence at all. The correct rates will be obtained later.

Lemma 5.5. *We have*

(5.38) $$\frac{1}{4} D_N \le \mathsf{E}\langle (R(\sigma, \sigma') - \langle R(\sigma, \sigma') \rangle)^2 \rangle \le 4 D_N.$$

In particular, (5.6) is equivalent to

$$(5.39) \qquad \lim_{N \to \infty} \mathsf{E}\langle (R(\sigma, \sigma') - \langle R(\sigma, \sigma') \rangle)^2 \rangle = 0.$$

What (5.39) means is that the function $(\sigma, \sigma') \mapsto R(\sigma, \sigma')$ on (Σ_N^2, G_N^2) is nearly constant. The value of the constant is quite naturally the integral of the function, namely $\langle R(\sigma, \sigma') \rangle$. This quantity might depend upon the disorder.

Proof. First, we write

$$D_N = \mathsf{E}\left\langle \left(\frac{\sigma^\sim \cdot \sigma^3}{N} \right)^2 \right\rangle$$
$$= \mathsf{E}((R(\sigma^2, \sigma^3) - R(\sigma^1, \sigma^3))^2)$$
$$\leq 2\,\mathsf{E}\langle (R(\sigma^2, \sigma^3) - \langle R(\sigma^2, \sigma^3) \rangle)^2 \rangle + 2\,\mathsf{E}\langle (R(\sigma^1, \sigma^3) - \langle R(\sigma^1, \sigma^3) \rangle)^2 \rangle$$
$$= 4\,\mathsf{E}\langle (R(\sigma^1, \sigma^2) - \langle R(\sigma^1, \sigma^2) \rangle)^2 \rangle.$$

Next we write

$$(5.40) \qquad b = \langle \sigma \rangle = (\langle \sigma_i \rangle)_{i \leq N}.$$

We observe that by Jensen's inequality

$$(5.41) \qquad \left\langle \left(\frac{(\sigma^1 - b) \cdot \sigma^3}{N} \right)^2 \right\rangle \leq \left\langle \left(\frac{\sigma^\sim \cdot \sigma^3}{N} \right)^2 \right\rangle.$$

This is because the left-hand side relates to the right-hand side by averaging σ^2 for G_N inside the square rather than outside. Similarly,

$$(5.42) \qquad \left\langle \left(\frac{(\sigma^1 - b) \cdot b}{N} \right)^2 \right\rangle \leq \left\langle \left(\frac{\sigma^\sim \cdot \sigma^3}{N} \right)^2 \right\rangle.$$

Let us also observe that

$$(5.43) \qquad b \cdot b = \sum_{i \leq N} \langle \sigma_i \rangle^2 = \sum_{i \leq N} \langle \sigma_i^1 \sigma_i^2 \rangle = \langle \sigma^1 \cdot \sigma^2 \rangle.$$

Thus

$$\mathsf{E}\left\langle \left(\frac{\sigma^1 \cdot \sigma^2}{N} - \left\langle \frac{\sigma^1 \cdot \sigma^2}{N} \right\rangle \right)^2 \right\rangle = \mathsf{E}\left\langle \left(\frac{(\sigma^1 - b) \cdot \sigma^2}{N} + \frac{(\sigma^2 - b) \cdot b}{N} \right)^2 \right\rangle$$
$$\leq 4D_N,$$

using (5.41), (5.42), and the fact that $(a + b)^2 \leq 2a^2 + 2b^2$. $\qquad \square$

In the use of the cavity method, we have met quantities such as

$$(5.44) \qquad \left\langle \exp \frac{t}{\sqrt{N}} \sum_{i \leq N} g_i \sigma_i \right\rangle$$

in (5.40) for $t = \beta \sigma_{N+1}$. These will occur often, so we write

$$(5.45) \qquad \sum_{i \leq N} g_i \sigma_i = g \cdot \sigma.$$

The quantities in (5.40) are averages with respect to Gibbs' measure, and as we do not yet understand Gibbs' measure, we might fear that we do not understand them either. A fundamental consequence of (5.6) is that the quantities (5.44) essentially depend upon Gibbs' measure only through the much simpler quantity $b = \langle \sigma \rangle$.

Proposition 5.6. *Under (5.7), if f is a bounded function on Σ_N, we have*

$$(5.46) \quad \lim_{N \to \infty} \mathsf{E} \left| \left\langle f(\sigma) \exp \frac{t}{\sqrt{N}} g \cdot \sigma \right\rangle - \langle f \rangle \exp\left(\frac{t}{\sqrt{N}} g \cdot b + \frac{t^2}{2}(1 - \bar{q}) \right) \right| \to 0$$

where

$$\bar{q} = \langle R(\sigma, \sigma') \rangle = \frac{\|b\|^2}{N}.$$

Strictly speaking the expression "f is a bounded function on Σ_N" is meaningless. This is a convenient way to express the following: for each N, we consider a function f_N on Σ_N, and we assume that $\sup_N |f_N| < \infty$.

Proof. First let

$$X = \left\langle f(\sigma) \exp \frac{t}{\sqrt{N}} g \cdot \sigma \right\rangle$$

$$Y = \left\langle f(\sigma) \exp\left(\frac{t}{\sqrt{N}} g \cdot b + \frac{t^2}{2}(1 - \bar{q}) \right) \right\rangle.$$

We will show that

$$(5.47) \qquad \mathsf{E}(X - Y)^2 = \mathsf{E}\big(\mathsf{E}_g(X^2 - 2XY + Y^2) \big) \to 0$$

where E_g denotes integration in g only. We use replicas to write

$$X^2 = \left\langle f(\sigma) f(\sigma') \exp \frac{t}{\sqrt{N}} g \cdot (\sigma + \sigma') \right\rangle,$$

$$XY = \left\langle f(\sigma) f(\sigma') \exp\left(\frac{t}{\sqrt{N}} g \cdot (\sigma + b) + \frac{t^2}{2}(1 - \bar{q}) \right) \right\rangle,$$

$$Y^2 = \exp\left(\frac{2t}{\sqrt{N}} g \cdot b + t^2(1 - \bar{q}) \right) \langle f(\sigma) f(\sigma') \rangle,$$

so that, since $\|\sigma\|^2 = \|\sigma'\|^2 = N$, using (2.14),

$$\mathsf{E}_g\, X^2 = \langle f(\sigma)f(\sigma')\exp t^2(1 + R(\sigma,\sigma'))\rangle,$$

$$\mathsf{E}_g\, XY = \Big\langle f(\sigma)f(\sigma')\exp\Big(\frac{t^2}{2}\Big(1 + 2\frac{\sigma\cdot b}{N} + \frac{\|b\|^2}{N}\Big) + \frac{t^2}{2}(1-\bar q)\Big)\Big\rangle$$

$$= \Big\langle f(\sigma)f(\sigma')\exp t^2\Big(1 + \frac{\sigma\cdot b}{N}\Big)\Big\rangle,$$

$$\mathsf{E}_g\, Y^2 = \exp\Big(2t^2\frac{\|b\|^2}{N} + t^2(1-\bar q)\Big)\langle f(\sigma)f(\sigma')\rangle$$

$$= \exp t^2(1+\bar q)\,\langle f(\sigma)f(\sigma')\rangle.$$

Now, it follows from (5.39) that

$$\mathsf{E}\langle (R(\sigma^1,\sigma^2) - \bar q)^2\rangle \to 0,$$

and from (5.42) and (5.43) that

$$\mathsf{E}\Big\langle\Big(\frac{\sigma^1\cdot b}{N} - \bar q\Big)^2\Big\rangle \to 0.$$

This implies that

$$\mathsf{E}\,|\mathsf{E}_g\, X^2 - \mathsf{E}_g\, Y^2| \to 0\,, \quad \mathsf{E}\,|\mathsf{E}_g\, XY - \mathsf{E}_g\, Y^2| \to 0.$$

This proves (5.43) and Proposition 5.6. $\qquad\square$

Not surprisingly, Proposition 5.6 will be very useful. But what is really behind it? Proposition 5.6 means that for the typical choice of g the law of $(g\cdot\sigma)/\sqrt{N}$ under G_N is approximately $\mathcal{N}((g\cdot b/\sqrt{N}),(1-\bar q))$. It is a kind of central limit theorem but it holds for the typical random coefficients g_i rather than for any choice. If it were true that G_N is a product measure (which it is certainly not) then for any sequence $(a_i)_{i\le N}$, with $\max a_i \ll (\sum a_i^2)^{\frac12}$, the law under G_N of $\sigma \mapsto \sum a_i\sigma_i$ would be approximatively

$$\mathcal{N}\Big(\sum_{i\le N} a_i\langle\sigma_i\rangle, \sum_{i\le N} a_i^2(1 - \langle\sigma_i\rangle^2)\Big).$$

When $a_i = g_i/\sqrt{N}$, one certainly expects that averaging takes place and that

$$\sum_{i\le N}\frac{g_i^2}{N}(1 - \langle\sigma_i\rangle^2) \simeq 1 - \frac{1}{N}\sum_{i\le N}\langle\sigma_i\rangle^2 = 1 - \bar q.$$

Proposition 5.6 is much less surprising as part of a general theme: under (5.6), Gibbs' measure somehow resembles a product measure. This is in fact the case as long as one considers only finitely many spins.

Proposition 5.7. *Under (5.6), for each p, we have*

(5.48) $$\lim_{N\to\infty} \mathsf{E}\big(\langle\sigma_1\dots\sigma_p\rangle - \langle\sigma_1\rangle\dots\langle\sigma_p\rangle\big)^2 = 0.$$

Thus, quite remarkably, the vanishing of the correlations of two spins implies a similar vanishing on any given number of spins.

Proof. We proceed by induction over p. This is true for $p = 2$. For the induction from $p - 1$ to p, since $|\langle \sigma_i \rangle| \leq 1$, it suffices to prove that

$$\lim_{N \to \infty} \mathsf{E}(\langle \sigma_1 \ldots \sigma_p \rangle - \langle \sigma_1 \ldots \sigma_{p-1} \rangle \langle \sigma_p \rangle)^2 = 0,$$

or, with the notation

(5.49) $$\dot{\sigma}_i = \sigma_i - \langle \sigma_i \rangle,$$

that

(5.50) $$\lim_{N \to \infty} \mathsf{E} \langle \sigma_1 \ldots \sigma_{p-1} \dot{\sigma}_p \rangle^2 = 0.$$

We observe that, by symmetry upon the sites, we have

(5.51) $$M \, \mathsf{E} \langle \sigma_1 \ldots \sigma_{p-1} \dot{\sigma}_p \rangle^2 \leq \mathsf{E} \sum_{i_1, \ldots, i_p} \langle \sigma_{i_1} \ldots \sigma_{i_{p-1}} \dot{\sigma}_{i_p} \rangle^2$$

where $M = N(N-1) \ldots (N - p + 1)$, and where the summation is over all choices $i_1, \ldots, i_p \leq N$. Now, using replicas,

(5.52) $$\sum_{i_1, \ldots, i_p} \langle \sigma_{i_1} \ldots \sigma_{i_{p-1}} \dot{\sigma}_{i_p} \rangle^2 = \Big\langle \sum_{i_1, \ldots, i_p} \sigma_{i_1} \ldots \sigma_{i_{p-1}} \dot{\sigma}_{i_p} \sigma'_{i_1} \ldots \sigma'_{i_{p-1}} \dot{\sigma}'_{i_p} \Big\rangle$$

$$= N^p \Big\langle \Big(\frac{\dot{\sigma} \cdot \dot{\sigma}'}{N} \Big) R(\sigma, \sigma')^{p-1} \Big\rangle.$$

Moreover, since $|R(\sigma, \sigma')| \leq 1$, by Cauchy-Schwarz,

(5.53) $$\mathsf{E} \Big\langle \Big(\frac{\dot{\sigma} \cdot \dot{\sigma}'}{N} \Big) R(\sigma, \sigma')^{p-1} \Big\rangle \leq \Big(\mathsf{E} \Big\langle \Big(\frac{\dot{\sigma} \cdot \dot{\sigma}'}{N} \Big)^2 \Big\rangle \Big)^{\frac{1}{2}} \leq C_N^{\frac{1}{2}},$$

where in the last inequality we have used Jensen's inequality as in (5.41). To finish the proof, we observe that (5.50) follows from (5.51) to (5.53).   □

Theorem 5.8. *Under (5.6), given any number p, $\sigma_1, \ldots, \sigma_p$ are asymptotically independent under G_N.*

Proof. The proof will provide a precise statement. Denote by $G_{N,p}$ the law of $\sigma_1, \ldots, \sigma_p$ under G_N, or equivalently, the projection of G_N onto $\{-1, 1\}^p$. Let us denote by μ the (random) product measure on $\{-1, 1\}^p$ such that

$$\int \sigma_i \, d\mu(\sigma_1, \ldots, \sigma_p) = \langle \sigma_i \rangle = \int \sigma_i \, dG_{N,p}(\sigma_1, \ldots, \sigma_p).$$

Then $\mathsf{E} |G_{N,p} - \mu| \to 0$, where $|G_{N,p} - \mu|$ is the total variation distance. To prove this, it suffices to show that given $\sigma_1, \ldots, \sigma_p$,

(5.54) $$\mathsf{E} |G_{N,p}(\{(\sigma_1, \ldots, \sigma_p)\}) - \mu(\{(\sigma_1, \ldots, \sigma_p)\})| \to 0.$$

Now, if $(\eta_1, \ldots, \eta_p) \in \{-1, 1\}^p$,

$$1_{\{(\sigma_1, \ldots, \sigma_p)\}}(\eta_1, \ldots, \eta_p) = 2^{-p} \prod_{i \leq p} (1 + \eta_i \sigma_i) = 2^{-p} \sum_I \eta_I \sigma_I$$

where the summation is over $I \subset \{1, \ldots, p\}$, and $\sigma_I = \prod_{i \in I} \sigma_i$, $\eta_I = \prod_{i \in I} \eta_i$. Thus, to prove (5.54), it suffices to prove that for each I

$$(5.55) \qquad E \left| \int \sigma_I \, dG_{N,p}(\sigma_1, \ldots, \sigma_p) - \int \sigma_I \, d\mu(\sigma_1, \ldots, \sigma_p) \right| \to 0,$$

i.e.

$$E \left| \langle \sigma_I \rangle - \prod_{i \in I} \langle \sigma_i \rangle \right| \to 0,$$

which follows from Proposition 5.7. $\qquad\square$

The remarkable feature of Theorem 5.8 is of course that the conclusion looks stronger than the hypothesis.

At this point the reader should be convinced of the importance of (5.6). With the exception of Proposition 5.2, the conclusions we have drawn from (5.6) are quite general, and rely only upon the symmetry between sites. The physicists say that a system satisfying (5.6) is in a *pure equilibrium state*. For all the systems where the author has been able to prove at high temperature the validity of the replica symmetric formulas, proving that the system is in a pure state has always been the first (and most difficult) step.

Lemma 5.5 tells us that the overlap of two generic configurations is independent of the particular choice of these configurations, although it might depend upon the disorder. Of course, we would like it to be in fact independent of the disorder. That is, we would like

$$\langle R(\sigma, \sigma') \rangle = \frac{1}{N} \sum_{i \leq N} \langle \sigma_i \sigma_i' \rangle = \frac{1}{N} \sum_{i \leq N} \langle \sigma_i \rangle^2 = \bar{q}$$

to be independent of the disorder. Before we prove this, we collect a few simple facts, the first of which, the integration by parts formula, is of fundamental importance.

Proposition 5.9. *If g is centered normal, for any smooth function f such that $(1 + x^2)^{-k} f(x)$ is bounded for some k then*

$$(5.56) \qquad E(gf(g)) = E(g^2) E(f'(g)).$$

Proof. This is, indeed, integration by parts. If $E(g^2) = \tau^2$,

$$\frac{1}{\tau\sqrt{2\pi}} \int_{-\infty}^{+\infty} t f(t) \exp\left(-\frac{t^2}{2\tau^2}\right) dt = \tau^2 \frac{1}{\tau\sqrt{2\pi}} \int_{-\infty}^{+\infty} f'(t) \exp\left(-\frac{t^2}{2\tau^2}\right) dt. \quad \square$$

We consider

$$(5.57) \quad \Psi(x) = \mathsf{E}\,\mathrm{th}^2(\beta g\sqrt{x}+h) = \frac{1}{\sqrt{2\pi}}\int \mathrm{th}^2(\beta g\sqrt{t}+h)\exp\left(-\frac{t^2}{2}\right)dt.$$

This notation will be used in Chapters 5 and 6. (The function ψ of Chapter 4 will be not used any more).

Lemma 5.10. *We have*

$$\Psi'(x) = \beta^2\,\mathsf{E}\,\frac{1-2\,\mathrm{sh}^2(\beta g\sqrt{x}+h)}{\mathrm{ch}^4(\beta g\sqrt{x}+h)}$$

and in particular $|\Psi'| \le 2\beta^2$, *and* $|\Psi'| \le \frac{1}{2}$ *if* $\beta \le \frac{1}{3}$.

Proof. We have

$$\Psi'(x) = \mathsf{E}\left(\frac{\beta g}{\sqrt{x}}\frac{\mathrm{th}}{\mathrm{ch}^2}(\beta g\sqrt{x}+h)\right)$$

and the result by (5.56), since

$$\left(\frac{\mathrm{th}(x)}{\mathrm{ch}^2(x)}\right)' = \frac{1-2\,\mathrm{sh}^2(x)}{\mathrm{ch}^4(x)}. \qquad \square$$

It follows, in particular, that (5.4) has indeed as unique solution q.

Theorem 5.11. *If* $\beta < \beta_0$, *for each h we have*

$$(5.58) \qquad \lim_{N\to\infty} \mathsf{E}(\bar{q}-q)^2 = 0.$$

Proof. We set

$$Q_N = Q_N(\beta,h) = \mathsf{E}(\bar{q}-q)^2 = \mathsf{E}\left(\frac{1}{N}\sum_{i\le N}\langle\sigma_i\rangle^2\right)^2 - 2q\,\mathsf{E}\,\frac{1}{N}\sum_{i\le N}\langle\sigma_i\rangle^2 + q^2.$$

By expanding the square and using symmetry between the sites, we have

$$(5.59) \qquad Q_N \le \frac{1}{N} + \mathsf{E}\langle\sigma_1\rangle^2\langle\sigma_N\rangle^2 - q\,\mathsf{E}\langle\sigma_1\rangle^2 - q\,\mathsf{E}\langle\sigma_N\rangle^2 + q^2,$$

and, changing N into $N+1$,

$$(5.60) \qquad Q_{N+1}(\beta',h) \le \frac{1}{N+1}$$
$$+ \mathsf{E}\langle\sigma_1\rangle'^2\langle\sigma_{N+1}\rangle'^2 - q'\,\mathsf{E}\langle\sigma_1\rangle'^2 - q'\,\mathsf{E}\langle\sigma_{N+1}\rangle'^2 + q'^2,$$

where q' is the solution of (5.4) when β' replaces β. It should be obvious that

$$(5.61) \qquad |q-q'| = o(1),$$

where $o(1)$ is a quantity that goes to zero as $N\to\infty$.

Lemma 5.12. *We have*

(5.62) $$\mathsf{E}\,|\langle\sigma_1\rangle'^2 - \langle\sigma_1\rangle^2| \to 0$$

(5.63) $$\mathsf{E}\left|\langle\sigma_{N+1}\rangle'^2 - \mathrm{th}^2\left(\frac{\beta}{\sqrt{N}}\,\boldsymbol{g}\cdot\boldsymbol{b}+h\right)\right| \to 0.$$

Proof. To prove (5.62), from (5.19) we have

$$\langle\sigma_1\rangle' = \frac{\langle \mathrm{Av}\,\sigma_1\mathcal{E}\rangle}{Z},$$

for

$$Z = \langle \mathrm{Av}\,\mathcal{E}\rangle$$

$$\mathcal{E} = \exp\sigma_{N+1}\left(\frac{\beta}{\sqrt{N}}\sum_{i\le N}g_i\sigma_i + h\right).$$

It follows from Proposition 5.6 that

$$\mathsf{E}\left|Z - \exp\frac{\beta^2}{2}(1-\bar{q})\,\mathrm{ch}\left(\frac{\beta}{\sqrt{N}}\,\boldsymbol{g}\cdot\boldsymbol{b}+h\right)\right| \to 0,$$

$$\mathsf{E}\left|\langle \mathrm{Av}\,\sigma_1\mathcal{E}\rangle - \langle\sigma_1\rangle\exp\frac{\beta^2}{2}(1-\bar{q})\,\mathrm{ch}\left(\frac{\beta}{\sqrt{N}}\,\boldsymbol{g}\cdot\boldsymbol{b}+h\right)\right| \to 0.$$

Since $Z \ge 1$, this implies

$$\mathsf{E}\,|\langle\sigma_1\rangle' - \langle\sigma_1\rangle| \to 0$$

from which (5.62) follows. The proof of (5.63) is similar. $\qquad\square$

End of the proof of Theorem 5.11. If we combine (5.60) to (5.63), we see that

(5.64) $$Q_{N+1}(\beta',h) \le \mathsf{E}\left(\langle\sigma_1\rangle^2\,\mathrm{th}^2\left(\frac{\beta}{\sqrt{N}}\,\boldsymbol{g}\cdot\boldsymbol{b}+h\right)\right)$$

$$- q\,\mathsf{E}\langle\sigma_1\rangle^2 - q\,\mathsf{E}\,\mathrm{th}^2\left(\frac{\beta}{\sqrt{N}}\,\boldsymbol{g}\cdot\boldsymbol{b}+h\right) + q^2 + o(1).$$

Since

$$\mathsf{E}_g\left(\frac{1}{\sqrt{N}}\,\boldsymbol{g}\cdot\boldsymbol{b}\right)^2 = \frac{1}{N}\|\boldsymbol{b}\|^2 = \bar{q},$$

we have

(5.65) $$\mathsf{E}_g\,\mathrm{th}^2\left(\frac{\beta}{\sqrt{N}}\,\boldsymbol{g}\cdot\boldsymbol{b}+h\right) = \Psi(\bar{q}).$$

Moreover, since $\langle\sigma_1\rangle^2$ does not depend upon g, we have

$$\mathsf{E}\left(\langle\sigma_1\rangle^2\,\mathrm{th}^2\left(\frac{\beta}{\sqrt{N}}\,\boldsymbol{g}\cdot\boldsymbol{b}+h\right)\right) = \mathsf{E}\left(\langle\sigma_1\rangle^2\,\mathsf{E}_g\,\mathrm{th}^2\left(\frac{\beta}{\sqrt{N}}\,\boldsymbol{g}\cdot\boldsymbol{b}+h\right)\right)$$

$$= \mathsf{E}(\langle\sigma_1\rangle^2\Psi(\bar{q})).$$

With (5.64) this implies

$$(5.66) \qquad Q_{N+1}(\beta', h) \le \mathsf{E}(\langle\sigma_1\rangle^2 \Psi(\bar{q})) - q\,\mathsf{E}\langle\sigma_1\rangle^2 - q\,\mathsf{E}\,\Psi(\bar{q}) + q^2 + o(1)$$

and, using the symmetry between sites,

$$
\begin{aligned}
(5.67) \qquad Q_{N+1}(\beta', h) &\le \mathsf{E}(\bar{q}\Psi(\bar{q})) - q\,\mathsf{E}(\bar{q}) - q\,\mathsf{E}\,\Psi(\bar{q}) + q^2 + o(1) \\
&= \mathsf{E}((\bar{q} - q)(\Psi(\bar{q}) - q)) + o(1) \\
&= \mathsf{E}((\bar{q} - q)(\Psi(\bar{q}) - \Psi(q))) + o(1) \\
&\le \frac{1}{2}\,\mathsf{E}((\bar{q} - q)^2) + o(1) \\
&= \frac{1}{2}Q_N(\beta, h) + o(1).
\end{aligned}
$$

Here, we have used Lemma 5.10 in the fourth line. Iteration of (5.67) yields (5.58). $\qquad\qquad\qquad\qquad\qquad\qquad\qquad\qquad\qquad\qquad\qquad\qquad\qquad\qquad$ $\square$

We have now improved (5.39) into

$$(5.68) \qquad\qquad \lim_{N\to\infty} \mathsf{E}\langle(R(\sigma, \sigma') - q)^2\rangle = 0$$

where q is given by (5.4).

We are now ready to start the proof of (5.3). We first remove some of the mystery of this formula. Fixing h once and for all, we denote by $\mathsf{SK}(\beta, q)$ the right-hand side of (5.3), when we think of β, q as independent variables. We then have

$$\frac{\partial\,\mathsf{SK}}{\partial q} = -\frac{\beta^2}{2}(1 - q) + \mathsf{E}\left(\frac{\beta g}{2\sqrt{q}}\,\mathrm{th}(\beta g\sqrt{q} + h)\right),$$

and using integration by parts (Proposition 5.9),

$$\frac{\partial\,\mathsf{SK}}{\partial q} = -\frac{\beta^2}{2}(1 - q) + \frac{\beta^2}{2}\,\mathsf{E}\,\frac{1}{\mathrm{ch}^2(\beta g\sqrt{q} + h)}.$$

Since $\mathrm{th}^2(x) = 1 - \dfrac{1}{\mathrm{ch}^2(x)}$, (5.4) means that $q = q(\beta, h)$ is such that

$$\frac{\partial\,\mathsf{SK}}{\partial q}(\beta, q) = 0.$$

It follows that

$$(5.69) \qquad\qquad \frac{d}{d\beta}\,\mathsf{SK}(\beta, q(\beta, h)) = \frac{\partial\,\mathsf{SK}}{\partial\beta}(\beta, q(\beta, h)).$$

Now,

$$\frac{\partial\,\mathrm{SK}}{\partial\beta} = \frac{\beta}{2}(1-q)^2 + \mathsf{E}\big(g\sqrt{q}\,\mathrm{th}(\beta g\sqrt{q}+h)\big)$$
$$= \frac{\beta}{2}(1-q)^2 + \beta q\,\mathsf{E}\,\frac{1}{\mathrm{ch}^2(\beta g\sqrt{q}+h)},$$

using Proposition 5.9 again. Now, by (5.4),

$$\mathsf{E}\,\frac{1}{\mathrm{ch}^2(\beta g\sqrt{q}+h)} = 1-q,$$

so that

(5.70)
$$\frac{\partial\,\mathrm{SK}}{\partial\beta} = \frac{\beta}{2}(1-q^2).$$

We note for $\beta = 0$

$$p_N(0,h) = \log(2\,\mathrm{ch}\,h) = \log(e^h + e^{-h})$$

so that (5.3) is obviously true. So to prove (5.3) we will show that

(5.71)
$$\lim_{N\to\infty}\left|\frac{\partial p_N}{\partial\beta}(\beta,h) - \frac{\beta}{2}(1-q^2)\right| = 0,$$

where q is given by (5.4). This, of course, removes much of the mystery of (5.3).

Lemma 5.13. *We have*

$$\frac{\partial p_N}{\partial\beta} = \frac{\beta}{2}\big(1 - \mathsf{E}\langle R(\sigma,\sigma')^2\rangle\big).$$

Proof. This fundamental formula is again a consequence of integration by parts. We write

(5.72)
$$p_N(\beta,h) = \frac{1}{N}\,\mathsf{E}\log Z_N = \frac{1}{N}\,\mathsf{E}\log\sum_\sigma B(\sigma),$$

where

$$B(\sigma) = B_N(\beta,h,\sigma) = \exp\Big(\frac{\beta}{\sqrt{N}}\sum_{i<j}g_{ij}\sigma_i\sigma_j + h\sum_{i\leq N}\sigma_i\Big)$$

and thus

(5.73)
$$\frac{\partial B(\sigma)}{\partial\beta} = \frac{1}{\sqrt{N}}\Big(\sum_{i<j}g_{ij}\sigma_i\sigma_j\Big)B(\sigma)$$

(5.74)
$$\frac{\partial B(\sigma)}{\partial g_{ij}} = \frac{\beta}{\sqrt{N}}\sigma_i\sigma_j B(\sigma).$$

Thus, from (5.71),

$$(5.75) \qquad \frac{\partial}{\partial \beta} p_N(\beta, h) = \frac{1}{N^{3/2}} \mathsf{E} \sum_{\sigma} \frac{\sum_{i<j} g_{ij} \sigma_i \sigma_j B(\sigma)}{Z_N}$$

$$= \frac{1}{N^{3/2}} \sum_{i<j} \sum_{\sigma} \sigma_i \sigma_j \mathsf{E} \frac{g_{ij} B(\sigma)}{Z_N}.$$

We now use integration by parts using the fact that the g_{ij} are independent, so that we can use Proposition 5.9 conditionally upon the $(g_{i'j'})$, $(i', j') \neq (i, j)$ and we get, using (5.74)

$$\mathsf{E}\Big(g_{ij} \frac{B(\sigma)}{Z_N}\Big) = \frac{\beta}{\sqrt{N}} \Big(\sigma_i \sigma_j \mathsf{E}\Big(\frac{B(\sigma)}{Z_N}\Big) - \mathsf{E}\frac{B(\sigma)(\sum_{\sigma'} \sigma_i' \sigma_j' B(\sigma'))}{Z_N^2}\Big)$$

and (5.75) yields

$$\frac{\partial}{\partial \beta} p_N(\beta, h) = \frac{\beta}{N^2} \Big(\frac{N(N-1)}{2} - \mathsf{E} \sum_{i<j} \langle \sigma_i \sigma_j \sigma_i' \sigma_j' \rangle\Big)$$

$$= \frac{\beta}{2N^2} \Big(N^2 - \mathsf{E} \sum_{i,j} \langle \sigma_i \sigma_j \sigma_i' \sigma_j' \rangle\Big)$$

$$= \frac{\beta}{2} \Big(1 - \frac{1}{N^2} \mathsf{E}\Big\langle \Big(\sum_i \sigma_i \sigma_i'\Big)^2 \Big\rangle\Big)$$

$$= \frac{\beta}{2} (1 - \mathsf{E} \langle R(\sigma, \sigma')^2 \rangle). \qquad \square$$

Combining Theorem 5.11, Lemma 5.13 and (5.70) proves (5.71), and hence Theorem 5.1.

Thus, it seems quite appropriate to say that the replica symmetric formula (5.3) is neither so mysterious nor quite central. The central fact is (5.68).

Theorem 5.8 means that "locally" (on finitely many spins), Gibbs' measure is determined by the sequence $b = (\langle \sigma_i \rangle)_{i \leq N}$, and Proposition 5.6 says that this is even true "globally", in some respects. So, how does this sequence $\langle \sigma_i \rangle_{i \leq N}$ behave?

Theorem 5.14. *For $\beta \leq \beta_0$, given any number p, $\langle \sigma_1 \rangle, \ldots, \langle \sigma_p \rangle$ are asymptotically i.i.d. Their common limit law is the law ν of $\mathrm{th}(\beta g \sqrt{q} + h)$ where g is $\mathcal{N}(0, 1)$ and q is given by (5.4).*

Proof. By induction over p, we show that if $f : \mathbb{R}^p \to \mathbb{R}$ is continuous with bounded support,

$$\mathsf{E} f(\langle \sigma_1 \rangle, \ldots, \langle \sigma_p \rangle) - \int f(x_1, \ldots, x_p) \, d\nu(x_1) \ldots d\nu(x_p) \to 0$$

uniformly over $\beta \leq \beta_0$. This is true for $p = 0$ (nothing to prove) so we proceed to the induction step from $p - 1$ to p. Since this requires use of the cavity method, we will change N into $N + 1$ and, using the symmetry among sites, we will prove that

(5.76)
$$\mathsf{E}\, f\big(\langle\sigma_1\rangle', \ldots, \langle\sigma_{p-1}\rangle', \langle\sigma_{N+1}\rangle'\big) - \int f(x_1, \ldots, x_p)\, d\nu'(x_1) \ldots d\nu'(x_p) \to 0$$

uniformly in $\beta' \leq \beta_0$, where β' and β are related by $\beta' = \beta\left(1 + \frac{1}{N}\right)^{1/2}$, and where ν' corresponds to β' rather than β. We observe from Lemma 5.12 that

(5.77)
$$\mathsf{E}\left|\langle\sigma_{N+1}\rangle' - \mathrm{th}\Big(\frac{\beta}{\sqrt{N}}\boldsymbol{g}\cdot\boldsymbol{b} + h\Big)\right| \to 0$$
$$\mathsf{E}\,|\langle\sigma_i\rangle' - \langle\sigma_i\rangle| \to 0, \quad \text{for all } i,$$

so that

(5.78)
$$\mathsf{E}\, f\big(\langle\sigma_1\rangle', \ldots, \langle\sigma_{p-1}\rangle', \langle\sigma_{N+1}\rangle'\big) - \mathsf{E}\, f\Big(\langle\sigma_1\rangle, \ldots, \langle\sigma_{p-1}\rangle, \mathrm{th}\Big(\frac{\beta}{\sqrt{N}}\boldsymbol{g}\cdot\boldsymbol{b} + h\Big)\Big) \to 0.$$

Consider the function
$$u(x_1, \ldots, x_{p-1}, y) = \mathsf{E}\, f\big(x_1, \ldots, x_{p-1}, \mathrm{th}(\beta g + h)\big),$$

where g is $\mathcal{N}(0, y)$. It is obviously continuous. In (5.78), g is independent of $\langle\,\cdot\,\rangle$, and the law of $(\boldsymbol{g}\cdot\boldsymbol{b})/\sqrt{N}$ given $\boldsymbol{b}$ is $\mathcal{N}(0, \bar{q})$. It follows that
$$\mathsf{E}\, f\Big(\langle\sigma_1\rangle, \ldots, \langle\sigma_{p-1}\rangle, \mathrm{th}\Big(\frac{\beta}{\sqrt{N}}\boldsymbol{g}\cdot\boldsymbol{b} + h\Big)\Big) = \mathsf{E}\, u\big(\langle\sigma_1\rangle, \ldots, \langle\sigma_{p-1}\rangle, \bar{q}\big).$$

Since we have shown that $\mathsf{E}\,|\bar{q} - q| \to 0$, we have
$$\mathsf{E}\, u\big(\langle\sigma_1\rangle, \ldots, \langle\sigma_{p-1}\rangle, \bar{q}\big) - \mathsf{E}\, u\big(\langle\sigma_1\rangle, \ldots, \langle\sigma_{p-1}\rangle, q\big) \to 0.$$

Now the function $u(x_1, \ldots, x_p, q)$ has bounded support, so by induction hypothesis (and since $\beta \leq \beta'$) we have
$$\mathsf{E}\, u\big(\langle\sigma_1\rangle, \ldots, \langle\sigma_{p-1}\rangle, q\big) - \int u(x_1, \ldots, x_{p-1}, q)\, d\nu(x_1) \ldots d\nu(x_{p-1}) \to 0.$$

Moreover
$$u(x_1, \ldots, x_{p-1}, q) = \int f(x_1, \ldots, x_{p-1}, x_p)\, d\nu(x_p).$$

Combining these estimates we have proved
$$\mathsf{E}\, f\big(\langle\sigma_1\rangle', \ldots, \langle\sigma_{p-1}\rangle', \langle\sigma_{N+1}\rangle'\big) - \int f(x_1, \ldots, x_p)\, d\nu(x_1) \ldots d\nu(x_p) \to 0$$

from which (5.76) follows. $\qquad\square$

6 Exponential inequalities

In the previous chapter, we have shown that for the SK-model, if $\beta \leq \frac{1}{3}$ we have

(6.1)
$$\lim_{N \to \infty} \mathsf{E}\langle (R(\sigma, \sigma') - q)^2 \rangle = 0.$$

We will prove here considerably more accurate results in the same direction, namely the following.

Theorem 6.1. *There exists a number L and a number β_0 such that if $\beta \leq \beta_0$, and $h \leq 1$, then*

(6.2)
$$\mathsf{E}\left\langle \exp \frac{N}{L} \left(R(\sigma, \sigma') - q \right)^2 \right\rangle \leq L.$$

Besides the fact that such an inequality is obviously a notable improvement on (6.1), it is also motivated by its possible relevance to the important open problem that will be discussed in the next chapter. The restriction $h \leq 1$ is not really needed. We impose it to avoid secondary complications. It can be removed with a few extra lines of work. In fact, large values of h improve the situation: one can show that (6.2) holds for $\beta \leq \beta(h)$ with $\beta(h) \to \infty$ as $h \to \infty$.

An exponential integrability condition such as (6.2) amounts to a uniform control of moments. We will prove the following

Proposition 6.2. *There exists a number β_0 with the following property. Set $A = 192$. If $\beta \leq \beta_0$, $h \leq 1$ then for each k, N we have*

(6.3)
$$\mathsf{E}\langle (\sigma^1 \cdot \sigma^2 - Nq)^{2k} \rangle \leq A^k N^k k!$$

Using that

$$e^{x^2} = \sum_{k \geq 0} \frac{x^{2k}}{k!},$$

we see that (6.3) implies (6.2). (These are in fact essentially equivalent).

An important observation before the proof starts is that to prove (6.3) we can assume $N \geq k$. Indeed, since $|\sigma^1 \cdot \sigma^2 - Nq| \leq 2N$, the left-hand side of (6.3) is at most $4^k N^{2k}$, so (6.3) holds provided

$$4^k N^k \leq A^k k!$$

so that if $N \leq k$, to prove (6.2) it suffices to see that $k^k \leq e^k k!$. This is a weak version of Stirling's formula which is very easy to prove by induction upon k.

To prove (6.3) we will proceed by double induction. Let us denote by $H(N, k)$ the validity of (6.3), so that $H(N, 0)$ is valid for all N. We will prove by induction over k that $H(N, k)$ is valid for all N. For a given k, we will prove the validity of $H(N, k)$ by induction over N.

Let us write

(6.4) $$U_{N,k}(\beta) = E\langle(\sigma^1 \cdot \sigma^2 - Nq)^{2k}\rangle.$$

To prove Proposition 6.2, it suffices to prove the following

Proposition 6.3. *If* $\beta' \leq \beta_0$ *and if*

(6.5) $$\forall l \leq k \quad U_{N,l}(\beta) \leq A^l N^l l!,$$

then

(6.6) $$U_{N+1,k}(\beta') \leq A^k (N+1)^k k!$$

Here, as elsewhere in this Chapter, β and β', q and q' as well as $\langle \cdot \rangle$ and $\langle \cdot \rangle'$ are related as in Chapter 5.

Lemma 6.4. *If* $\beta \leq \beta_0$, *we have*

$$(N+1)|q - q'| \leq 1$$

Proof. We write as $\Psi(x, \beta)$ the quantity (5.57). Thus if $q = \Psi(q, \beta)$, then

$$\frac{dq}{d\beta} = \frac{\frac{\partial \Psi}{\partial \beta}(q, \beta)}{1 - \frac{\partial \Psi}{\partial q}(q, \beta)}.$$

Now $\partial \Psi / \partial \beta = 2q/\beta \, \partial \Psi / \partial q$, and the result follows from Lemma 5.10 and the fact that $|\beta' - \beta| \leq \beta/2N$. $\square$

Let us start the proof of Proposition 6.3. (The reader is suggested to assume $k = 1$ at first reading). To simplify notation, we write

$$V = \sigma^1 \cdot \sigma^2 - Nq, \quad v_{N+1} = \sigma^1_{N+1}\sigma^2_{N+1} - q, \quad \delta = (N+1)(q - q').$$

We have

$$(6.7) \qquad U_{N+1,k}(\beta') = E\langle(\sigma^1 \cdot \sigma^2 + \sigma^1_{N+1}\sigma^2_{N+1} - (N+1)q')^{2k}\rangle'$$

$$= E\langle(V + v_{N+1} + \delta)^{2k}\rangle'$$

$$= \sum_{0 \le l \le 2k} \binom{2k}{l} \delta^l E\langle(V + v_{N+1})^{2k-l}\rangle'$$

$$\le \sum_{0 \le l \le 2k} \binom{2k}{l} W^{\frac{2k-l}{2k}}$$

$$= (1 + W^{\frac{1}{2k}})^{2k}$$

where

$$(6.8) \qquad W = E\langle(V + v_{N+1})^{2k}\rangle',$$

and where we have used Lemma 6.4 and Hölder's inequality in the fourth line. To prove (6.6), it suffices to prove that

$$(6.9) \qquad W \le \frac{1}{2}(N+1)^k A^k k!$$

Indeed, by (6.7), we have

$$(6.10) \qquad U_{N+1,k}(\beta') \le (1 + W^{\frac{1}{2k}})^{2k} \le \left(1 + \left(\frac{1}{2}(N+1)^k A^k k!\right)^{\frac{1}{2k}}\right)^{2k}$$

$$\le \frac{1}{2}(N+1)^k A^k k! \exp\left(\frac{2k}{\left(\frac{1}{2}(N+1)^k A^k k!\right)^{\frac{1}{2k}}}\right)$$

since, for $x > 0$,

$$(1 + x)^{2k} = x^{2k}(1 + \frac{1}{x})^{2k} \le x^{2k} \exp(\frac{2k}{x}).$$

Now, $N + 1 \ge k$ and $A^k k! \ge 4^{2k+1}k^k$, so that

$$\left(\frac{1}{2}(N+1)^k A^k k!\right)^{\frac{1}{2k}} \ge 4k$$

and (6.10) implies (6.6).

We turn to the proof of (6.9). Using symmetry between sites, we get

$$(6.11) \qquad W \le (N+1) E\langle v_{N+1}(V + v_{N+1})^{2k-1}\rangle'.$$

Of course in the bracket, average is integration of $\varrho^l = (\sigma_1^l, \ldots, \sigma_N^l, \sigma_{N+1}^l)$ ($l \le 2$) in G_{N+1}. We write

$$(V + v_{N+1})^{2k-1} = \sum_{0 \le l \le 2k-1} \binom{2k-1}{l} V^{2k-1-l} v_{N+1}^l.$$

The most important term is for $l = 0$. For $l \ge 1$, we bound $|v_{N+1}|^{l+1}$ by 2^{l+1}, and we obtain from (6.11) that

(6.12) $$W \le (N+1) \, \mathsf{E} \langle v_{N+1} V^{2k-1} \rangle' + S$$

where

(6.13) $$S = (N+1) \sum_{1 \le l \le 2k-1} \binom{2k-1}{l} 2^{l+1} \, \mathsf{E} \langle |V|^{2k-1-l} \rangle'.$$

We first perform the easy task, the control of S.

Lemma 6.5. If $N + 1 \ge k$, we have

(6.14) $$S \le \frac{1}{4}(N+1)^k A^k k!$$

Proof. We use (5.22), and the fact that $Z \ge 1$ to get

$$\mathsf{E} \langle |V|^{2k-1-l} \rangle' \le \mathsf{E} \langle \mathrm{Av} |V|^{2k-1-l} \mathcal{E}_2 \rangle$$
$$\le \exp(2\beta^2 + 2h) \, \mathsf{E} \langle |V|^{2k-1-l} \rangle,$$

by integrating first in $g = (g_i)_{i \le N}$.

Next, we use Hölder's inequality,

$$\mathsf{E} \langle |V|^{2k-l-1} \rangle \le (\mathsf{E} \langle V^{2k-2} \rangle)^{\frac{2k-1-l}{2k-2}}$$
$$= (U_{N,k-1}(\beta))^{\frac{2k-1-l}{2k-2}}.$$

We use the trivial bound

$$\binom{2k-1}{l} \le 2k(2(k-1))^{l-1},$$

to obtain from (6.13) that

(6.15) $$S \le 8k(N+1) \exp(2\beta^2 + 2h) \sum_{1 \le l \le 2k-1} (4(k-1))^{l-1} (U_{N,k-1}(\beta))^{\frac{2k-l-1}{2k-2}}.$$

Now, we use that $p^p \le 3^p p!$ for $p = k - 1$ to see that, since we assume $N \ge k - 1$, we have

$$8^{2(k-1)}(k-1)^{2(k-1)} \le A^{k-1} N^{k-1}(k-1)!,$$

so that

(6.16) $$(4(k-1))^{l-1} \le 2^{1-l}(A^{k-1}N^{k-1}(k-1)!)^{\frac{l-1}{2k-2}}.$$

Next we use (6.5) to see that

$$(6.17) \qquad U_{N,k-1}(\beta) \leq A^{k-1} N^{k-1}(k-1)!.$$

and substitution of (6.16), (6.17) in (6.15) gives

$$S \leq 16k(N+1)\exp(2\beta^2 + 2h)A^{k-1}N^{k-1}(k-1)!$$
$$\leq \left(\frac{16}{A}\exp(2\beta^2 + 2h)\right)(N+1)^k A^k k!$$

But since $\beta \leq \frac{1}{3}, h \leq 1$, we have $\frac{16}{A}\exp(2\beta^2 + 2h) \leq \frac{1}{4}$. $\qquad\square$

Now we turn to the main issue, the study of $E\langle v_{N+1} V^{2k-1}\rangle'$. We use (5.22) to get

$$(6.18) \qquad E\langle v_{N+1} V^{2k-1}\rangle' = E\frac{1}{Z^2}\langle V^{2k-1} \operatorname{Av}(\varepsilon_1\varepsilon_2 - q)\mathcal{E}_2\rangle,$$

where

$$\mathcal{E}_2 = \exp\sum_{l\leq 2}\varepsilon_l(\frac{\beta}{\sqrt{N}}g\cdot\sigma^l + h),$$

$$Z = \left\langle\operatorname{Av}\exp\varepsilon(\frac{\beta}{\sqrt{N}}g\cdot\sigma + h)\right\rangle.$$

To study (6.18), we will use the following principle.

Proposition 6.6. *There exists a number L with the following property. Consider a function $f : \Sigma_N^2 \to \mathbb{R}$ and p, p' with $1/p + 1/p' = 1$. Consider q in $[0, 1]$. Then if $\beta \leq 1, h \leq 1$, and if J is a subset of $\{1, 2\}$ we have*

$$(6.19) \qquad \left|E\frac{1}{Z^2}\langle f \operatorname{Av}\prod_{l\in J}\varepsilon_l\mathcal{E}_2\rangle - E\operatorname{th}^m(Y)E\langle f\rangle\right|$$

$$\leq \beta^2 L(E\langle|f|^p\rangle)^{\frac{1}{p}}\left(E\langle|R(\sigma^1, \sigma^2) - q|^{p'}\rangle\right)^{\frac{1}{p'}},$$

where $Y = \beta g\sqrt{q} + h$, and where g is $N(0, 1)$.

The ideas behind Proposition 6.6 are of fundamental importance, so we delay their discussion until after the proof of Theorem 6.1.

We will use (6.19) when q is given by (5.4). This simply means that $q = E\operatorname{th}^2 Y$. We apply (6.19) to $f = V^{2k-1}, p = 2k/(2k-1), p' = 2k$, and for two choices of $J : J = \{1, 2\}$ and $J = \emptyset$. Since

$$|f|^p = N^{2k}|R(\sigma^1, \sigma^2) - q|^{2k} = |V|^{2k},$$

we get

$$(6.20) \qquad E\langle v_{N+1}V^{2k-1}\rangle' \leq \frac{L\beta^2}{N}U_{N,k}(\beta)$$
$$\leq L\beta^2 N^{k-1}A^k k!,$$

using (6.5). Going back to (6.12), and using (6.4), we get

$$W \leq (N+1)^k A^k k!\left(\frac{1}{4} + L\beta^2\right)$$

and this proves (6.9) if $L\beta^2 \leq 1/4$. This proves Proposition 6.2 and hence Theorem 6.1.

We now turn to the proof of Proposition 6.6. Even though it will require some abstraction to see it, the true nature of the problem we face is as follows. Consider two centered jointly Gaussian families $(u_i)_{i\leq n}$, $(v_i)_{i\leq n}$ and a smooth function F on $\mathbb{R}^n$. How do we compare

$$E F(u_1,\ldots,u_n) \text{ and } E F(v_1,\ldots,v_n)?$$

The circle of ideas we will use to do this will be referred to as Kahane's principle (in honor of [K]). As in all great mathematics, the starting point is almost absurdly simple. It is to find a smooth path $w(t)$ from the family u to the family v so that $w(0) = u$, $w(1) = v$, and to write

$$(6.21) \qquad E F(v) - E F(u) = \int_0^1 \theta'(t)\,dt,$$

where $\theta(t) = E F(w(t))$. The basic tool for computation of $\theta'(t)$ is integration by parts, and we first state the form we need.

Proposition 6.7. *Consider a (centered) jointly Gaussian family $u, u_1, \ldots, u_n$ of r.v. Then*

$$(6.22) \qquad E uF(u_1,\ldots,u_n) = \sum_{i\leq n} E(uu_i)\, E\frac{\partial F}{\partial x_i}(u_1,\ldots,u_n).$$

Proof. This is an application of (5.56). We write

$$u_i' = u_i - \frac{E(uu_i)}{E u^2}u$$

so that $E(u_i'u) = 0$. The family $(u_i')_{i\leq n}$ is independent of u. Setting $c_i = E(uu_i)/E(u^2)$, we write

$$E uF(u_1,\ldots,u_n) = E uF(u_1' + c_1u, \ldots, u_n' + c_nu)$$

and we use (5.56) at (u_i') fixed. $\qquad\square$

Here is a particularly elegant formulation.

Proposition 6.8. *Consider two centered jointly Gaussian families of r.v.*
$u = (u_i)_{i\leq n}$, $v = (v_i)_{i\leq n}$. *We assume that these families are independent of each other (but we do not assume that $(u_i)_{i\leq n}$ or $(v_i)_{i\leq n}$ are independent). For $0 \leq t \leq 1$ consider*

$$\theta(t) = E\,F(\sqrt{1-t}\,u + \sqrt{t}\,v). \tag{6.23}$$

Then for $0 < t < 1$ we have

$$\theta'(t) = \frac{1}{2}\sum_{i,j}\big(E(v_iv_j) - E(u_iu_j)\big)\,E\,\frac{\partial^2 F}{\partial x_i\partial x_j}(\sqrt{1-t}\,u + \sqrt{t}\,v). \tag{6.24}$$

Proof. We have

$$\theta'(t) = \frac{1}{2}E\sum_{i\leq n}\Big(\frac{1}{\sqrt{t}}v_i - \frac{1}{\sqrt{1-t}}u_i\Big)\frac{\partial F}{\partial x_i}(\sqrt{1-t}\,u + \sqrt{t}\,v).$$

By independence of u, v, we have

$$E\Big(\Big(\frac{1}{\sqrt{t}}v_i - \frac{1}{\sqrt{1-t}}u_i\Big)\big(\sqrt{1-t}\,u_j + \sqrt{t}\,v_j\big)\Big) = E(v_iv_j - u_iu_j),$$

so that the result follows from Proposition 6.7. ☐

The way Kahane uses this is that if

$$\forall i,j\ \ E(v_iv_j) \geq E(u_iv_j) \text{ and } \frac{\partial^2 F}{\partial x_i\partial x_j} \geq 0,$$

then $E\,F(v) \geq E\,F(u)$. (Slepian's Lemma)

Proof of Proposition 6.6. For further purposes, we will consider a more general setting, namely replicas of order s rather than 2 and we will try to approximate

$$E\,\frac{\langle f\,\mathrm{Av}\,\varepsilon_J\mathcal{E}_s\rangle}{Z^s} = E\,\frac{\langle f\,\mathrm{Av}\,\varepsilon_J\mathcal{E}_s\rangle}{\langle\mathrm{Av}\,\mathcal{E}_s\rangle} \tag{6.25}$$

where $J \subset \{1,\ldots,s\}$ and $\varepsilon_J = \prod_{l\in J}\varepsilon_l$. We will work conditionally upon Gibbs' measure on Σ_N, that is, we integrate first in g. Rather than Gibbs' measure, it turns out to be very convenient to consider points $(\sigma(j))_{j\leq M}$ in Σ_N and the probability μ that gives mass $1/M$ to each $\sigma(j)$, i.e. $\mu = M^{-1}\sum_{j\leq M}\delta_{\sigma(j)}$. The points $\sigma(j)$ are not assumed to be distinct. Fixing N, we will take $M \to \infty$ to approximate Gibbs' measure. Thus, we want to understand

$$E_g\,\frac{\sum f\,\mathrm{Av}\,\varepsilon_J\exp\sum_{l\leq s}\varepsilon_l(\frac{\beta}{\sqrt{N}}g\cdot\sigma(j_l) + h)}{\sum\mathrm{Av}\exp\sum_{l\leq s}\varepsilon_l(\frac{\beta}{\sqrt{N}}g\cdot\sigma(j_l) + h)}. \tag{6.26}$$

Here

$$f = f\big(\sigma(j_1),\ldots,\sigma(j_s)\big), \tag{6.27}$$

and the sum $\sum$ are over all possible choices $j_1, \ldots, j_s \leq M$. It turns out to be very useful to think to the quantities $g \cdot \sigma(j_l)$ as Ms different variables (s for $l \leq s$, M for the M possible values of j). That is, we consider $x = (x(j,l))_{j \leq M, l \leq s}$ and we consider

(6.28)
$$F_1(x) = \sum_{j_1, \ldots, j_s} f \, \mathrm{Av} \, \varepsilon_J \exp\left(\sum_{l \leq s} \varepsilon_l(\beta x(j_l, l) + h)\right)$$

(6.29)
$$F_2(x) = \sum_{j_1, \ldots, j_s} \mathrm{Av} \exp\left(\sum_{l \leq s} \varepsilon_l(\beta x(j_l, l) + h)\right),$$

where f is as in (6.27) and the summation is over all possible values of $j_1, \ldots, j_s$. We set

(6.30)
$$F(x) = \frac{F_1(x)}{F_2(x)}.$$

We consider the Gaussian family

$$v = (v(j,l))_{j \leq M, l \leq s}$$

where

$$v(j,l) = \frac{1}{\sqrt{N}} g \cdot \sigma(j),$$

so that (6.26) is $\mathrm{E}_g F(v)$.

We consider now independent Gaussian variables u, $w(j,l)$ such that

(6.31)
$$\mathrm{E}\, u^2 = q$$

(6.32)
$$\mathrm{E}\, w^2(j,l) = 1 - q.$$

We set

(6.33)
$$u(j,l) = u + w(j,l).$$

Thus, if $(j,l) \neq (j',l')$, we have

(6.34)
$$\mathrm{E}\, u(j,l) u(j',l') = q.$$

The idea of considering u is as follows. We have

(6.35) $M^{-s} F_1(u)$
$$= M^{-s} \, \mathrm{Av}\left(\varepsilon_J \exp\left(\sum_{l \leq s} \varepsilon_l(\beta u + h)\right) \sum f \exp\left(\sum_{l \leq s} \beta \varepsilon_l w(j_l, l)\right)\right).$$

Now, the variables $w(j_l, l)$ are all independent, so by a trivial computation

(6.36) $\mathrm{E}\left| M^{-s} \sum f \exp \sum_{l \leq s} \beta \varepsilon_l w(j_l, l) - M^{-s} \sum f \exp \frac{\beta^2}{2} s(1-q) \right|^2 \leq \frac{K}{M},$

where K depends upon f, β, s but not upon M. Proceeding as in (6.35) for $F_2(u)$, we see that

$$(6.37) \quad \lim \mathsf{E}_g \frac{F_1(u)}{F_2(u)} = \langle f(\sigma^1, \ldots, \sigma^s) \rangle \, \mathsf{E}_g \frac{\mathrm{Av} \prod_{l \in J} \varepsilon_l \exp \sum_{l \leq s} \varepsilon_l(\beta u + h)}{\mathrm{Av} \exp \sum_{l \leq s} \varepsilon_l(\beta u + h)}$$

$$= \mathsf{E}_g \left(\mathrm{th}(\beta u + h) \right)^{\mathrm{card}\, J} \langle f(\sigma^1, \ldots, \sigma^s) \rangle.$$

In (6.37), the limit is taken as $M \to \infty$ and as $\nu = M^{-1} \sum_{j \leq M} \delta_{\sigma(j)}$ converges to Gibbs' measure. The reader has by now certainly understood that the purpose of introducing ν rather than working with Gibbs' measure is not to use equal weights (this makes little difference) but to consider the atoms of Gibbs' measure as made up from many different small pieces in order to use (6.37) as $M \to \infty$. We consider

$$\theta(t) = \mathsf{E}_g F(\sqrt{1-t}\, u + \sqrt{t}\, v)$$

so that

$$(6.38) \quad \mathsf{E}_g F(v) - \mathsf{E}_g F(u) = \theta(1) - \theta(0) = \int_0^1 \theta'(t)\, \mathrm{d}t,$$

and we compute $\theta'(t)$ through Proposition 6.8

Lemma 6.9. *We have*

$$(6.39) \qquad \theta'(t) = \sum_{1 \leq l_1 < l_2 \leq s} \mathsf{E}_g F^1_{l_1,l_2}(\sqrt{1-t}\, u + \sqrt{t}\, v)$$

$$- \sum_{1 \leq l_1 < l_2 \leq s} \mathsf{E}_g F^2_{l_1,l_2}(\sqrt{1-t}\, u + \sqrt{t}\, v)$$

$$- \sum_{1 \leq l_1, l_2 \leq s} \mathsf{E}_g F^3_{l_1,l_2}(\sqrt{1-t}\, u + \sqrt{t}\, v)$$

$$+ \sum_{1 \leq l_1, l_2 \leq s} \mathsf{E}_g F^4_{l_1,l_2}(\sqrt{1-t}\, u + \sqrt{t}\, v),$$

where the functions $F^r_{l_1,l_2}$, $r = 1, 2, 3, 4$ are defined as follows:

$$(6.40) \qquad F^1_{l_1,l_2}(x) = \frac{\beta^2}{F_2(x)} \sum_{j_1, \ldots, j_s} f A_{l_1,l_2} \, \mathrm{Av}\, \varepsilon_{l_1} \varepsilon_{l_2} \varepsilon_J \mathcal{E}(x)$$

for

$$\mathcal{E}(x) = \exp \sum_{l \leq s} \varepsilon_l(\beta x(j_l, l) + h) \; ;$$

$$(6.41)$$

$$A_{l_1,l_2} = \frac{\sigma(j_{l_1}) \cdot \sigma(j_{l_2})}{N} - q$$

$$(6.42) \qquad F^2_{l_1,l_2}(x) = \frac{\beta^2 F_1(x)}{F_2(x)^2} \sum_{j_1, \ldots, j_s} A_{l_1,l_2} \, \mathrm{Av}\, \varepsilon_{l_1} \varepsilon_{l_2} \mathcal{E}(x).$$

Moreover,

$$(6.43) \qquad F^3_{l_1,l_2}(x) = \frac{\beta^2}{F_2(x)^2} \sum_{j_1,\dots,j_{2s}} f A_{l_1,l_2+s} \operatorname{Av} \varepsilon_{l_1}\varepsilon_{l_2+s}\varepsilon_J \mathcal{E}(x)\mathcal{E}'(x).$$

Here $j_1,\dots,j_{2s}$ take all values $\leq M$, with $(j_{l_1},l_1) \neq (j_{l_2+s},l_2)$, the average Av is over $\varepsilon_1,\dots,\varepsilon_{2s} = \pm 1$, and

$$\mathcal{E}'(x) = \exp \sum_{1\leq l\leq s} \varepsilon_{l+s}\big(\beta x(j_{l+s},l) + h\big) \; ;$$

$$A_{l_1,l_2+s} = \frac{\sigma(j_{l_1}) \cdot \sigma(j_{l_2+s})}{N} - q,$$

and finally

$$(6.44) \qquad F^4_{l_1,l_2}(x) = \beta^2 \frac{F_1(x)}{F_2(x)^3} \sum_{j_1,\dots,j_{2s}} A_{l_1,l_2+s} \operatorname{Av} \varepsilon_{l_1}\varepsilon_{l_2+s}\mathcal{E}(x)\mathcal{E}'(x),$$

where $j_1,\dots,j_{2s}$ take all values $\leq M$ with $(j_{l_1},l_1) \neq (j_{l_2+s},l_2)$.

This of course looks totally overwhelming! To prove that this is not the case, we finish the proof of Proposition 6.6.

We use that $F_2(x) \geq M^s$. Taking absolute values, and expectations, we see that all these terms on the right of (6.39) are bounded by

$$\beta^2 K(\beta,h,s)M^{-3s} \sum_{j_1,\dots,j_{3s}} |f(\sigma(j_1),\dots,\sigma(j_s))|\left|\frac{\sigma(j_l) \cdot \sigma(j_{l'})}{N} - q\right|$$

where $l,l' \leq 3s,\ l \neq l'$. Using Hölder's inequality, this is at most

$$\beta^2 K(\beta,h,s)\left(M^{-s} \sum_{j_1,\dots,j_s} |f(\sigma(j_1),\dots,\sigma(j_s))|^p\right)^{\frac{1}{p}}$$

$$\times \left(M^{-2} \sum_{j,j'}\left|\frac{\sigma(j) \cdot \sigma(j')}{N} - q\right|^{p'}\right)^{\frac{1}{p'}}.$$

This is a bound for the left-hand side of (6.38). As $M \to \infty$ and $M^{-1} \sum_{j\leq M} \delta_{\sigma(j)}$ converges to Gibbs' measure, this bound becomes

$$\beta^2 K(\beta,h,s)\langle|f|^p\rangle^{\frac{1}{p}}\langle|R(\sigma,\sigma') - q|^{p'}\rangle^{\frac{1}{p'}},$$

where $K(\beta,h,s)$ remains bounded with β,h,s, and we conclude by Hölder's inequality. $\qquad\square$

Proof of Lemma 6.9. This is a mere computation, although of course not a pleasant one. We write, with obvious notation,

$$
(6.45) \qquad \frac{\partial^2 F}{\partial x_a \partial x_b} = \frac{\partial^2 F_1}{\partial x_a \partial x_b} \frac{1}{F_2} - \frac{\partial^2 F_2}{\partial x_a \partial x_b} \frac{F_1}{F_2^2}
$$
$$
- \left(\frac{\partial F_1}{\partial x_b} \frac{\partial F_2}{\partial x_a} + \frac{\partial F_1}{\partial x_a} \frac{\partial F_2}{\partial x_b} \right) \frac{1}{F_2^2} + 2 \frac{\partial F_2}{\partial x_a} \frac{\partial F_2}{\partial x_b} \frac{F_1}{F_2^3}
$$

so that

$$
(6.46) \qquad \frac{1}{2} \sum \mathsf{E}(v_a v_b - u_a u_b) \frac{\partial^2 F}{\partial x_a \partial x_b}
$$

is the sum of four pieces, each of which creates one of the sums of (6.39). The indexes a, b in the sum (6.46) are of the type

$$
(6.47) \qquad a = (k_1, l_1), \quad b = (k_2, l_2).
$$

Restricting the summation to l_1, l_2 fixed will create the individual terms in the sums of (6.39). Let us observe that $\mathsf{E} v_a^2 = 1 = \mathsf{E} u_a^2$ so that in (6.46) we can restrict the sum to $a \neq b$. If $l_1 = l_2$, no term of the sum (6.28) can contain both variables $x(k_1, l_1)$, $x(k_2, l_2)$ if $k_1 \neq k_2$ so that

$$
\frac{\partial^2 F_1}{\partial x_a \partial x_b} = 0.
$$

If $l_1 \neq l_2$,

$$
\frac{\partial^2 F_1}{\partial x_a \partial x_b} = \beta^2 \sum f \, \mathrm{Av} \, \varepsilon_{l_1} \varepsilon_{l_2} \varepsilon_J \exp \sum_{l \leq s} \varepsilon_l (\beta x(j_l, l) + h)
$$

where the sum is restricted over those families $j_1, \ldots, j_s$ such that $j_{l_1} = k_1$, $j_{l_2} = k_2$. Now if $a \neq b$, we have seen (using again the notation (6.47)) that

$$
(6.48) \qquad \mathsf{E}(v_a v_b - u_a u_b) = \frac{\sigma(k_1) \cdot \sigma(k_2)}{N} - q.
$$

Thus

$$
\sum \mathsf{E}(v_a v_b - u_a u_b) \frac{\partial^2 F_1}{\partial x_a \partial x_b} = \beta^2 \sum f(\sigma(j_1), \ldots, \sigma(j_s)) \left(\frac{\sigma(j_{l_1}) \cdot \sigma(j_{l_2})}{N} - q \right)
$$
$$
\times \mathrm{Av} \, \varepsilon_{l_1} \varepsilon_{l_2} \varepsilon_J \exp \sum_{l \leq s} \varepsilon_l (\beta x(j_l, l) + h),
$$

where the sum on the left is over $k_1, k_2 \leq M$, $a = (k_1, l_1)$, $b = (k_2, l_2)$, l_1, l_2 fixed. This is how (6.40) occurs. The restriction $l_1 < l_2$ creates a factor 2. The term (6.42) occurs for the same reason, so we turn toward (6.43), that occur from the term $\dfrac{\partial F_1}{\partial x_b}\dfrac{\partial F_2}{\partial x_b}$. Using again (6.47), we have

$$\frac{\partial F_1}{\partial x_a}(x)\frac{\partial F_2}{\partial x_b}(x) = \beta^2\left(\sum f\,\mathrm{Av}\,\varepsilon_{l_1}\varepsilon_J\exp\sum_{l\le s}\varepsilon_l\big(\beta x(j_l,l)+h\big)\right)$$
$$\times\left(\sum\mathrm{Av}\,\varepsilon_{l_2}\exp\sum_{l\le s}\varepsilon_l\big(\beta x(j_l,l)+h\big)\right),$$

where the first sum is restricted to $j_{l_1}=k_1$, while the second sum is restricted to $j_{l_2}=k_2$. We write this product of two sums as a single sum (replicas) to obtain

$$\frac{\partial F_1}{\partial x_a}(x)\frac{\partial F_2}{\partial x_b}(x) = \beta^2\sum f\,\mathrm{Av}\,\varepsilon_{l_1}\varepsilon_{l_2+s}\varepsilon_J$$
$$\times\exp\left(\sum_{l\le s}\varepsilon_l\big(\beta x(j_l,l)+h\big)+\sum_{l\le s}\varepsilon_{l+s}\big(\beta x(j_{l+s},l)+h\big)\right).$$

There the sum is over $j_1,\ldots,j_{2s}\le M$, such that $j_{l_1}=k_1$, $j_{l_2+s}=k_2$, the average is over $\varepsilon_1,\ldots,\varepsilon_{2s}$. If we recall (6.48), computation of the sum

$$\sum\mathsf{E}(v_av_b-u_au_b)\frac{\partial F_1}{\partial x_a}(x)\frac{\partial F_2}{\partial x_b}(x)$$

yields (6.43). The restriction $(j_{l_1},l_1)\ne(j_{l_2+s},l_2)$ is because (6.48) requires $a\ne b$. The term (6.44) is similar.   □

Challenge problem 6.10. a) After reading Chapter 7, prove that

$$|q-\mathsf{E}\,\bar{q}|\le\frac{L}{N}.$$

b) Prove that $\lim\limits_{N\to\infty} N(q-\mathsf{E}\,\bar{q})$ exists.

We now end this chapter by the following weaker version of Proposition 2.1. It was discovered around 1985 by G. Pisier[Pi].

Proposition 6.11. *Consider a function f on $\mathbb{R}^M$ and assume that*

(6.49) $|f(x)-f(y)|\le L\|x-y\|.$

Then, for all $t>0$,

$$\mathsf{P}\big(|f(g)-\mathsf{E}\,f(g)|\ge t\big)\le 2\exp\left(-\frac{t^2}{4L^2}\right).$$

Proof. We assume first f differentiable. Consider s, and the function F on R^{2M} given by

(6.50) $F(z)=\exp s\big(f((z_i)_{i\le M})-f((z_{i+M})_{i\le M})\big).$

Consider the Gaussian family u given by

$$\begin{cases} u_i = g_i \\ u_{i+M} = g_i \end{cases} \quad \text{for } i \leq M.$$

Then $F(u) = 1$. Consider now i.i.d. $\mathcal{N}(0,1)$ variables $(v_i)_{i \leq 2M}$. Thus, for $i \leq j \leq 2M$ we have $E(v_i v_j - u_i u_j) = 0$ unless $j = i + M$, in which case $E(v_i v_j - u_i u_j) = -1$. Consider now $\theta(t) = E\, F(\sqrt{1-t}\, u + \sqrt{t}\, v)$ so that, by Proposition 6.8,

$$(6.51) \qquad \theta'(t) = \sum_{i<j} E(v_i v_j - u_i u_j)\, E\, \frac{\partial^2 F}{\partial x_i \partial x_j}(\sqrt{1-t}\, u + \sqrt{t}\, v)$$

$$= -\sum_{i \leq M} E\, \frac{\partial^2 F}{\partial x_i \partial x_{i+M}}(\sqrt{1-t}\, u + \sqrt{t}\, v).$$

Now,

$$(6.52) \qquad \frac{\partial^2 F}{\partial x_i \partial x_{i+M}}(x) = -\frac{s^2 \partial f}{\partial x_i}((x_i)_{i \leq M})\frac{\partial f}{\partial x_i}((x_{i+M})_{i \leq M})F(x).$$

Since (6.49) implies

$$\forall x, \quad \|\nabla f(x)\| \leq L,$$

i.e.

$$\sum_{i \leq M}\left(\frac{\partial f^2}{\partial x_i}(x)\right)^2 \leq L^2,$$

we see from (6.52) and Cauchy-Schwarz that

$$\left|\sum_{i \leq M}\frac{\partial^2 F}{\partial x_i \partial x_{i+M}}(x))\right| \leq L^2 s^2 F(x),$$

so that (6.52) implies

$$\theta'(t) \leq s^2 L^2 \theta(t)$$

and since $\theta(0) = 1$ this implies

$$\theta(1) = E\, F(v) \leq \exp s^2 L^2.$$

Now by Jensen's inequality (integration in $(v_i)_{i \geq M}$) we have

$$E \exp s\big(f(g) - E f(g)\big) \leq E\, F(v) \leq \exp s^2 L^2,$$

so that

$$P\left(f(g) - E f(g) \geq t\right) \leq \exp(s^2 L^2 - st),$$

and

$$P\left(f(g) - E f(g) \geq t\right) \leq \exp\left(-\frac{t^2}{4L^2}\right)$$

by optimization over s. Changing F into $-F$ finishes the proof when F is differentiable. The general case follows by approximation (e.g. convolution with a smooth function). □

7 Central limit theorems and the Almeida-Thouless line

In the previous chapter we have seen how to obtain sharp bounds on certain moments when $\beta \leq \beta_0$. But how do we perform exact computation ? For example, we know since Chapter 5 that for $\beta \leq 1/3$

$$NC_N = N \, \mathsf{E}\left\langle \left(\frac{\sigma^{\sim} \cdot \sigma^*}{N}\right)^2 \right\rangle$$

remains bounded. We surely expect that this quantity has a limit, and we would like to compute it. Using symmetries between the sites, we write as usual

$$NC_N = \mathsf{E}\langle \sigma_N^{\sim} \sigma_N^* \sigma^{\sim} \cdot \sigma^* \rangle,$$

and changing N into $N+1$ and β in β',

(7.1) $\quad (N+1)C_{N+1}(\beta') = \mathsf{E}\langle \sigma_{N+1}^{\sim} \sigma_{N+1}^* \sigma^{\sim} \cdot \sigma^* \rangle' + \mathsf{E}\left\langle (\sigma_{N+1}^{\sim} \sigma_{N+1}^*)^2 \right\rangle'.$

It should be obvious that the last term has a limit, as we will detail later. In view of what we did in Chapter 5, or even in Chapter 6, it is not sufficient to give a bound for the middle term of (7.1). We need an exact expression, or at least an expression with error $o(1)$. We will as usual use (5.22), so we need to evaluate

(7.2) $$\mathsf{E}\frac{1}{Z^4}\langle \sigma^{\sim} \cdot \sigma^* \, \mathsf{Av}\, \varepsilon\tilde{\varepsilon}^*\mathcal{E}_4 \rangle.$$

The closest we can do for the computation of such an expression is in Proposition 6.6. But certainly the proposition (or its obvious extensions to more replicas) is not accurate enough. Since $f = \sigma^\sim \cdot \sigma^*$ satisfies $\langle f \rangle = 0$, it would suggest 0 as approximation of (7.2), which is not really satisfactory (the error term in (6.19) is of course of order 1). Our first task will then be to obtain a result in the spirit of (6.19), but with higher accuracy. It will be a "second order expansion" rather than a "first order expansion". The reader who has not followed in detail the computations of the second half of Chapter 6 should at this point skip directly to Corollary 7.4 below. This is the key result that allows to prove Proposition 7.5. This result in turn plays an important role in our discussion of the A-T line, the conjectured separation between high and low temperature behavior. We could perform the main computation in the case of Corollary 7.4 only. Rather, we have decided to give a more general result. There are several good reasons to do this. First, the extra complication is minimal, and purely algebraic. Second, this general result opens the door to interesting computations (Proposition 7.7) which show how far the high temperature phase of the SK model is from being trivial despite what the physicists say.

Our general result will involve s replicas and a given subset $J \subset \{1, \ldots, s\}$ of these replicas. We set $n = \operatorname{card} J$. For $l_1 < l_2 \leq s$, we set

$$n(J, l_1, l_2) = \begin{cases} n+2 & \text{if } l_1, l_2 \notin J \\ n & \text{if } l_1 \notin J, l_2 \in J \text{ or } l_1 \in J, l_2 \notin J \\ n-2 & \text{if } l_1, l_2 \in J. \end{cases}$$

If $\varepsilon_l \in \{-1, 1\}$, $n(J, l_1, l_2)$ is the number of terms that actually occur in the product

$$\varepsilon_{l_1} \varepsilon_{l_2} \prod_{l \in J} \varepsilon_l$$

after cancelation ($\varepsilon_l^2 = 1$). In a similar manner, we define

$$n(J, l') = \begin{cases} n+1 & \text{if } l' \notin J \\ n-1 & \text{if } l' \in J. \end{cases}$$

We recall our usual notation

$$\mathcal{E}_s = \exp \sum_{l \leq s} \varepsilon_l \left(\frac{\beta}{\sqrt{N}} g \cdot \sigma^l + h \right)$$

$$Z = \left\langle \operatorname{Av} \exp \varepsilon \left(\frac{\beta}{\sqrt{N}} g \cdot \sigma + h \right) \right\rangle$$

$$\bar{q} = \langle R(\sigma, \sigma') \rangle,$$

and we recall that q is the solution of (5.4).

To lighten the notation, we write

$$(7.3) \qquad\qquad Y = \beta g \sqrt{\bar{q}} + h$$

where g is $N(0,1)$.

Proposition 7.1. *Given s, β_1, h_1, there is a number $K = K(s, \beta_1, h_1)$ such that if $\beta \leq \beta_1$, $h \leq h_1$, for each function f on Σ_N^s, each p, p' with $\dfrac{1}{p} + \dfrac{1}{p'} = 1$, each subset J of $\{1, \ldots, s\}$ of cardinal n we have*

$$(7.4) \qquad \mathsf{E} \frac{1}{Z^s} \left\langle f \operatorname{Av}\left(\prod_{l \in J} \varepsilon_l\right) \mathcal{E}_s \right\rangle$$

$$= \mathsf{E} \operatorname{th}^n(Y) \, \mathsf{E}\langle f \rangle$$

$$+ \beta^2 \sum_{1 \leq l_1 < l_2 \leq s} \mathsf{E} \operatorname{th}^{n(J,l_1,l_2)}(Y) \, \mathsf{E}\left\langle f\left(\frac{\sigma^{l_1} \cdot \sigma^{l_2}}{N} - q\right)\right\rangle$$

$$- s\beta^2 \sum_{1 \leq l \leq s} \mathsf{E} \operatorname{th}^{1+n(J,l)}(Y) \, \mathsf{E}\left\langle f\left(\frac{\sigma^l \cdot b}{N} - q\right)\right\rangle$$

$$+ \frac{s(s+1)}{2} \beta^2 \, \mathsf{E} \operatorname{th}^{n+2}(Y) \, \mathsf{E}\Big(\langle f \rangle (\bar{q} - q)\Big)$$

$$+ R,$$

where

$$(7.5) \qquad |R| \leq K \, \mathsf{E}\langle |f|^p \rangle^{\frac{1}{p}} \, \mathsf{E}\left\langle \left|R(\sigma, \sigma') - q\right|^{2p'}\right\rangle^{\frac{1}{p'}}.$$

Compared with Proposition 6.6, the gain is that we have a better error term, due to the exponent $2p'$ rather than p'. But, rather than approximating the left-hand side of (7.4) only by $\mathsf{E}(\operatorname{th}^n(Y)\langle f \rangle)$ we now have three extra terms, each of which is a sum. Probably the reader will find that this is awfully complicated. But the model (rather than the author) should be blamed: when $\langle f \rangle = 0$ these extra terms are what matters. What will make (7.4) usable is that we use it for combinations of terms such as the left-hand side of (7.4) with various choices of J. These will be chosen such that there is abundant cancelation among the complicated terms. This will be done in Proposition 7.2 and most notably in Proposition 7.3.

Proof of Proposition 7.1. The proof is a refinement of the proof of Proposition 6.6, so that the reader who has not studied this should go directly to Corollary 7.4. We use (6.30), and the same families u, v as in the proof of Proposition 6.6. Rather than (6.38), however, we use a second order expansion

$$(7.6) \qquad \qquad \theta(1) = \theta(0) + \theta'(0) + R$$

where

$$|R| \leq \left|\int_0^1 \theta''(t)(1-t)\,dt\right| \leq \sup_t |\theta''(t)|.$$

It was shown in (6.37) that as $M \to \infty$ and $M^{-1}\sum_{j \leq M} \delta_{\sigma(j)}$ converges to Gibbs' measure, $\theta(0)$ contributes as the first term in the right-hand side of (7.4). We compute the contribution of $\theta'(0)$. For each $r = 1, \ldots, 4$, as $M \to \infty$, we examine the limit of $\mathsf{E}_g\, F^r_{l_1,l_2}(u)$ as $M \to \infty$. For $r = 1$, this creates the second term on the right-hand side of (7.4). For $r = 2$, the term A_{l_1,l_2} creates an average $\langle \frac{\sigma^1 \cdot \sigma^2}{N} - q \rangle = \overline{q} - q$, so this term contributes as $\frac{s(s-1)}{2}\beta^2\, \mathsf{E}\,\mathrm{th}^{n+2}(Y)\, \mathsf{E}\left(\langle f\rangle(\overline{q}-q)\right)$. For the same reason, the term for $r = 4$ contributes as $s^2\beta^2\, \mathsf{E}\,\mathrm{th}^{n+2}(Y)\, \mathsf{E}\left(\langle f\rangle(\overline{q}-q)\right)$, and regrouping these gives the penultimate term of (7.4). The term $\mathsf{E}_g\, F^3_{l_1,l_2}(u)$ has a limit

$$\beta^2 \left\langle f\left(\frac{\sigma^{l_1} \cdot \sigma^{l_2+s}}{N} - q\right)\right\rangle \mathsf{E}\,\mathrm{th}^m(Y),$$

where m is the number of terms ε_l occurring in $\varepsilon_J \varepsilon_{l_1} \varepsilon_{l_2+s}$, that is (since $l_1 \leq s$) $1 + n(J, l_1)$. Since f depends only upon $\sigma^1, \ldots, \sigma^s$, we have

$$\left\langle f\left(\frac{\sigma^{l_1} \cdot \sigma^{l_2+s}}{N} - q\right)\right\rangle = \left\langle f\left(\frac{\sigma^{l_1} \cdot b}{N} - q\right)\right\rangle,$$

so that the terms arising from $\mathsf{E}_g\, F^3_{l_1,l_2}(u)$ contribute as the third term on the right-hand side of (7.4). Thus, to prove Proposition 7.1, it remains to control the error term, that is to show that $\mathsf{E}|\theta''(t)|$ is bounded as in (7.5). What one would like to say is that Lemma 6.9 proves that $\theta'(t)$ is the sum of pieces of the same nature as $\theta(t)$, except that they involve more replicas (up to $3s$ replicas in (6.44)) and that one has replaced f by $f(\frac{\sigma^{l'} \cdot \sigma^{l''}}{N} - q)$ for certain values $l' \neq l''$. Thus, applying Lemma 6.9 once more will express $\theta''(t)$ as a sum of similar terms, but now involving yet more replicas, and functions

$$f\left(\frac{\sigma^{l'} \cdot \sigma^{l''}}{N} - q\right)\left(\frac{\sigma^{m'} \cdot \sigma^{m''}}{N} - q\right).$$

Bounding these terms as in the case of $\theta'(t)$, but using Hölder's inequality with coefficients $p, 2p', 2p'$ will then yield the required error term

$$\langle |f|^p\rangle^{\frac{1}{p}} \left\langle \left|\frac{\sigma \cdot \sigma'}{N} - q\right|^{2p'}\right\rangle^{\frac{1}{p'}}.$$

It is however not exactly true that $F^r_{l_1,l_2}(x)$ is of the same nature as $F(x)$, because $F(x)$ involves different variables $x(\sigma, l)$ in each replica while this is not the case e.g. for $F^4_{l_1,l_2}$. Thus, before saying that we can iterate Lemma 6.9, we must investigate how to modify this lemma in a situation where different replicas need not have different variables; that is, we consider different variables $y(\sigma, m)$, $m = 1, \ldots$ and we take $x(\sigma, l) = y(\sigma, m(l))$ where the map $l \mapsto m(l)$ need not be one to one. We then see that (6.39) remains valid, provided that in the first two terms in the right-hand side of (6.39) we restrict the summation to the pairs $l_1 < l_2$ with $m(l_1) \neq m(l_2)$. This completes the proof.   □

Proposition 7.2. *We have*

(7.7)
$$\mathsf{E} \frac{1}{Z^s} \langle f \operatorname{Av} \varepsilon \tilde{\varepsilon}^3 \mathcal{E}_s \rangle = \beta^2 \mathsf{E} \frac{1}{\operatorname{ch}^2 Y} \mathsf{E} \Big\langle f \frac{\boldsymbol{\sigma}^{\sim} \cdot \boldsymbol{\sigma}^3}{N} \Big\rangle$$
$$+ \beta^2 \sum_{l \geq 4} \mathsf{E} \frac{\operatorname{th}^2 Y}{\operatorname{ch}^2 Y} \mathsf{E} \Big\langle f \frac{\boldsymbol{\sigma}^{\sim} \cdot \boldsymbol{\sigma}^l}{N} \Big\rangle$$
$$- s\beta^2 \mathsf{E} \frac{\operatorname{th}^2 Y}{\operatorname{ch}^2 Y} \mathsf{E} \Big\langle f \frac{\boldsymbol{\sigma}^{\sim} \cdot \boldsymbol{b}}{N} \Big\rangle + R,$$

where R is as in (7.5).

Proof. Let $J_1 = \{1, 3\}$, $J_2 = \{2, 3\}$. From (7.4) we get

(7.8)
$$\mathsf{E} \frac{1}{Z^s} \langle f \operatorname{Av} \varepsilon \tilde{\varepsilon}^3 \mathcal{E}_s \rangle$$
$$= \beta^2 \sum_{1 \leq l_1 < l_2 \leq s} \mathsf{E} \big((\operatorname{th} Y)^{n(J_1, l_1, l_2)} - (\operatorname{th} Y)^{n(J_2, l_1, l_2)} \big) \mathsf{E} \Big\langle f \Big(\frac{\boldsymbol{\sigma}^{l_1} \cdot \boldsymbol{\sigma}^{l_2}}{N} - q \Big) \Big\rangle$$
$$- s\beta^2 \sum_{1 \leq l \leq s} \mathsf{E} \big((\operatorname{th} Y)^{1 + n(J_1, l)} - (\operatorname{th} Y)^{1 + n(J_2, l)} \big) \mathsf{E} \Big\langle f \Big(\frac{\boldsymbol{\sigma}^l \cdot \boldsymbol{b}}{N} - q \Big) \Big\rangle + R.$$

Now $n(J_1, l_1, l_2) = n(J_2, l_1, l_2)$ unless $l_1 \leq 2 < l_2$; and $n(J_1, l) = n(J_2, l)$ unless $l = 1, 2$. To obtain the result, we regroup the contributions of $l_1 = 1$, $l_1 = 2$ (treating separately the case $l_2 = 3$ or $l_2 > 3$) and the contributions of $l = 1, 2$. We then use that

$$1 - \operatorname{th}^2 Y = \frac{1}{\operatorname{ch}^2 Y}.$$ □

Proposition 7.3. *We have*

(7.9)
$$\mathsf{E} \frac{1}{Z^s} \langle f \operatorname{Av} \varepsilon \tilde{\varepsilon}^* \mathcal{E}_s \rangle = \beta^2 \mathsf{E} \frac{1}{\operatorname{ch}^4 Y} \mathsf{E} \Big\langle f \frac{\boldsymbol{\sigma}^{\sim} \cdot \boldsymbol{\sigma}^*}{N} \Big\rangle + R,$$

where R is as in (7.5).

Proof. Use (7.7) twice on $\varepsilon \tilde{\varepsilon}^3$ and $\varepsilon \tilde{\varepsilon}^4$, and subtract.   □

Corollary 7.4. *Under the conditions of Proposition 7.1, we have*

(7.10)
$$\mathsf{E} \frac{1}{Z^s} \langle f \operatorname{Av} \varepsilon \tilde{\varepsilon}^* \mathcal{E}_s \rangle = \beta^2 \mathsf{E} \frac{1}{\operatorname{ch}^4 Y} \mathsf{E} \Big\langle f \frac{\boldsymbol{\sigma}^{\sim} \cdot \boldsymbol{\sigma}^*}{N} \Big\rangle + R,$$

where

$$R \leq K \frac{p'}{N} (\mathsf{E} \langle |f|^p \rangle)^{\frac{1}{p}}.$$

Proof. We have shown in Proposition 6.2 that

$$E\left\langle \left|\frac{\sigma \cdot \sigma'}{N} - q\right|^{2k}\right\rangle \leq \left(\frac{Lk}{N}\right)^k.$$

If k is the smallest integer such that $p' \leq k$ we write

$$E\left\langle \left|\frac{\sigma \cdot \sigma'}{N} - q\right|^{2p'}\right\rangle^{\frac{1}{p'}} \leq E\left\langle \left|\frac{\sigma \cdot \sigma'}{N} - q\right|^{2k}\right\rangle^{\frac{1}{k}} \leq \frac{Lk}{N}$$

to control the remainder (7.5). □

We now go back to the study of (7.1), when $\beta \leq \beta_0$, $h \leq 1$ (so that we can use Corollary 7.4). We use (7.10) for $s = 4$, $f = \sigma^{\sim} \cdot \sigma^*$, $p = p' = 2$. We know that $E\langle f^2\rangle \leq LN$ by Theorem 6.1, so that (7.1) yields

$$(7.11) \quad (N+1)C_{N+1}(\beta') = \beta^2 E\frac{1}{\text{ch}^4 Y}NC_N(\beta) + E\langle(\sigma_{\tilde{N}+1}\sigma^*_{N+1})^2\rangle' + R,$$

where $|R| \leq K/\sqrt{N}$. Let us compute the last term. With the notation $\varepsilon_l = \sigma^l_{N+1}$, and since $\varepsilon^{\sim 2} = 2(1 - \varepsilon_1\varepsilon_2)$, it is

$$E\frac{1}{Z^4}\langle\text{Av}(\varepsilon^{\sim}\tilde{\varepsilon}^*)^2\mathcal{E}_4\rangle = 4E\frac{1}{Z^4}\langle\text{Av}(1 - \varepsilon_1\varepsilon_2 - \varepsilon_3\varepsilon_4 + \varepsilon_1\varepsilon_2\varepsilon_3\varepsilon_4)\mathcal{E}_4\rangle,$$

and using (7.4) when $f = 1$, we see that

$$(7.12) \qquad E\langle(\sigma_{\tilde{N}+1}\sigma^*_{N+1})^2\rangle' = 4E\frac{1}{\text{ch}^4 Y} + R,$$

where $|R| \leq K/\sqrt{N}$ (in fact, $|R| \leq K/N$). Now we see that we are on the right track, but (7.11) alone would require the use of iteration. To avoid this, we will show that $(N+1)C_{N+1}(\beta') \simeq NC_N(\beta)$. We write

$$(7.13) \quad NC_N = E\left\langle\frac{1}{N}\sum_{i\leq N}\sigma_i\sigma^*_i\sigma^{\sim}\cdot\sigma^*\right\rangle = E\left\langle\frac{1}{N-1}\sum_{i\leq N-1}\sigma_i\sigma^*_i\sigma^{\sim}\cdot\sigma^*\right\rangle,$$

using symmetry between sites. Changing N into $N+1$, we get

$$(7.14) \quad (N+1)C_{N+1}(\beta') = E\left\langle\frac{1}{N}\sigma^{\sim}\cdot\sigma^*(\sigma^{\sim}\cdot\sigma^* + \sigma_{\tilde{N}+1}\sigma^*_{N+1})\right\rangle'$$

$$= NE\left\langle\left(\frac{\sigma^{\sim}\cdot\sigma^*}{N}\right)^2\right\rangle' + \frac{1}{N}E\langle\sigma_{\tilde{N}+1}\sigma^*_{N+1}\sigma^{\sim}\cdot\sigma^*\rangle'.$$

We leave it to the reader to use (7.4), with $n = 0$, to obtain

$$(7.15) \qquad |(N+1)C_{N+1}(\beta') - NC_N(\beta)| \leq KNE\left\langle\left|\frac{\sigma \cdot \sigma'}{N} - q\right|^3\right\rangle,$$

so that, using Proposition 6.2, we have

$$(7.16) \qquad |(N+1)C_{N+1}(\beta') - NC_N(\beta)| \leq \frac{K}{\sqrt{N}}.$$

We combine with (7.11), (7.12), to see that

$$(7.17) \qquad NC_N = \beta^2 \,\mathrm{E}\, \frac{1}{\mathrm{ch}^4 Y} NC_N + 4\,\mathrm{E}\, \frac{1}{\mathrm{ch}^4 Y} + R,$$

with $|R| \le K/\sqrt{N}$. We have proved the following:

Proposition 7.5. *If $\beta \le \beta_0$, $h \le 1$, we have*

$$(7.18) \qquad \lim_{N \to \infty} N \,\mathrm{E}\!\left\langle \left(\frac{\sigma^{\tilde{}} \cdot \sigma^*}{N}\right)^2 \right\rangle = \frac{4A}{1 - \beta^2 A},$$

where

$$A = \mathrm{E}\, \frac{1}{\mathrm{ch}^4 Y} = \mathrm{E}\, \frac{1}{\mathrm{ch}^4(\beta g \sqrt{q} + h)}.$$

The importance of this result is that it provides an identification of the so-called Almeida-Thouless line. This line is given by the relation

$$(7.19) \qquad \beta^2 \,\mathrm{E}\, \frac{1}{\mathrm{ch}^4(\beta g \sqrt{q} + h)} = 1,$$

where q is "the" root of (5.4). It is certainly not obvious that (5.4) has a unique root; This was recently proved by R. Latala (private communication). Since the left-hand side of (7.18) is a square, (7.18) can hold only if the right-hand side is positive, that is if

$$(7.20) \qquad \beta^2 \,\mathrm{E}\, \frac{1}{\mathrm{ch}^4(\beta g \sqrt{q} + h)} < 1.$$

The physicists believe that (5.4) holds under (7.20) ("the high temperature region"), but fails if

$$(7.21) \qquad \beta^2 \,\mathrm{E}\, \frac{1}{\mathrm{ch}^4(\beta g \sqrt{q} + h)} > 1,$$

("the low temperature region"). It is certain that under (7.21), (7.18) cannot hold, so one of the facts we have used in its derivation must be wrong. Using (7.17), that we read as

$$(7.22) \qquad NC_N = BNC_N + 4\,\mathrm{E}\, \frac{1}{\mathrm{ch}^4 Y} + R,$$

where $B > 1$, we see that we can take advantage of this only if $R \ll NC_N$, or, rather, that (7.22) implies that R is of the same order as NC_N. This error term has several sources. One of them is (7.5), where we have used $f = \sigma^{\tilde{}} \cdot \sigma^*$. The natural choice of $p = 2 = p'$ there gives an error term

$$(7.23) \qquad \sqrt{C_N} \left(\mathrm{E}\!\left\langle \left(\frac{\sigma \cdot \sigma'}{N} - q\right)^4 \right\rangle \right)^{\frac{1}{2}}.$$

There is also an error term from (7.15), but we can argue that this is less important. What we would really like to show is that under (7.21) C_N cannot be small, because the fact "C_N small" is really the main reason behind (5.3).

So we try to argue by contradiction. If C_N is small ($\lim C_N = 0$ as $N \to \infty$) does it follow that the quantities (7.23) are much smaller than NC_N? A first difficulty is that a priori it is not clear how to relate

$$(7.24) \qquad E\left\langle \left(\frac{\sigma \cdot \sigma'}{N} - q\right)^4 \right\rangle$$

and

$$(7.25) \qquad E\left\langle \left(\frac{\tilde{\sigma} \cdot \sigma^*}{N}\right)^4 \right\rangle.$$

(See Lemma 5.3). But even if we know that the quantities (7.24), (7.25) are of the same order, it is not clear why we should have

$$(7.26) \quad E\left\langle \left(\frac{\tilde{\sigma} \cdot \sigma^*}{N}\right)^2 \right\rangle \text{ small} \implies E\left\langle \left(\frac{\tilde{\sigma} \cdot \sigma^*}{N}\right)^4 \right\rangle \ll E\left\langle \left(\frac{\tilde{\sigma} \cdot \sigma^*}{N}\right)^2 \right\rangle.$$

In other words, are terms that look second order terms really second order terms? This at first looks like a purely technical point. Rather, as will be explained in detail later in this chapter, this appears to be a question of fundamental importance.

Let us now turn to the discussion of the following important issue: is it true that (5.3) holds under (7.20)? The following is probably more relevant

Problem 7.6. Is it true that in the region (7.20) we have

$$\lim_{N\to\infty} E\langle (R(\sigma, \sigma') - q)^2 \rangle = 0?$$

This problem, to a large extend, was recently solved in [**T8**]. Nonetheless, it seems worthwhile to discuss in detail the underlying issues. The question of controlling the entire "high temperature" region is of fundamental interest for many models, and the solution given in [**T8**] is unfortunately very specific to the SK model.

We have found condition (7.20) through the computation (7.18). A first obvious question is: couldn't we find a more stringent condition through another computation? Apparently this is not the case. It seems desirable that the function Ψ of (5.57) be a contraction near q, and Lemma 5.10 shows that this amounts to

$$\beta^2 E \frac{2\,\text{sh}^2(\beta\sqrt{q} + h) - 1}{\text{ch}^4(\beta g\sqrt{q} + h)} > -1.$$

The physicists say that this holds under (7.20). They have probably checked this numerically.

To illustrate the fact that not everything is so simple, we now start another interesting computation, that of

$$(7.27) \qquad \lim_{N\to\infty} N E\left\langle \left(\frac{\tilde{\sigma} \cdot \sigma^3}{N}\right)^2 \right\rangle = \lim_{N\to\infty} N D_N.$$

The computation will be valid when $\beta \le \beta_0$, $h \le 1$. The error terms are controlled as in the computation (7.18), so we will not explain again how this is done each time. It always follows from Proposition 6.2 and Hölder's inequality. We will denote by R a quantity such that $|R| \le K/\sqrt{N}$. We start as usual with the relation

$$ND_N = \mathsf{E}\left\langle \left(\frac{1}{N}\sum_{i \le N}\sigma_i^{\tilde{}}\sigma_i^3\right)\sigma^{\tilde{}}\cdot\sigma^3\right\rangle = \mathsf{E}\langle\sigma_N^{\tilde{}}\sigma_N^3\sigma^{\tilde{}}\cdot\sigma^3\rangle,$$

so that

(7.28) $(N+1)D_{N+1}(\beta') = \mathsf{E}\langle\sigma_{\tilde{N}+1}\sigma_{N+1}^3\sigma^{\tilde{}}\cdot\sigma^3\rangle' + \mathsf{E}\langle(\sigma_{\tilde{N}+1}\sigma_{N+1}^3)^2\rangle'.$

Now, $(\sigma_{\tilde{N}+1}\sigma_{N+1}^3)^2 = 2(1 - \sigma_{N+1}^1\sigma_{N+1}^2)$, and as in (7.12) we obtain

(7.29) $\mathsf{E}\langle(\sigma_{\tilde{N}+1}\sigma_{N+1}^3)^2\rangle' = 2\,\mathsf{E}\dfrac{1}{\mathrm{ch}^2 Y} + R.$

Next, we use Proposition 7.2 to write

(7.30) $\mathsf{E}\langle\sigma_{\tilde{N}+1}\sigma_{N+1}^3\sigma^{\tilde{}}\cdot\sigma^3\rangle' = \mathsf{E}\dfrac{1}{Z^3}\langle\sigma^{\tilde{}}\cdot\sigma^3\,\mathrm{Av}\,\varepsilon\,\tilde{\varepsilon}^3\mathcal{E}_3\rangle$

$$= \beta^2\,\mathsf{E}\dfrac{1}{\mathrm{ch}^2 Y}ND_N$$

$$- 3\beta^2\,\mathsf{E}\dfrac{\mathrm{th}^2 Y}{\mathrm{ch}^2 Y}\mathsf{E}\left\langle\sigma^{\tilde{}}\cdot\sigma^3\dfrac{\sigma^{\tilde{}}\cdot b}{N}\right\rangle + R.$$

Now, as in (7.15) we find

(7.31) $(N+1)D_{N+1}(\beta') = ND_N + R,$

and combining with (7.28) we have

(7.32) $ND_N = \left(\beta^2\,\mathsf{E}\dfrac{1}{\mathrm{ch}^2 Y}\right)ND_N$

$$- \dfrac{3\beta^2}{N}\mathsf{E}\dfrac{\mathrm{th}^2 Y}{\mathrm{ch}^2 Y}\mathsf{E}\langle\sigma^{\tilde{}}\cdot\sigma^3\,\sigma^{\tilde{}}\cdot b\rangle + 2\,\mathsf{E}\dfrac{1}{\mathrm{ch}^2 Y} + R.$$

Now we observe that

$$\langle(\sigma^{\tilde{}}\cdot\sigma^*)^2\rangle = \langle(\sigma^{\tilde{}}\cdot\sigma^3 - \sigma^{\tilde{}}\cdot\sigma^4)^2\rangle$$

$$= 2\langle(\sigma^{\tilde{}}\cdot\sigma^3)^2\rangle - 2\langle\sigma^{\tilde{}}\cdot\sigma^3\,\sigma^{\tilde{}}\cdot\sigma^4\rangle,$$

so that

$$\mathsf{E}\langle\sigma^{\tilde{}}\cdot\sigma^3\sigma^{\tilde{}}\cdot b\rangle = \mathsf{E}\langle\sigma^{\tilde{}}\cdot\sigma^3\,\sigma^{\tilde{}}\cdot\sigma^4\rangle$$

$$= N^2 D_N - \dfrac{1}{2}N^2 C_N,$$

and combining with (7.32) we have

(7.33) $ND_N\left(1 - \beta^2\,\mathsf{E}\left(\dfrac{1 - 2\,\mathrm{sh}^2 Y}{\mathrm{ch}^4 Y}\right)\right) = \dfrac{3}{2}\beta^2\,\mathsf{E}\dfrac{\mathrm{th}^2 Y}{\mathrm{ch}^2 Y}NC_N + 2\,\mathsf{E}\dfrac{1}{\mathrm{ch}^2 Y} + R.$

In order to make the point that not everything is very simple, we state the following conclusion, that follows from (7.33) and (7.18).

Proposition 7.7. *If $\beta \leq 1/10$ and $h \leq 1$, we have*

$$(7.34) \qquad \lim_{N \to \infty} N \, \mathsf{E}\left\langle \left(\frac{\sigma^{\sim} \cdot \sigma^3}{N}\right)^2 \right\rangle = \frac{2B - 6\beta^2 A^2 + 4\beta^2 AB}{(1 - \beta^2 A)(1 - \beta^2 (3A - 2B))}$$

where A is as in Proposition 7.5 and

$$B = \mathsf{E}\,\frac{1}{\mathrm{ch}^2 Y}.$$

Challenge problem 7.8. Compute

$$\lim_{N \to \infty} N \, \mathsf{E}\left\langle \left(\frac{\sigma^1 \cdot \sigma^2 - \langle \sigma^1 \cdot \sigma^2 \rangle}{N}\right)^2 \right\rangle.$$

And what about

$$\lim_{N \to \infty} N \, \mathsf{E}\left\langle \left(\frac{\sigma^1 \cdot \sigma^2 - \mathsf{E}\langle \sigma^1 \cdot \sigma^2 \rangle}{N}\right)^2 \right\rangle?$$

Do you find a new constraint other than (7.20)?
Hint: [**T4**], Section 6.

The main difficulty in proving that Problem 7.6 has a positive answer is that it seems very difficult to do any kind of computation about C_N unless one also has a control about

$$\mathsf{E}\left\langle \left(\frac{\sigma^{\sim} \cdot \sigma^*}{N}\right)^4 \right\rangle,$$

or, even better, about

$$\mathsf{E}\left\langle \left(R(\sigma, \sigma') - q\right)^4 \right\rangle.$$

In turn, it seems possible to control these only by controlling quantities where 4 is replaced by 6, etc. This was one motivation behind the work of Chapter 6.

Everything would be easy if it were true that (7.26) holds. At first sight the failure of (7.26) looks like a rather remote possibility, one of these pathologies that are "in principle" possible, and that only weird people (such as mathematicians) care about. This is however not the case. The failure of (7.26) is not only possible. It is also natural. It does represent a different type of transition from high to low temperature than that predicted for the SK model. To understand what happens, we will compare the behavior of the SK model and the p-spin interaction model as β increases to the critical value β_p, and p is even. We will assume that $h = 0$, because then all the claims we make can be proved rather than conjectured. We expect that if h is small the situation is similar.

When $h = 0$ we have $b = \langle \sigma \rangle = 0$ by symmetry. There is no "centering problem" here and the natural quantity to measure the correlation of spins is

$$E\langle R(\sigma, \sigma')^2 \rangle = E\left\langle \left(\frac{\sigma \cdot \sigma'}{N}\right)^2 \right\rangle$$

$$= \frac{1}{N} + \frac{N-1}{N} E(\langle \sigma_i \sigma_j \rangle^2).$$

Proposition 7.9. *If $h = 0$, $\beta < 1$, for the typical value of the disorder, and N large, the map*

$$(\sigma, \sigma') \mapsto \frac{\sigma \cdot \sigma'}{\sqrt{N}}$$

resembles a $\mathcal{N}\left(0, \dfrac{1}{1 - \beta^2}\right)$ r.v. More precisely, for each $k \geq 0$, the r.v.

$$\left\langle \left(\frac{\sigma \cdot \sigma'}{\sqrt{N}}\right)^{2k} \right\rangle$$

converges in probability to $(1 - \beta^2)^{-2k} E g^{2k}$ where g is $\mathcal{N}(0,1)$.

This is proved in [**C-N**]. When $h \neq 0$, we will prove a similar result (Theorem 7.12 below). The difference is that Theorem 7.12 is proved only for $\beta < 1/10$, while Proposition 7.9 holds for $\beta < 1$. (See Challenge problem 7.15 to understand better the relationship).

In other words, what Proposition 7.9 says is that $N^{-1/2}\sigma \cdot \sigma'$ is distributed like $C(\beta)g$ (where g is $\mathcal{N}(0,1)$), where $C(\beta) \to \infty$ as β gets close to the critical value β_c. What happens for large p is totally different.

Proposition 7.10. *For the p-spin interaction model with $h = 0$, there exist $L > 0$, p_0 and $x > 0$ such that if $p \geq p_0$ and $\beta \leq 2^{p/2-6}$, we have*

(7.35) $$E(R(\sigma, \sigma')^2 1_{\{|R(\sigma, \sigma')| \leq x\}}) \leq \frac{L}{N}.$$

If moreover $\beta > 2\sqrt{\log 2}$, we have

(7.36) $$\liminf_{N \to \infty} E(\langle 1_{\{|R(\sigma, \sigma')| \geq x\}}\rangle) > 0.$$

Comment. In particular if p is large, there are situations where (7.35) and (7.36) occur simultaneously.

Proposition 7.10 says that if we do not look at the values of $R(\sigma, \sigma')$ that are of order 1 we do not see any increase in the fluctuations of this quantity as β increases beyond the critical value β_c. On the other hand, values of $R(\sigma, \sigma')$ of that order exist for $\beta > \beta_c$. We will prove Proposition 7.10 in Chapter 8.

Proposition 7.11. *For the 4-spin interaction model there exist arbitrarily large values of N for which for certain values of β,*

$$\mathsf{E}(\langle R(\sigma,\sigma')^2\rangle) \leq \frac{L}{N},$$

but

(7.37)
$$\mathsf{E}(\langle R(\sigma,\sigma')^2\rangle) \leq L\,\mathsf{E}(\langle R(\sigma,\sigma')^4\rangle).$$

This in particular shows that (7.26) cannot follow from general principles.

If we do not insist on $p = 4$, but allow p to be large, this is a consequence of Proposition 7.10 as we now show. Let us assume x irrational, so that we never have $R(\sigma,\sigma') = Nx$. Let us consider N and β_0 with $0 \leq \beta_0 < \beta_p + \frac{1}{2}$ and

(7.38)
$$\mathsf{E}(\langle 1_{\{|R(\sigma,\sigma')|\geq x\}}\rangle_{\beta_0}) \geq \frac{1}{N}.$$

Here, the subscript β_0 indicates of course that Gibbs' measure is taken at this value of the parameter β.

Since N is fixed, the quantity

(7.39)
$$\mathsf{E}(\langle 1_{\{|R(\sigma,\sigma')|\geq x\}}\rangle_{\beta})$$

is a continuous function of β. It is exponentially small at $\beta = 0$, so we can fix $\beta \leq \beta_0 < \beta_p + \frac{1}{2}$ for which

(7.40)
$$\mu := \mathsf{E}(\langle 1_{\{|R(\sigma,\sigma')|\geq x\}}\rangle) = \frac{1}{N}.$$

(As β is now fixed, it is no longer indicated in the notation). We then deduce (7.37) from the fact that

$$\mathsf{E}(\langle R(\sigma,\sigma')^p\rangle) \geq x^p \mu \geq \frac{x^p}{N},$$

while

$$\mathsf{E}(\langle R(\sigma,\sigma')^2\rangle) \leq \mathsf{E}(\langle R(\sigma,\sigma')^2 1_{\{|R(\sigma,\sigma')|^2\leq x\}}\rangle) + \mu \leq \frac{L+1}{N}. \qquad \square$$

It is nonetheless very instructive to give a direct proof of Proposition 7.11. This will also be done in Chapter 8.

The preceding discussion has shown that, as β increase, there are (at least) two very distinct ways in which the structure described in Chapter 6 can deteriorate. Condition (7.21) identifies the value of β where deterioration will occur following the pattern of Proposition 7.5. The challenge of Problem 7.6 is to show that the other type of deterioration, following the pattern of Proposition 7.10 will not occur first.

We now give an example of computation of higher moments.

Theorem 7.12. *If $\beta \leq \beta_0$, $h \leq 1$, for $n = 1, 2$ we have*

$$(7.41) \qquad \lim_{N \to \infty} N^{nk} \, \mathsf{E} \Big\langle \Big(\frac{\tilde{\sigma} \cdot \sigma^*}{N} \Big)^{2k} \Big\rangle^n = \Big(\frac{4A}{1 - \beta^2 A} \Big)^{nk} (\mathsf{E} \, g^{2k})^n$$

where g is $\mathcal{N}(0, 1)$ and A is as in Proposition 7.5.

This implies that for each k, the r.v.

$$M_k = N^k \Big\langle \Big(\frac{\tilde{\sigma} \cdot \sigma^*}{N} \Big)^{2k} \Big\rangle$$

satisfies $\mathsf{E} \, M_k^2 \simeq (\mathsf{E} \, M_k)^2$ so that M_k converges in probability to $\mathsf{E} \, M_k$. Given k_0, for large N and the typical value of the disorder, the moments of order $k \leq k_0$ of the function

$$(\sigma^1, \ldots, \sigma^4) \mapsto \frac{\tilde{\sigma} \cdot \sigma^*}{\sqrt{N}}$$

are approximately those of a $\mathcal{N}(0, 4A/(1 - \beta^2 A))$ r.v. Thus, Theorem 7.12 is indeed a central limit theorem.

Proof of Theorem 7.12. For $k_1, k_2 \geq 0$, we prove that

$$(7.42) \qquad \lim_{N \to \infty} N^{k_1 + k_2} \, \mathsf{E} \Big\langle \Big(\frac{\tilde{\sigma} \cdot \sigma^*}{N} \Big)^{2k_1} \Big\rangle \Big\langle \Big(\frac{\tilde{\sigma} \cdot \sigma^*}{N} \Big)^{2k_2} \Big\rangle$$

$$= \Big(\frac{4A}{1 - \beta^2 A} \Big)^{k_1 + k_2} \mathsf{E} \, g^{2k_1} \, \mathsf{E} \, g^{2k_2}.$$

The proof is by induction over $k = k_1 + k_2$. For $k = 1$, this is Proposition 7.5. We write

$$a_{k_1, k_2, N} = a_{k_1, k_2, N}(\beta) = \mathsf{E} \langle (\tilde{\sigma} \cdot \sigma^*)^{2k_1} \rangle \langle (\tilde{\sigma} \cdot \sigma^*)^{2k_2} \rangle.$$

We use the symmetry between sites to write

$$a_{k_1 + 1, k_2, N} = N \, \mathsf{E} \langle \tilde{\sigma}_N \sigma_N^* (\tilde{\sigma} \cdot \sigma^*)^{2k_1 + 1} \rangle \langle (\tilde{\sigma} \cdot \sigma^*)^{2k_2} \rangle,$$

and, changing N into $N + 1$,

$$a_{k_1 + 1, k_2, N+1}(\beta') = (N + 1) \, \mathsf{E} \, \langle \tilde{\sigma}_{N+1} \sigma_{N+1}^* (\tilde{\sigma} \cdot \sigma^* + \tilde{\sigma}_{N+1} \sigma_{N+1}^*)^{2k_1 + 1} \rangle'$$

$$(7.43) \qquad \langle (\tilde{\sigma} \cdot \sigma^* + \tilde{\sigma}_{N+1} \sigma_{N+1}^*)^{2k_2} \rangle'. \qquad \square$$

Lemma 7.13. *We have*

$$\mathsf{E} \langle |\tilde{\sigma} \cdot \sigma^*|^a \rangle' \langle |\tilde{\sigma} \cdot \sigma^*|^b \rangle' \leq K N^{\frac{(a+b)}{2}}.$$

Proof. We use Cauchy-Schwarz to reduce this to the proof of

$$\mathsf{E} \langle |\tilde{\sigma} \cdot \sigma^*|^{2a} \rangle' \leq K N^a.$$

We use (5.22) and that $Z \geq 1$. We first integrate w.r. to g to reduce the proof to the fact that $\mathsf{E} \langle |\tilde{\sigma} \cdot \sigma^*|^{2a} \rangle \leq K N^a$, which follows from Proposition 6.2. $\qquad \square$

We want to study

$$\lim_{N\to\infty}(N+1)^{-k_1-k_2-1}a_{k_1+1,k_2,N+1}.$$

We expand the powers in (7.43) and we use that $|\sigma_{N+1}^{\sim}\sigma_{N+1}^{*}|\le 4$ and Lemma 7.13 to see that

$$(7.44)\quad a_{k_1+1,k_2,N+1}(\beta')$$
$$= (N+1)\,\mathsf{E}\langle\sigma_{N+1}^{\sim}\sigma_{N+1}^{*}(\sigma^{\sim}\cdot\sigma^{*})^{2k_1+1}\rangle'\langle(\sigma^{\sim}\cdot\sigma^{*})^{2k_2}\rangle'$$
$$+ (2k_1+1)(N+1)\,\mathsf{E}\langle(\sigma_{N+1}^{\sim}\sigma_{N+1}^{*})^2(\sigma^{\sim}\cdot\sigma^{*})^{2k_1}\rangle'\langle(\sigma^{\sim}\cdot\sigma^{*})^{2k_2}\rangle'$$
$$+ 2k_2(N+1)\,\mathsf{E}\langle\sigma_{N+1}^{\sim}\sigma_{N+1}^{*}(\sigma^{\sim}\cdot\sigma^{*})^{2k_1+1}\rangle'$$
$$\langle\sigma_{N+1}^{\sim}\sigma_{N+1}^{*}(\sigma^{\sim}\cdot\sigma^{*})^{2k_2-1}\rangle'$$
$$+ R$$
$$= \mathrm{I}+\mathrm{II}+\mathrm{III}+R,$$

where

$$(7.45)\qquad\qquad |R|\le K N^{k_1+k_2+\frac{1}{2}}.$$

We will study each of the terms I to III. To reduce the product of brackets to a single bracket we set

$$U = \big((\sigma^5-\sigma^6)\cdot(\sigma^7-\sigma^8)\big)^{2k_2},$$

so that

$$\mathrm{I} = (N+1)\,\mathsf{E}\langle\sigma_{N+1}^{\sim}\sigma_{N+1}^{*}(\sigma^{\sim}\cdot\sigma^{*})^{2k_1+1}U\rangle'.$$

We use the cavity method and Proposition 7.3 to obtain

$$(7.46)\qquad \mathrm{I} = \frac{N+1}{N}\beta^2\,\mathsf{E}\,\frac{1}{\mathrm{ch}^4 Y}E\langle(\sigma^{\sim}\cdot\sigma^{*})^{2k_1+2}U\rangle + R,$$

where R is as in (7.5), i.e.

$$(7.47)\quad |R|\le L(N+1)\,\mathsf{E}\big(\langle|(\sigma^{\sim}\cdot\sigma^{*})^{2k_1+1}U|^p\rangle\big)^{1/p}\,\mathsf{E}\Big\langle\Big|\frac{\sigma\cdot\sigma'}{N}-q\Big|^{2p'}\Big\rangle^{1/p'}.$$

Taking $p=p'=2$, and using Hölder's inequality, we see that $|R|$ is as in (7.45). It is essential here (as was already essential in Proposition 7.5) to have $2p'$ rather than p' in (7.47).

The term II is less dangerous, because $\sigma^{\sim}\cdot\sigma^{*}$ occurs with a lower power. It is enough there to use Proposition 6.6 to get

$$(7.48)\qquad \mathrm{II} = (2k_1+1)(N+1)\,\mathsf{E}\,\frac{1}{\mathrm{ch}^4 Y}E\langle(\sigma^{\sim}\cdot\sigma^{*})^{2k_1}U\rangle + R,$$

where R is as before.

To study the term III, we also use Proposition 6.6. The difference with the term II is that the main contributions cancel out. Writing ε_l for σ_{N+1}^l, the dependence in the variables ε_l is as in

$$(\varepsilon_1 - \varepsilon_2)(\varepsilon_3 - \varepsilon_4)(\varepsilon_5 - \varepsilon_6)(\varepsilon_7 - \varepsilon_8)$$

and out of the 16 resulting terms, each involving a product of 4 different ε_l, 8 of them have a negative sign.

Now we have obtained

$$a_{k_1+1,k_2,N+1}(\beta') = \frac{N+1}{N}\, \mathsf{E}\, \frac{\beta^2}{\mathrm{ch}^4 Y} E\langle(\sigma^\sim \cdot \sigma^*)^{2k_1+1}\rangle\langle(\sigma^\sim \cdot \sigma^*)^{2k_2}\rangle$$

$$+ (N+1)(2k_1+1)\, \mathsf{E}\, \frac{4}{\mathrm{ch}^4 Y} E\langle(\sigma^\sim \cdot \sigma^*)^{2k_1+1}\rangle\langle(\sigma^\sim \cdot \sigma^*)^{2k_2}\rangle$$

$$+ R.$$

(7.49) $$\qquad a_{k_1+1,k_2,N+1}(\beta') = \beta^2\, \mathsf{E}\, \frac{1}{\mathrm{ch}^4 Y} a_{k_1+1,k_2,N}(\beta)$$

$$+ (2k_1+1)N\, \mathsf{E}\, \frac{4}{\mathrm{ch}^4 Y} a_{k_1,k_2,N}(\beta) + R.$$

Now we claim that

(7.50) $$\qquad a_{k_1+1,k_2,N+1}(\beta') = a_{k_1+1,k_2,N}(\beta) + R.$$

Indeed, we write

(7.51) $$\quad a_{k_1+1,k_2,N}(\beta') = \frac{N}{N-1}\, \mathsf{E}\langle \sum_{i\leq N-1} \sigma_i^\sim \sigma_i^* (\sigma^\sim \cdot \sigma^*)^{2k_1+1}\rangle\langle(\sigma^\sim \cdot \sigma^*)^{2k_2}\rangle,$$

we replace N by $N+1$; we use Lemma 7.13 to see that only the term

$$\frac{N+1}{N} E\langle(\sigma^\sim \cdot \sigma^*)^{2k_1+2}\rangle\langle(\sigma^\sim \cdot \sigma^*)^{2k_2}\rangle$$

matters, and we conclude with Proposition 6.6 again.

Combining (7.49), (7.50), we have reached the conclusion that

(7.52) $$\qquad (N+1)^{-k_1-k_2-1} a_{k_1+1,k_2,N}\left(1 - \beta^2\, \mathsf{E}\, \frac{1}{\mathrm{ch}^4 Y}\right)$$

$$= (2k_1+1)\, \mathsf{E}\, \frac{4}{\mathrm{ch}^4 Y} N^{-k_1-k_2} a_{k_1,k_2,N} + R,$$

where

$$|R| \leq K/\sqrt{N}.$$

We also observe that by integration by parts,

$$E(g^{2k_1+2}) = E(gg^{2k_1+1}) = (2k_1+1)\, E(g^{2k_1}).$$

Thus (7.52) finishes the induction step. $\qquad\qquad\qquad\qquad\qquad$ □

Challenge problem 7.14. Prove Proposition 7.9.
Hint: Use the results of Chapter 3.

Challenge problem 7.15. Compute the following, when $\beta < \beta_0$:

$$\lim_{N \to \infty} N \, \mathsf{E}\langle (\frac{\sigma^1 \cdot \sigma^2}{N} - q)^2 \rangle \; ; \; \lim_{N \to \infty} N \, \mathsf{E}\langle (\frac{\sigma^1 \cdot b}{N} - q)^2 \rangle$$

$$\lim_{N \to \infty} N \, \mathsf{E}(\bar{q} - q)^2 .$$

If you don't have the stamina to compute these limits, at least prove their existence.

8 Emergence and separation of the lumps in the p-spin interaction model

We go back to the study of the p-spin interaction model with no external field, and our efforts will focus on the low temperature region.

Theorem 8.1. *There exists a number p_0 such that, if $p \geq p_0$, then for $\beta \leq 2^{\frac{p}{2}-6}$, with overwhelming probability we have*

$$
(8.1) \qquad G_N^{\otimes 2}(\{|R(\sigma,\sigma')| \in [2^{-\frac{p}{4}}, 1 - 2^{-\frac{p}{2}}]\}) \leq \exp\left(-\frac{N}{K}\right).
$$

Proof. This follows from Theorem 4.13 and Fubini's theorem. $\qquad\square$

Thus, for large p, $|R(\sigma,\sigma')|$ is typically close to 0 or 1. This fact has strong consequences about Gibbs' measure.

It is convenient and will somewhat clarify matters to consider a more general setting. Let

$$
S_N = \Big\{\sigma \in \mathbb{R}^N \;\Big|\; \sum_{i \leq N} \sigma_i^2 = N\Big\},
$$

be the sphere of radius $\sqrt{N}$ (so that $\Sigma_N \subset S_N$), and for σ, σ' in S_N, we write

$$(8.2) \qquad R(\sigma, \sigma') = \frac{\sigma \cdot \sigma'}{N} = \frac{1}{N} \sum_{i \leq N} \sigma_i \sigma_i'.$$

Consider now a probability measure μ on S_N, and assume that for a certain (small) number a,

$$(8.3) \qquad \mu \otimes \mu(T) = \varepsilon$$

is very small, where

$$T = \{(\sigma, \sigma') \mid |R(\sigma, \sigma')| \in [a, 1 - a]\}.$$

What does (8.3) tell us about μ? First it could happen that (8.3) is true simply because it is already true that

$$(8.4) \qquad \mu \otimes \mu(\{(\sigma, \sigma') \mid |R(\sigma, \sigma')| \geq a\})$$

is very small. This is for example the case if μ is uniform on Σ_N. It could also happen that μ is concentrated on a single point, or more generally on a set A such that $R(\sigma, \sigma') \geq 1 - a$ if $(\sigma, \sigma') \in A$. We will show that the general situation is a mixture of the previous two cases.

The construction will involve thinking of S_N as a metric space, equipped with the distance

$$d(\sigma, \sigma') = \arccos(R(\sigma, \sigma')).$$

We consider δ given by

$$(8.5) \qquad \delta = \arccos(1 - a)$$

so that, if a is small enough, we have

$$(8.6) \qquad a < \cos 7\delta,$$

and for σ in S_N, we consider the sets

$$C_1(\sigma) = \{\sigma' \in S_N \mid |R(\sigma, \sigma')| \geq \cos \delta\}$$
$$C(\sigma) = \{\sigma' \in S_N \mid |R(\sigma, \sigma')| \geq \cos 3\delta\}$$
$$D(\sigma) = \{\sigma' \in S_N \mid \cos 6\delta \leq |R(\sigma, \sigma')| < \cos 2\delta\}.$$

Thus $C_1(\sigma)$ (resp. $C(\sigma)$) is the union of two balls of S_N for the metric d, of radius δ (resp. 3δ) that are centered at σ and $-\sigma$. One should keep in mind that μ might (and, in the case of G_N, will) be invariant by the symmetry $\sigma \mapsto -\sigma$, so that all our constructions should respect this symmetry.

Definition 8.2. A set $C(\sigma)$ such that

$$(8.7) \qquad \mu(C_1(\sigma)) \geq \varepsilon^{\frac{1}{3}}$$

is called a *lump*.

Thus, a lump is a small set (its diameter is $\leq 6\delta$) and it carries a significant mass. The motivation of $D(\sigma)$ is that this is "nearly empty space" around a lump $C(\sigma)$.

Lemma 8.3. *For each lump $C(\sigma)$ we have*

$$(8.8) \qquad \mu(D(\sigma)) \leq \varepsilon^{\frac{2}{3}}.$$

Proof. By (8.3), (8.7) it suffices to prove that

$$C_1(\sigma) \times D(\sigma) \subset T.$$

Consider $\sigma' \in C_1(\sigma)$, $\sigma'' \in D(\sigma)$. Without loss of generality we can assume $R(\sigma, \sigma') \geq 0$, $R(\sigma, \sigma'') \geq 0$. Thus $d(\sigma', \sigma) \leq \delta$, $2\delta \leq d(\sigma, \sigma'') \leq 6\delta$, so that $\delta < d(\sigma', \sigma'') \leq 7\delta$, i.e. $\cos 7\delta \leq R(\sigma', \sigma'') \leq \cos \delta$, and $(\sigma', \sigma'') \in T$ using (8.6). $\qquad\square$

Theorem 8.4. *Under (8.3), (8.6), we can find a disjoint family $(C_\alpha)_{\alpha \leq M}$ of lumps such that*

$$(8.9) \qquad \mu \otimes \mu\Big(\{(\sigma, \sigma') \mid |R(\sigma', \sigma'')| \geq a\} \setminus \bigcup_{\alpha \leq M} C_\alpha^2\Big) \leq 3\varepsilon^{\frac{1}{3}}.$$

Comment. Here, and below, α is an integer. The content of the theorem is that essentially the only way one can have $|R(\sigma, \sigma')| \geq a$ is that σ, σ' belong to the same lump.

This of course contains information only if

$$(8.10) \qquad \mu \otimes \mu(\{(\sigma, \sigma') \mid |R(\sigma, \sigma')| \geq a\})$$

is not too small, or equivalently under (8.3), if

$$\mu \otimes \mu(\{(\sigma, \sigma') \mid |R(\sigma, \sigma')| \geq 1 - a\})$$

is not too small.

Proof of Theorem 8.4. We consider a maximal disjoint family of lumps $(C_\alpha)_{\alpha \leq M}$. (It might very well happen that $M = 0$) and we prove (8.9). First, by (8.7) and since the lumps C_α are disjoint, we have $M\varepsilon^{\frac{1}{3}} \leq 1$, so that $M \leq \varepsilon^{-\frac{1}{3}}$. Next, if $C_\alpha = C(\sigma_\alpha)$, by Lemma 8.3 we have $\mu(D(\sigma_\alpha)) \leq \varepsilon^{\frac{2}{3}}$, so that if $D = \bigcup_{\alpha \leq M} D(\sigma_\alpha)$ we have $\mu(D) \leq M\varepsilon^{\frac{2}{3}} \leq \varepsilon^{\frac{1}{3}}$. Thus to prove (8.9), it suffices to prove that

$$(8.11) \quad \{(\sigma, \sigma') \in S_N^2 \mid |R(\sigma, \sigma')| \geq 1 - a\} \subset S' \cup (D \times S_N) \cup \Big(\bigcup_{\alpha \leq M} C_\alpha^2\Big),$$

where

$$S' = \{(\sigma, \sigma') \mid \sigma' \in C_1(\sigma), \ \mu(C_1(\sigma)) \leq \varepsilon^{\frac{1}{3}}\}.$$

This implies (8.9) because $\mu \otimes \mu(S') \leq \varepsilon^{\frac{1}{3}}$ by Fubini's theorem. To prove (8.11), consider $(\sigma, \sigma') \in S_N^2$, with $|R(\sigma, \sigma')| \geq 1 - a$ and assume that

$(\sigma, \sigma') \notin S'$, that is, $\mu(C_1(\sigma)) \geq \varepsilon^{\frac{1}{3}}$. The maximality of the family $(C_\alpha)_{\alpha \leq M}$ implies that for some α we have

$$(8.12) \qquad C(\sigma) \cap C(\sigma_\alpha) \neq \emptyset.$$

Let us assume for clarity $R(\sigma, \sigma_\alpha) > 0$, $R(\sigma, \sigma') > 0$. Then by (8.12) we have $d(\sigma, \sigma_\alpha) \leq 6\delta$. If $d(\sigma, \sigma_\alpha) \geq 2\delta$, we are done because $\sigma \in D(\sigma_\alpha) \subset D$. If $d(\sigma, \sigma_\alpha) \leq 2\delta$, then, since $R(\sigma, \sigma') \geq 1 - a$, we have $d(\sigma, \sigma') \leq \delta$, so that $d(\sigma', \sigma_\alpha) \leq 3\delta$ and $(\sigma, \sigma') \in C_\alpha^2$. $\qquad \square$

Challenge problem 8.5. Formulate and prove a statement showing that the lumps are essentially determined.

We now go back to the case of Gibbs' measure. When (8.3) occurs for $\mu = G_N$, under (8.6), we can construct the lumps $(C_\alpha)_{\alpha \leq M}$ where ε is the left-hand side of (8.3). It is perfectly possible a priori that $M = 0$, that there exist no lumps, or that there exist few of them, and that they are far from exhausting all of G_N. To keep notations simple, it is better not to distinguish these various cases, and to adopt a procedure that exhausts all of the configuration space Σ_N. This will be done in a trivial way (this does not bring more information than Theorem 8.4). After having constructed the lumps $(C_\alpha)_{\alpha \leq M}$ we continue the construction by enumerating all the remaining pairs $\{\sigma, -\sigma\}$ as $C_{M+1}, \dots$ until Σ_N is exhausted. This happens at a certain index α_0, and for $\alpha > \alpha_0$ we set $C_\alpha = \emptyset$; we thus define a sequence of lumps $(C_\alpha)_{\alpha \geq 1}$. (Only the $(C_\alpha)_{\alpha \leq M}$ actually deserve the name of lumps).

Theorem 8.6. *There exists a number p_0 such that if $p \geq p_0$, then for $\beta \leq 2^{\frac{p}{2}-6}$, we can find a decomposition $(C_\alpha)_{\alpha \geq 1}$ of Σ_N in disjoint sets with the following properties:*

$(8.13) \qquad$ *Each set C_α is symmetric, i.e. $\sigma \in C_\alpha \implies -\sigma \in C_\alpha$.*

$(8.14) \qquad \sigma, \sigma' \in C_\alpha \implies |R(\sigma, \sigma')| \geq 1 - 2^{-\frac{p}{4}+6}.$

$(8.15) \qquad \mathsf{E}\big(G_N^2(\{(\sigma, \sigma') \mid |R(\sigma, \sigma')| \geq 2^{-\frac{p}{4}}\} \setminus \bigcup_{\alpha \geq 1} C_\alpha^2)\big) \leq \exp\big(-\frac{N}{K}\big).$

Proof. We apply the construction of Theorem 8.4 for each value of the disorder with $a = 2^{-\frac{p}{4}}$. To prove (8.14), we observe that $\sigma, \sigma' \in C_\alpha \implies R(\sigma, \sigma') \geq \cos 6\delta$ where $\cos \delta = 1 - a$ and that $\cos 6\delta \geq 1 - 2^6(1 - \cos \delta)$ for small δ. $\qquad \square$

Of course one should repeat that *all* the information contained in this statement is contained in (8.15); so for example if it is already true that

$$(8.16) \qquad \mathsf{E}\, G_N^2(\{\sigma, \sigma' \mid |R(\sigma, \sigma')| \geq 2^{-\frac{p}{4}}\}) \leq \exp\big(-\frac{N}{K}\big)$$

(as in the case where $\beta < \beta_p$) a decomposition such as that of Theorem 8.6 is uninteresting. On the other hand, things become interesting as soon a (8.16) fails. Indeed, in that case, since the lumps are disjoint, we have $\sum_{\alpha \geq 1} G_N(C_\alpha) = 1$, so that

$$\mathsf{E} \max_\alpha G_N(C_\alpha) \geq \mathsf{E}\left(\sum_{\alpha \geq 1} G_N(C_\alpha)^2\right)$$

$$= \mathsf{E}\left(G_N \otimes G_N\left(\bigcup C_\alpha^2\right)\right)$$

$$\geq \mathsf{E}\left(G_N \otimes G_N(\{(\sigma, \sigma') \mid |R(\sigma, \sigma')| \geq 2^{-\frac{p}{4}}\})\right) - \exp\left(-\frac{N}{K}\right),$$

where we have used (8.15) in the last line. The ultimate goal (for which we must refer to [T7]) is to show that the "total mass of the macroscopic lumps is one".

We can reformulate (8.15) by saying that if σ, σ' do not belong to the same lump, then (generically) we have $|R(\sigma, \sigma')| \leq 2^{-\frac{p}{4}}$. The rest of the chapter is devoted to show that in fact not only do we have $|R(\sigma, \sigma')| \leq 2^{-\frac{p}{4}}$, but in fact $R(\sigma, \sigma') \simeq 0$. The proof will rely upon the cavity method, and our first task is to set up notation to do this. Given σ in Σ_N, we set

$$T(\sigma) = (\sigma_{i_1} \ldots \sigma_{i_{p-1}}),$$

where the index ranges over all families $i_1 < i_2 < \cdots < i_{p-1}$. We consider a family of independent $\mathcal{N}(0,1)$ variables,

(8.17)
$$g = (g_{i_1 \ldots i_{p-1}}),$$

with the same range of indexes as above, and we define

$$g \cdot T(\sigma) = \sum g_{i_1 \ldots i_{p-1}} \sigma_{i_1} \ldots \sigma_{i_{p-1}}$$

where the range of the index is as above.

We recall the Hamiltonian $H_N(\sigma)$ of (4.1). Consider $\sigma_{N+1} \in \{-1, 1\}$. Then we have

(8.18)
$$-\beta H_N(\sigma) + \sigma_{N+1}\beta\left(\frac{p!}{2N^{p-1}}\right)^{\frac{1}{2}} T(g) \cdot \sigma$$

$$= \beta\left(\frac{p!}{2N^{p-1}}\right)^{\frac{1}{2}} \sum g_{i_1 \ldots i_p} \sigma_{i_1} \ldots \sigma_{i_p},$$

where the sum is over $1 \leq i_1 < \cdots < i_p \leq N + 1$, and where

$$g_{i_1 \ldots i_{p-1} i_p} = g_{i_1 \ldots i_{p-1}}$$

when $i_p = N + 1$. If we recall the notation

$$\varrho = (\sigma_i)_{i \leq N+1} \in \Sigma_{N+1},$$

we see that the right-hand side of (8.18) is

$$-\beta' H_{N+1}(\varrho),$$

where

(8.19)
$$\beta' = \left(\frac{N+1}{N}\right)^{\frac{p-1}{2}}.$$

That is, (8.18) is the Hamiltonian of an $(N+1)$-spin system at a slightly lower temperature. We denote by $G_{N+1} = G_{N+1}(\beta')$ the corresponding Gibbs' measure and by $\langle \cdot \rangle'$ average with respect to this measure: we will use the same conventions as for the SK model. Then (5.19) holds, except that now

(8.20)
$$\mathcal{E} = \mathcal{E}(\sigma, \sigma_{N+1}) = \exp \sigma_{N+1} \beta \left(\frac{p!}{2N^{p-1}}\right)^{\frac{1}{2}} g \cdot T(\sigma)$$

and in (5.22) we now have

(8.21)
$$\mathcal{E}_k = \exp\left(\sum_{l \le k} \sigma^l_{N+1} \beta \left(\frac{p!}{2N^{p-1}}\right)^{\frac{1}{2}} g \cdot T(\sigma^l)\right).$$

Even though everything looks pretty much like in the case $p = 2$, it is good to observe that

(8.22)
$$\mathsf{E}\left(\left(\left(\frac{p!}{2N^{p-1}}\right)^{\frac{1}{2}} g \cdot T(\sigma)\right)^2\right) = \frac{p!}{2N^{p-1}}\binom{N}{p-1}$$
$$= \frac{p!}{2N^{p-1}} \frac{N(N-1)\ldots(N-p+1)}{(p-1)!}$$
$$\simeq \frac{p}{2}.$$

Thus as p increases, the exponent in $\mathcal{E}_k$ gets larger, and is harder to control. This is unavoidable: as p increases, there is less relationship between the structure of the $(N+1)$-spin system and the structure of the N-spin system. The good news is that, on the other hand, the correlation between the variables $g \cdot T(\sigma)$ decreases.

Lemma 8.7. *We have*

(8.23)
$$\left| T(\sigma) \cdot T(\sigma') - \frac{(\sigma \cdot \sigma')^{p-1}}{(p-1)!}\right| \le K(p)N^{p-2}.$$

Proof. We have, setting $a_i = \sigma_i \sigma_i'$

$$T(\sigma) \cdot T(\sigma') = \sum_{i_1 < \cdots < i_{p-1}} \sigma_{i_1} \cdots \sigma_{i_{p-1}} \sigma_{i_1}' \cdots \sigma_{i_{p-1}}'$$

$$= \sum_{i_1 < \cdots < i_{p-1}} a_{i_1} \cdots a_{i_{p-1}}$$

$$= \frac{1}{(p-1)!} \sum_d a_{i_1} \cdots a_{i_{p-1}},$$

where $\sum_d$ means that the summation is over indices $i_1, \ldots i_{p-1}$ all different (but not necessarily ordered). Now

$$(\sigma \cdot \sigma')^{p-1} = \left(\sum_{i \leq N} a_i \right)^{p-1} = \sum a_{i_1} \cdots a_{i_{p-1}},$$

where the summation is over all choices of $i_1 \leq N, \ldots, i_p \leq N$. For each i, we have $|a_i| \leq 1$ and there are at most $K(p) N^{p-2}$ choices of $i_1, \ldots, i_{p-1}$ that are not all different. $\qquad \square$

Theorem 8.8. *There exists a number p_0 such that if $p \geq p_0$ and $\beta \leq 2^{\frac{p}{2}-6}$, we have*

$$(8.24) \qquad \mathsf{E}\langle R(\sigma, \sigma')^2 1_{\{|R(\sigma, \sigma')| \leq 1/2\}} \rangle \leq \frac{K(p)}{N}.$$

We will show that

$$(8.25) \qquad \beta \leq 2^{\frac{p}{2}-6} \implies A_N(\beta) \leq \frac{K(p)}{N}$$

where

$$(8.26) \qquad A_N(\beta) = \mathsf{E}\langle R(\sigma^1, \sigma^2)^2 1_{\{|R(\sigma^1, \sigma^2)| \leq c\}} \rangle,$$

and where $c = 2^{-\frac{p}{4}+1}$. This implies (8.24) by Theorem 4.13.

We use symmetry between sites to see that

$$(8.27) \qquad A_N(\beta) = \mathsf{E}\langle \sigma_N^1 \sigma_N^2 R(\sigma^1, \sigma^2) 1_{\{|R(\sigma^1, \sigma^2)| \leq c\}} \rangle,$$

and, changing N into $N+1$ and β into β', we get

$$(8.28) \qquad A_{N+1}(\beta') = \mathsf{E}\langle \sigma_{N+1}^1 \sigma_{N+1}^2 R(\varrho^1, \varrho^2) 1_{\{|R(\varrho^1, \varrho^2)| \leq c\}} \rangle.$$

We have set

$$(8.29) \qquad \varrho^l = (\sigma_1^l, \ldots, \sigma_{N+1}^l)$$

so that

$$(8.30) \qquad R(\varrho^1, \varrho^2) = \frac{N}{N+1} R(\sigma^1, \sigma^2) + \frac{\sigma_{N+1}^1 \sigma_{N+1}^2}{N+1}.$$

This implies that

$$(8.31) \quad A_{N+1}(\beta') \leq \frac{1}{N+1} + \frac{N}{N+1} \, \mathsf{E} \langle \sigma_{N+1}^1 \sigma_{N+1}^2 R(\sigma^1, \sigma^2) 1_{\{|R(\varrho^1, \varrho^2)| \leq c\}} \rangle'.$$

Now (8.30) implies

$$|R(\varrho^1, \varrho^2) - R(\sigma^1, \sigma^2)| \leq \frac{2}{N+1}.$$

Thus

$$\frac{2}{N+1} \leq \frac{c}{2} \implies |1_{\{|R(\varrho^1, \varrho^2)| \leq c\}} - 1_{\{|R(\sigma^1, \sigma^2)| \leq c\}}| \leq 1_{\{\frac{c}{2} \leq |R(\varrho^1, \varrho^2)| \leq \frac{3c}{2}\}},$$

and thus, if $\beta' \leq 2^{\frac{p}{2}-6}$, we deduce from Theorem 4.13 that

$$(8.32) \quad A_{N+1}(\beta') \leq \frac{1}{N+1} + \exp\left(-\frac{N}{K}\right)$$

$$+ \frac{N}{N+1} \, \mathsf{E} \langle \sigma_{N+1}^1 \sigma_{N+2}^2 R(\sigma^1, \sigma^2) 1_{\{|R(\sigma^1, \sigma^2)| \leq c\}} \rangle.$$

Setting

$$f = f(\sigma^1, \sigma^2) = R(\sigma^1, \sigma^2) 1_{\{|R(\sigma^1, \sigma^2)| \leq c\}},$$

we are now in a position to use the cavity method to write

$$(8.33) \quad \mathsf{E} \langle \sigma_{N+1}^1 \sigma_{N+1}^2 f \rangle' = \mathsf{E} \frac{\langle f \, \mathrm{Av} \, \varepsilon_1 \varepsilon_2 \mathcal{E}_2 \rangle}{Z^2},$$

where, as usual, $\varepsilon_l = \sigma_{N+1}^l$. The family g of (8.17) is independent of the disorder involved in G_N (the variables $g_{i_1, \ldots, i_p}$) so we will try to evaluate

$$(8.34) \quad \mathsf{E}_g \frac{\langle f \, \mathrm{Av} \, \varepsilon_1 \varepsilon_2 \mathcal{E}_2 \rangle}{Z^2}.$$

This will be done using Kahane's principle. The proof will resemble that of Proposition 6.6. Let us write

$$\mu(\sigma) = G_N(\{\sigma\}).$$

For $x = (x(\sigma)) \in \mathbb{R}^{\Sigma_N}$ we write

$$F_1(x) = \sum \mu(\sigma^1)\mu(\sigma^2) f(\sigma^1, \sigma^2) \, \mathrm{Av} \, \varepsilon_1 \varepsilon_2 \exp \sum_{l \leq 2} \varepsilon_l x(\sigma^l)$$

$$= \sum \mu(\sigma^1)\mu(\sigma^2) f(\sigma^1, \sigma^2) \, \mathrm{sh} \, x(\sigma^1) \, \mathrm{sh} \, x(\sigma^2)$$

$$F_2(x) = \sum \mu(\sigma^1)\mu(\sigma^2) \, \mathrm{Av} \exp \sum_{l \leq 2} \varepsilon_l x(\sigma^l)$$

$$= \left(\sum \mu(\sigma) \, \mathrm{ch} \, x(\sigma) \right)^2$$

where the summations are over all values of σ^1, σ^2. We consider

$$F(x) = \frac{F_1(x)}{F_2(x)}.$$

Consider the Gaussian family $v = (v(\sigma))$ where

(8.35)
$$v(\sigma) = \beta \left(\frac{p!}{2N^{p-1}} \right)^{\frac{1}{2}} g \cdot T(\sigma),$$

so that the quantity (8.34) is $E_g F(v)$.

We consider a centered Gaussian family u such that

(8.36) $\qquad (\sigma, \sigma') \in \bigcup_\alpha C_\alpha^2 \implies E\, u(\sigma) u(\sigma') = E\, v(\sigma) v(\sigma')$

(8.37) $\qquad (\sigma, \sigma') \notin \bigcup_\alpha C_\alpha^2 \implies E\, u(\sigma) u(\sigma') = 0.$

The first condition means that u and v have the same covariance inside a lump, while the second condition means that different lumps are independent for u.

It should be obvious that such a family exists. The idea of (8.36) is that we do not disturb anything inside a lump, while making the lumps independent. In this manner, it is not an obstacle to the proof that we do not know what happens inside the lumps.

Lemma 8.9. $E_g F(u) = 0.$

Proof. We observe that we have

(8.38) $\qquad F_1(x) = \sum f(\sigma^1, \sigma^2) \mu(\sigma^1) \mu(\sigma^2)\, \mathrm{sh}\, x(\sigma^1)\, \mathrm{sh}\, x(\sigma^2)$

(8.39) $\qquad F_2(x) = \sum \mu(\sigma^1) \mu(\sigma^2)\, \mathrm{ch}\, x(\sigma^1)\, \mathrm{ch}\, x(\sigma^2).$

We will prove that, given σ^1, σ^2 we have $E_g A(u, \sigma^1, \sigma^2) = 0$ where

$$A(u, \sigma^1, \sigma^2) = \frac{f(\sigma^1, \sigma^2)\, \mathrm{sh}\, u(\sigma^1)\, \mathrm{sh}\, u(\sigma^2)}{F_2(u)}.$$

Certainly we can assume $|R(\sigma^1, \sigma^2)| \le \frac{1}{2}$, for otherwise $f(\sigma^1, \sigma^2) = 0$. Consider α such that $\sigma^1 \in C_\alpha$. Then $\sigma^2 \notin C_\alpha$, for otherwise $|R(\sigma^1, \sigma^2)| > 1/2$. Consider the Gaussian family u_α given by

$$u_\alpha(\sigma) = u(\sigma) \qquad \text{if } \sigma \notin C_\alpha$$
$$u_\alpha(\sigma) = -u(\sigma) \quad \text{if } \sigma \in C_\alpha.$$

Then u_α and u have the same distribution by (8.36), (8.37) so that

(8.40) $\qquad E_g A(u, \sigma^1, \sigma^2) = E_g A(u_\alpha, \sigma^1, \sigma^2).$

But we have $u_\alpha(\sigma^1) = -u(\sigma^1)$, $u_\alpha(\sigma^2) = u(\sigma^2)$, $F_2(u_\alpha) = F_2(u)$ so that

$$A(u, \sigma^1, \sigma^2) = -A(u_\alpha, \sigma^1, \sigma^2),$$

and combining with (8.40) this proves the lemma. $\qquad\qquad \square$

To estimate $\mathsf{E}_g F(v) - \mathsf{E}_g F(u)$, we will use (6.38). That is, we set

$$u_t = \sqrt{1-t}\,u + \sqrt{t}\,v$$
$$\theta(t) = \mathsf{E}_g F(u_t)$$

so that

$$(8.41) \qquad \mathsf{E}_g F(v) = \mathsf{E}_g F(v) - \mathsf{E}_g F(u) = \int_0^1 \theta'(t)\,dt.$$

To compute $\theta'(t)$, it will be convenient to introduce the quantity $B(\sigma^1, \sigma^2)$ given by

$$B(\sigma^1, \sigma^2) = \frac{\beta^2 p!}{2N^{p-1}} T(\sigma^1) \cdot T(\sigma^2)$$

when σ^1, σ^2 do not belong to the same lump, and $B(\sigma^1, \sigma^2) = 0$ otherwise.

Lemma 8.10. *We have*

$$(8.42) \qquad \theta'(t) = \sum_{r=3}^{6} \mathsf{E}_g F_r(u_t)$$

where
(8.43)
$$F_3(x) = \frac{1}{F_2(x)} \sum f(\sigma^1, \sigma^2)\mu(\sigma^1)\mu(\sigma^2)B(\sigma^1, \sigma^2)\,\mathrm{ch}\,x(\sigma^1)\,\mathrm{ch}\,x(\sigma^2)$$

(the summation is over σ^1, σ^2 in Σ_N),

$$(8.44) \qquad F_4(x) = -\frac{F_1(x)}{F_2^2(x)} \sum \mu(\sigma^1)\mu(\sigma^2)B(\sigma^1, \sigma^2)\,\mathrm{sh}\,x(\sigma^1)\,\mathrm{sh}\,x(\sigma^2)$$

$$(8.45) \qquad F_5(x) = -\frac{4}{F_2^2(x)} \sum \mu(\sigma^1)\mu(\sigma^2)\mu(\sigma^3)\mu(\sigma^4)f(\sigma^1, \sigma^2)$$
$$\times B(\sigma^2, \sigma^3)\,\mathrm{sh}\,x(\sigma^1)\,\mathrm{ch}\,x(\sigma^2)\,\mathrm{sh}\,x(\sigma^3)\,\mathrm{ch}\,x(\sigma^4)$$

$$(8.46) \qquad F_6(x) = \frac{4F_1(x)}{F_2^3(x)} \sum \mu(\sigma^1)\mu(\sigma^2)B(\sigma^1, \sigma^2)\,\mathrm{sh}\,x(\sigma^1)\,\mathrm{sh}\,x(\sigma^2).$$

Proof. We use (6.46). Now the indexes a, b belong to Σ_N. If $a = \eta^1$, $b = \eta^2$, we have

$$(8.47) \qquad \mathsf{E}(v_a v_b - u_a u_b) = B(\eta^1, \eta^2).$$

Since F is a ratio, the quantity

$$\frac{1}{2}\sum \mathsf{E}(v_a v_b - u_a u_b)\frac{\partial^2 F}{\partial x_a \partial x_b}$$

is the sum of four terms, as in the formula

$$\frac{\partial^2}{\partial x_a \partial x_b}\left(\frac{F_1}{F_2}\right) = \frac{\partial^2 F_1}{\partial x_a \partial x_b}\frac{1}{F_2} - F_1 \frac{\partial^2 F_2}{\partial x_a \partial x_b}\frac{1}{F_2^2}$$
$$- \left(\frac{\partial F_1}{\partial x_a}\frac{\partial F_2}{\partial x_b} + \frac{\partial F_1}{\partial x_b}\frac{\partial F_2}{\partial x_a}\right)\frac{1}{F_2^2} + 2F_1\left(\frac{\partial F_2}{\partial x_a}\frac{\partial F_2}{\partial x_b}\right)\frac{1}{F_2^3}.$$

These terms are F_3 to F_6 respectively. Let us see how (8.45) arises. For $a = \eta^1$, we have

$$\frac{\partial F_1}{\partial x_a} = \sum_{\sigma^2} \mu(\eta^1)\mu(\sigma^2)f(\eta^1,\sigma^2)\operatorname{ch} x(\eta^1)\operatorname{sh} x(\sigma^2)$$
$$+ \sum_{\sigma^1} \mu(\sigma^1)\mu(\eta^1)f(\sigma^1,\eta^1)\operatorname{sh} x(\sigma^1)\operatorname{ch} x(\eta^1)$$
$$= 2\sum_{\sigma^2} \mu(\eta^1)\mu(\sigma^2)f(\eta^1,\sigma^2)\operatorname{ch} x(\eta^1)\operatorname{sh} x(\sigma^2)$$

by symmetry. Similarly, if $b = \eta^2$,

$$\frac{\partial F_2}{\partial x_b} = 2\sum_{\sigma^1} \mu(\sigma^1)\mu(\eta^2)\operatorname{ch} x(\sigma^1)\operatorname{sh} x(\eta^2).$$

Writing $\dfrac{\partial F_1}{\partial x_a}\dfrac{\partial F_2}{\partial x_b}$ as a double sum, summing over η^1, η^2 we get (8.45). $\square$

Corollary 8.11. *We have*

(8.48)
$$|\theta'(t)| \le 10\beta^2 p 2^{-(p-3)(\frac{p}{4}-1)}\, \mathsf{E}_g \frac{F_7(u_t)}{F_2(u_t)} + S$$

where

$$F_7(x) = \sum \mu(\sigma^1)\mu(\sigma^2)f^2(\sigma^1,\sigma^2)\operatorname{ch} x(\sigma^1)\operatorname{ch} x(\sigma^2)$$

and

$$\mathsf{E}\,S \le \frac{K(p)}{N}.$$

Proof. We use the inequalities $|\operatorname{sh} x| \le \operatorname{ch} x$ and $|f| \le 1$ to see that

(8.49)
$$\sum_{r=3}^{6} |F_r(x)| \le 10\frac{F_8(x)}{F_2(x)},$$

with

$$F_8(x) = \sum \mu(\sigma^1)\mu(\sigma^2)B^2(\sigma^1,\sigma^2)\operatorname{ch} x(\sigma^1)\operatorname{ch} x(\sigma^2).$$

Now, from (8.23) we see that

$$|B(\sigma^1,\sigma^2)| \le \frac{\beta^2 p}{2}|R^{p-1}(\sigma^1,\sigma^2)| + \frac{K(p)}{N}.$$

Since $B(\sigma^1, \sigma^2) = 0$ when $(\sigma^1, \sigma^2) \in \cup C_\alpha^2$ we have

(8.50) $\displaystyle |B(\sigma^1, \sigma^2)| \le \frac{\beta^2 p}{2} |R(\sigma^1, \sigma^2)|^{p-1} 1_{\{|R| \le c\}} + \frac{\beta^2 p}{2} 1_C + \frac{K(p)}{N}$

where

$$C = \{(\sigma^1, \sigma^2) \mid |R(\sigma^1, \sigma^2)| \ge c; \; (\sigma^1, \sigma^2) \notin \cup C_\alpha^2\}.$$

Thus,

(8.51) $\displaystyle |B(\sigma^1, \sigma^2)| \le \frac{\beta^2 p}{2} c^{p-3} R(\sigma^1, \sigma^2)^2 1_{\{|R| \le c\}} + \frac{\beta^2 p}{2} 1_C + \frac{K(p)}{N}.$

Using elementary inequalities such as $(a+b)^2 \le 2(a^2 + b^2)$, $\sqrt{a+b} \le \sqrt{a} + \sqrt{b}$, etc., from (8.49), (8.51) we see that (since $F_2 \ge 1$)

(8.52) $\displaystyle \sum_{r=3}^{6} |F_r(x)| \le 10 \beta^2 p c^{p-2} \frac{F_7(x)}{F_2(x)} + \frac{K(p)}{N}$

$\displaystyle \qquad\qquad + \beta^2 p \sum \mu(\sigma^1) \mu(\sigma^2) 1_C(\sigma^1, \sigma^2) \operatorname{ch} x(\sigma^1) \operatorname{ch} x(\sigma^2).$

We now appeal to (8.39) to obtain (8.48) with

$$S = \frac{K(p)}{N} + \beta^2 p \, \mathsf{E}_g \Big(\sum \mu(\sigma^1) \mu(\sigma^2) 1_C(\sigma^1, \sigma^2) \operatorname{ch} u_t(\sigma^1) \operatorname{ch} u_t(\sigma^2) \Big).$$

It follows from (8.15) that the expected value of S (in the disorder) is $\le K(p)/N$. $\square$

Proposition 8.12. *If $p \ge p_0$, $\beta \le 2^{\frac{p}{2}-6}$ we have*

(8.53) $\theta_1'(t) \ge -\theta_1(t) - S_1,$

where

(8.54) $\displaystyle \theta_1(t) = \mathsf{E}_g \frac{F_7(u_t)}{F_2(u_t)}$

and

$$\mathsf{E}\, S_1 \le \frac{K(p)}{N}.$$

Proof. It is similar to the proof of Corollary 8.11. $\square$

Proof of Theorem 8.8. First, we claim that (8.53) implies

(8.55) $\displaystyle 0 \le t \le 1 \implies \theta_1(t) \le e^{1-t} \theta_1(t) + S_1(1-t).$

Indeed, if $u = 1 - t$

$$\xi(u) = \theta_1(1-u) - e^u \theta_1(1) - S_1 u$$

satisfies $\xi(0) = 0$, $\xi'(u) \le \xi(u)$, so that $\xi(u) \le 0$.

If we combine (8.55) with (8.54), (8.48), we see that for $p \geq p_0$, $\beta \leq 2^{\frac{p}{4}-6}$ we have

$$(8.56) \qquad |\theta'(t)| \leq \frac{1}{2}\theta_1(1) + S + S_1,$$

and, combining with (8.38), we have

$$\left|\mathsf{E}_g \frac{\langle f \operatorname{Av}\varepsilon_1\varepsilon_2\mathcal{E}_2\rangle}{Z^2}\right| = |\mathsf{E}_g F(v)| \leq \frac{1}{2}\mathsf{E}_g \frac{F_7(v)}{F_1(v)} + S + S_1.$$

Recalling (8.33), and taking expectation in the disorder, we have

$$(8.57) \qquad |\mathsf{E}\langle \sigma_{N+1}^1 \sigma_{N+1}^2 f\rangle'| \leq \frac{1}{2}\mathsf{E}\frac{F_7(v)}{F_1(v)} + \frac{K(p)}{N}$$

$$= \frac{1}{2}\mathsf{E}\langle R(\sigma^1,\sigma^2)^2 1_{\{R \leq c\}}\rangle' + \frac{K(p)}{N}.$$

By an argument similar to that in the proof of (8.33), we see that $\mathsf{E}\langle \sigma_{N+1}^1 \sigma_{N+1}^2 f\rangle$ and $\mathsf{E}\langle R(\sigma^1,\sigma^2)^2 1_{\{R \leq c\}}\rangle'$ differ from $A_{N+1}(\beta')$ by at most $K(p)/N$, so that (8.57) yields

$$A_{N+1}(\beta') \leq \frac{K(p)}{N}. \qquad \square$$

Challenge problem 8.13. Improve Theorem 8.8 to an exponential inequality.

If we combine Theorems 4.13 and 8.8, we see that, if

$$B = \bigcup_\alpha C_\alpha^2,$$

then

$$(8.58) \qquad |\mathsf{E}\langle R(\sigma^1,\sigma^2)^p 1_B\rangle - \mathsf{E}\langle R(\sigma^1,\sigma^2)^p\rangle| \leq \frac{K(p)}{N}.$$

Also, since $(1 - 2^{-\frac{p}{4}})^p \geq \frac{1}{2}$ for large p, we have

$$\sigma, \sigma' \in C_\alpha \implies \frac{1}{2} \leq R(\sigma^1,\sigma^2)^p \leq 1$$

so that, if

$$(8.59) \qquad w_\alpha = G_N(C_\alpha),$$

$$\frac{1}{2}\sum_\alpha w_\alpha^2 \leq \langle R(\sigma^1,\sigma^2)^p\rangle \leq \sum_\alpha w_\alpha^2,$$

and, combining with (8.58),

$$(8.60) \qquad \frac{1}{2}\mathsf{E}\sum w_\alpha^2 - \frac{K(p)}{N} \leq \mathsf{E}\langle R(\sigma^1,\sigma^2)^p\rangle \leq \mathsf{E}\sum w_\alpha^2 + \frac{K(p)}{N}.$$

This allows us to obtain information about the weights (8.59). The following is a generalization of Lemma 5.13.

Lemma 8.14. *We have*

$$(8.61) \qquad \left| \frac{d}{d\beta} p_N(\beta) - \frac{\beta}{2}(1 - \mathsf{E}\langle R(\sigma, \sigma')^p \rangle) \right| \leq \frac{K}{N}.$$

Proof. The same computation as in Lemma 5.13 gives

$$\frac{dp_N}{d\beta} = \frac{\beta p!}{2N^p} \sum_{i_1 < \cdots < i_p} (1 - \mathsf{E}\langle \sigma_{i_1}\sigma'_{i_1} \ldots \sigma_{i_p}\sigma'_{i_p} \rangle).$$

Changing p into $p+1$, (8.23) shows that

$$\left| \frac{p!}{N^p} \sum_{i_1 < \cdots < i_p} \sigma_{i_1}\sigma'_{i_1} \ldots \sigma_{i_p}\sigma'_{i_p} - R(\sigma, \sigma')^p \right| \leq \frac{K}{N}. \qquad \square$$

Corollary 8.15. *If $\beta > 2\sqrt{\log 2}$, then*

$$(8.62) \qquad \liminf_{N \to \infty} \mathsf{E}\langle R(\sigma, \sigma')^p \rangle > 0,$$

and consequently, if $\beta \leq 2^{\frac{p}{4} - 6}$,

$$(8.63) \qquad \liminf_{N \to \infty} \mathsf{E} \sum_{\alpha \geq 1} w_\alpha^2 > 0.$$

Proof. (8.62) follows from (8.61) and Lemma 4.14 and this implies (8.63) by (8.60). $\qquad \square$

Comment. Observe that (8.62) and Theorem 8.6 prove (7.43).

Since $\sum w_\alpha = 1$, a consequence of (8.63) is that $\liminf_{N \to \infty} \mathsf{E} \max w_\alpha > 0$: For large N, with positive probability, there exists at least one lump of mass of order 1. Our theory of lumps is not empty. (It is in fact shown in [**T7**] that the lumps do carry all the mass).

Proof of Proposition 7.11. We write

$$A_{N,k}(\beta) = \mathsf{E}\langle R(\sigma, \sigma')^k \rangle.$$

Proceeding as in (8.37), we obtain

$$(8.64) \qquad A_{N+1,2}(\beta') \leq \frac{1}{N+1} + \frac{N}{N+1} \mathsf{E} \frac{1}{Z^2} \langle R(\sigma^1, \sigma^2) \operatorname{Av} \varepsilon_1 \varepsilon_2 \mathcal{E}_2 \rangle.$$

We note that $\langle R(\sigma^1, \sigma^2) \operatorname{Av} \varepsilon_1 \varepsilon_2 \mathcal{E}_2 \rangle$ is a square (as in the proof of Proposition 5.2) so that, since $Z \geq 0$,

$$(8.65) \qquad A_{N+1,2}(\beta') \leq \frac{1}{N+1} + \mathsf{E}\langle R(\sigma^1, \sigma^2) \operatorname{Av} \varepsilon_1 \varepsilon_2 \mathcal{E}_2 \rangle.$$

We recall that

$$\mathcal{E}_2 = \exp \beta \left(\frac{4!}{2N^3} \right)^{\frac{1}{2}} \sum_{l \leq 2} \varepsilon_l \boldsymbol{g} \cdot \boldsymbol{T}(\sigma^l),$$

so that

$$E_g \operatorname{Av} \varepsilon_1 \varepsilon_2 \mathcal{E}_2 = B \operatorname{sh}\left(\frac{\beta^2 4!}{2N^3} T(\sigma^1) \cdot T(\sigma^2)\right)$$

for

$$B = \exp \frac{\beta^2 4!}{2N^3} \frac{N(N-1)(N-2)}{3!} \le \exp 2\beta^2.$$

Assuming $\beta \le 3$, and using Lemma 8.7, we then have

$$(8.66) \qquad A_{N+1,2}(\beta') \le \frac{L_1}{N+1} + L_1 A_{N,4}(\beta).$$

To prove Proposition 7.11, we now argue by contradiction. If this proposition was wrong, we could say that

$$\exists N_0, \ \forall N \ge N_0, \ \forall \beta < 3,$$

$$(8.67) \qquad A_{N,2}(\beta) \le \frac{2L_1}{N} \implies A_{N,4}(\beta) \le \frac{1}{4} A_{N,2}(\beta).$$

Let us define

$$B(N) = \sup\left\{\beta > 0 \ \Big|\ A_{N,2}(\beta) \le \frac{2L_1}{N}\right\}.$$

The function $A_{N,2}(\cdot)$ is continuous, and $A_{N,2}(0) = 1/N$, so that

$$A_{N,2}(B(N)) = \frac{2L_1}{N}.$$

Thus, if $\beta \le B(N)$, $\beta \le 3$ by (8.67) we have

$$A_{N,4}(\beta) \le \frac{1}{4} A_{N,2}(\beta) \le \frac{1}{2}\frac{L_1}{N},$$

so that by (8.66)

$$A_{N+1,2}(\beta') \le \frac{L_1}{N+1} + \frac{1}{2}\frac{L_1}{N} \le \frac{2L_1}{N+1},$$

and this means that $\beta' \le B(N+1)$. Thus we have proved that

$$B(N) \le 3 \implies B(N+1) \ge \left(\frac{N+1}{N}\right)^{\frac{3}{2}} B(N).$$

This implies that there are arbitrarily large values of N for which $B(N) \ge 3$. This however contradicts (8.57). $\qquad\qquad\square$

Bibliography

[A-L-R] M. Aizenman, J.L. Lebowitz, D. Ruelle, Some rigorous results on the Sherrington-Kirkpatrick model, *Commun. Math. Phys.*, **112** (1987) 3–20.

[B-P] A. Bovier, P. Picco (editors), *Mathematical Aspects of Spin Glasses and Neural Networks*, Progress in Probability, Vol. 41, Birkhäuser, Boston, 1997.

[C] F. Comets, A spherical bound for the Sherrington-Kirkpatrick model, Hommage à P.-A. Meyer et J. Neveu, *Astérisque*, **236**, (1996) 103–108.

[C-N] F. Comets, J. Neveu, The Sherrington-Kirkpatrick model of spin glasses and stochastic calculus: the high temperature case, *Comm. Math. Phys.*, **166**, 3 (1995) 549–564.

[D] B. Derrida, Random energy model: An exactly solvable model of disordered systems, *Phys. Rev. B*, **24**, #5 (1981) 2613–2626.

[F-Z] J. Fröhlich, B. Zegarlinski, Some comments on the Sherrington-Kirkpatrick model of spin glasses, *Commun. Math. Phys.*, **112** (1987) 553–566.

[G] E. Gardner, Spin glasses with p-spin interactions, *Nuclear Phys. B*, **257**, #6 (1985) 747–765.

[G-G] S. Ghirlanda, F. Guerra, General properties of overlap probability distributions in disordered spin systems. Towards Parisi ultrametricity, *J. Phys. A*, **31**, #46 (1998) 9149–9155.

[I-S-T] I.A. Ibragimov, V. Sudakov, B.S. Tsirelson, Norms of Gaussian sample functions, "Proceedings of the Third Japan-USSR Symposium on Probability", Tashkent, 1975, *Lecture Notes in Math.*, **550**, Springer Verlag, Berlin, 1976, 20–41.

[K] J.-P. Kahane, Une inegalité du type de Slepian et Gordon sur les processus gaussiens, *Israel J. Math.*, **55**, 1 (1986) 109–110.

[K-T-J] J. Kosterlitz, D. Thouless, R. Jones, Spherical model of spin glass, *Phys. Rev. Lett.*, **36** (1976) 1217–1220.

[M-P-V] M. Mézard, G. Parisi, M. Virasoro, *Spin glass Theory and beyond*, World Scientific, Singapore, 1987.

[P] G. Parisi, *Field Theory, Disorder, Simulation*, World Scientific Lecture Notes in Physics 45, World Scientific, Singapore, 1992.

[Pi]   G. Pisier, Probabilistic methods in the geometry of Banach Spaces, Probability and analysis, Varenna 1985, *Springer Verlag Lecture Notes in Math.* n° **1206** (1996) 167–241.

[P-Y]   J. Pitman, M. Yor, The two-parameter Poisson-Dirichlet distribution derived from a stable subordinator, *Ann. Probab.*, **25** (1997) 855–900.

[Sh]   M. Shcherbina, On the replica-symmetric solution of the Sherrington-Kirkpatrick model, *Helv. Phys. Acta*, **70** (1997) 838–853.

[S-K]   D. Sherrington, S. Kirkpatrick, Solvable model of spin glass, *Phys. Rev. Lett.*, **35** (1972) 1792–1796.

[T1]   M. Talagrand, Concentration of measure and isoperimetric inequalities in product spaces, *Publ. Math. I.H.E.S.*, **81** (1995) 73–205.

[T2]   ———, The Sherrington-Kirkpatrick model: a challenge to mathematicians, *Probab. Theory Related Fields*, **110** (1998) 109–176.

[T3]   ———, Rigorous low temperature results for the p-spin interaction model, *Probab. Theory Related Fields*, **117** (2000) 303–360.

[T4]   ———, Exponential inequalities and replica-symmetry breaking for the Sherrington-Kirkpatrick model, *Ann. Probab.*, **28** (2000) 1018–1062.

[T5]   ———, Huge random structures and mean field models for spin glasses, in "Proceedings of the International Congress of Mathematicians, Vol. I (Berlin 1998)", *Documenta Math.*, Extra Vol. I (1998) 507–536.

[T6]   ———, Verres de spin et optimisation combinatoire, Séminaire Bourbaki, March 1999, *Astérisque*, **266** (2000) Exp. n° 859, 287–317.

[T7]   ———, Self organization in a spin glass model, in preparation.

[T8]   ———, On the high temperature region of the Sherrington-Kirkpatrick model, *Ann. Probab.*, to appear.

LIST OF OTHER TALKS

AIT OUAHRA Mohamed
Limit theorems for some functionals of stable processes in Hölder space

AYACHE Antoine
Processus Multifractionnaires

BARRAL Julien
Uniform convergence results for matinales in the BRW

BELOPOLSKAYA Yana
Nonlinear PDEs in diffusion theory

BERARD Jean
Asymptotics of a genetic algorithm

BEZNEA Lucian
Potential kernels, smooth measures and the Revuz correspondence

BOUCHERON Stéphane
A concentration inequality with applications

BUICULESCU Mioara
Exponential decay parameters associated with excessive measures

CALKA Pierre
On the spectral function of the Johnson-Mehl and Poisson-Voronoi

CARASSUS Laurence
Portfolio optmization for piecearse concave criteria functions

CONT Rama
Stochastic PDEs, infinite dimensional diffusions and interest rate dynamics

DEELSTRA Griselda
Optimal investment strategies in a Cir framework

DELMOTTE Thierry
Random walks on graphs of fractal nature

DURY Marie-Eliette
Estimation du paramètre de Hurst pour certains processus stables autosimilaires à accroissements stationnaires

ES-SAKY Elhassan
Backward stochastic differential equations and homogenization of partial differential equations

GAUBERT Stéphane
Iterates of monotone homogeneous maps

GUILLIN Arnaud
Moderate deviations of inhomogeneous functionals of Markov Process and Averaging

HERNANDEZ Daniel
Risk sensitive contraol of Markov processes with applications to Portfolio Management

HERMANN Samuel
A singular large deviations phenomenon

LAKHEL Elhassan
Un résultat d'approximation pour les processus stochastiques à deux paramètres dans les espaces de Besov

LATALA Rafal
Exponential inequalities for U-statistics

LE NY Arnaud
Mesures de Gibbs sur un réseau, groupe de renormalization et non-gibbsiannité: une présentation de la presque quasilocalité, de la gibbsiannité faible et de la quasilocalité fractale

LEOBACHER Gunther
The exact solution fo an optimization problem in long-term hedging

MARQUEZ-CARRERAS David
On stochastic partial differential equations with spatially correlated noïse

MARTIN Andreas
Small balls for the stochastic wave equation

MORATO Laura M.
Stochastic mechanics and Dirichlet forms

OLESZKIEWICZ Krzysztof
Between Sobolev and Poincaré

RAIC Martin
Stein's method

RUEDIGER Barbara
Non local Dirichlet forms and processes with jumps obtained by subordination
on general state spaces

SOOS Anna
Invariant sets of random variables

SORTAIS Michel
Dynamique de Langevin du modèle d'Edwards-Anderson et du "Random Field
Ising Model"

STOICA Lucretiu
A probabilistic interpretation of divergence and backward stochastic differential
equations

TARRES Pierre
Vertex-reinforced random walks and stochastic approximation algorithms

TEICHMANN Josef
Regularity of infinite-dimensional lie groups by metric space methods

TUDOR Ciprian
Tanaka formula for the fractional brownian motion

VERSCHUERE Michel
Entropy production for particle systems

WINKEL Matthias
Burgers turbulence initialized by a regenerative impulse

ZAMBOTTI Lorenzo
A reflected stochastic heat equation as symmetric dynamics with respect to the
3-d Bessel Bridge

ZERNER Martin P.W.
A zero-one law for planar random walks in random environment

LIST OF PARTICIPANTS

Mr.	ADARVE Sergio	Université de Bogotà (Colombie)
Mr.	AIT OUAHRA Mohamed	Université Cadi Ayyad, Marrakech (Maroc)
Mlle	AKIAN Marianne	INRIA, Domaine de Voluceau, Rocquencourt, Le Chesnay
Mr.	AYACHE Antoine	Université Paul Sabatier, Toulouse
Mr.	BARRAL Julien	Université de Montpellier II
Mlle	BELOPOLSKAYA Yana	State University for Architecture and Civil Engineering, St. Petersbourg, (Russie)
Mr.	BERARD Jean	Université Claude Bernard Lyon 1
Mr.	BERNARD Pierre	LMA, Université Blaise Pascal, Clermont-Ferrand
Mr.	BERTOLDI Marcello	Dipartimento di Matematica Université de Trento (Italie)
Mr.	BERTRAND Pierre	U.F.R. Psychologie Université Blaise Pascal, Clermont-Ferrand
Mr.	BEZNEA Lucian	Institute of Mathematics, Bucarest (Roumanie)
Mlle	BIAGINI Sara	Scuola Normale Superiore, Pise (Italie)
Mr.	BOUCHERON Stéphane	Maths, Université PARIS XI (Orsay)
Mr.	BOUFOUSSI Brahim	Mathématiques, Université Cadi Ayyad, Marrakech (Maroc)
Mr.	BUICULESCU Mioara Lucia	Centre for Mathematical Statistics Bucarest (Roumanie)
Mr.	CALKA Pierre	Université Claude Bernard, Lyon 1
Mr.	CAMPI Luciano	Dipartimento Matematica Pura ed Applicata Universitat di Padova, Padova (ITALIE)
Mme	CARASSUS Laurence	Université PARIS VII
Mr.	CARDONA Alexander	LMA, Université Blaise Pascal, Clermont-Ferrand
Mr.	CARMONA Philippe	LSP, Université Paul Sabatier, Toulouse
Mr.	COMETS Francis	Mathématiques, Université PARIS VII
Mr.	CONT Rama	CMAP, Ecole Polytechnique, Palaiseau
Mme	DEELSTRA Griselda	ENSAE, Malakoff
Mr.	DELAHAUT Thierry	LSP, Université Paul Sabatier, Toulouse
Mr.	DELMOTTE Thierry	LSP, Université Paul Sabatier, Toulouse
Mr.	DJELLOUT Hacène	LMA, Université Blaise Pascal, Clermont-Ferrand
Mme	DONATI-MARTIN Catherine	LSP, Université Paul Sabatier, Toulouse
Mlle	DURY Marie-Eliette	LMA, Université Blaise Pascal, Clermont-Ferrand
Mr.	DZIWISZ Artur	Mathematical Institute University of Wroclaw (Pologne)
Mr.	EMERY Michel	IRMA, Université René Descartes, Strasbourg
Mr.	ENGELBERT Hans-Jürgen	Institute for Stochastics University of Jen (Allemagne)
Mr.	ESSAKY Elhassan	Université Cadi Ayyad, Marrakech (Maroc)
Mr.	FAWCETT Thomas	M.C.R., St Anne's College Oxford (Royaume Uni)
Mr.	FLEURY Gérard	LMA, Université Blaise Pascal, Clermont-Ferrand
Mr.	FOUGERES Pierre	LSP, Université Paul Sabatier, Toulouse
Mr.	FRANZ Uwe	Institut für Mathematik und Informatik Universität Greifswald (Allemagne)

Mlle	GAIER Johanna	Department for Financial and Actuarial Mathematics
Mr.	GAUBERT Stéphane	Unité de Mathématiques Appliquées ENSTA, Paris
Mr.	GEISS Stefan	Department of Mathematics University of Jyväskylä (Finlande)
Mr.	GIROUX Gaston	Mathématiques,Université de Sherbrooke (Canada)
Mr.	GRORUD Axel	Maths et Info, Université de Marseille
Mr.	GROSSET Luca	Dipartimento di Matematica Pura et Applicata Università degli studi, Padova (Italie)
Mr.	GUILLIN Arnaud	LMA, Université Blaise Pascal, Clermont-Ferrand
Mlle	HANIG Kristina	Institut für Mathematik und Informatik Universität Greifswald (Allemagne)
Mr.	HERNANDEZ Daniel	Centro de Investigacion en Matematicas Guanajuato Gto (Mexique)
Mr.	HERRMANN Samuel	Institut Elie Cartan,Université H.Poincaré, Nancy 1
Mr.	ISHIKAWA Yasushi	Department of Mathematics Ehime University (Japon)
Mlle	KINZ Mélanie	Institut für Mathematik und Informatik Universität Greifswald (Allemagne)
Mr.	LAKHEL Elhassan	Université Cadi Ayyad, Marrakech (Maroc)
Mr.	LATALA Rafal	Institute of Mathematics Warsaw University (Pologne)
Mr.	LE NY Arnaud	IREM de Rennes Université de Rennes 1
Mr.	LEOBACHER Gunther	Institut fuer Mathematik Salzburg (Autriche)
Mr.	LEONARD Christian	Département de Mathématiques Université PARIS X, Nanterre
Mr.	MACHRAFI Hatim	LSP, Université de Sciences et Technologies, Lille
Mr.	MARDIN Arif	Département Signal & Image Institut National des Télécommunications, Evry
Mr.	MARQUEZ-CARRERAS David	Facultat de Matemàtiques Universitat de Barcelona (Espagne)
Mr.	MARTIN Andreas	Département de Mathématiques Ecole Polytechnique Fédérale, Lausanne (Suisse)
Mlle	MAZZOCCHI Sonia	Riga Technical University Riga, Lettonie (Russie)
Mme	MORATO Laura M.	Facultà di Scienze, Università di Verona (Italie)
Mr.	MOUTSINGA Octave	LPS, Université des Sciences et Technologies,Lille
Mr.	MYTNIK Leonid	Faculty of Industrial Engineering and Management, Technion, Haifa (Israël)
Mr.	OLESZKIEWICZ Krzysztof	Institute of Mathematics Warsaw University (Pologne)
Mme	PAYCHA Sylvie	LMA, Université Blaise Pascal, Clermont-Ferrand
Mr.	PEREZ PEREZ Aroldo	Centro de Investigacion en Matematicas Guanajuato, Gto (Mexique)
Mme	PETIT Frédérique	Probabilités et Modèles Aléatoires Université Paris VI
Mr.	PRATELLI Maurizio	Dipartimento di Matematica Universita di Pisa (Italie)

Mr.   RAIC Martin   Institute of Mathematics
                                           Ljubljana (Slovénie)
Mr.   ROUX Daniel   LMA, Université Blaise Pascal, Clermont-Ferrand
Mme RUEDIGER Barbara   Institut Angewandte Mathematik
                                           Universität Bonn (Allemagne)
Mr.   SAINT LOUBERT BIE Erwan   LMA, Université Blaise Pascal, Clermont-Ferrand
Mlle SARRA ROVIRA Monica   Facultat de Matemàtiques,
                                           Universitat de Barcelona (Espagne)
Mr.   SCHIED Alexander   Institut für Mathematik/stochastik
                                           Humboldt-Universität, Berlin (Allemagne)
Mr.   SCHILTZ Jang   Mathématiques, Centre Universitaire de
                                           Luxembourg
Mlle SOOS Anna   Babes Bolyai University
                                           Cluj-Napoca (Roumanie)
Mr.   SORTAIS Michel   DMA, Ecole Polytechnique Fédérale
                                           Lausanne (Suisse)
Mr.   STOICA Lucretiu   Institut de Mathématiques
                                           Bucarest (Roumanie)
Mr.   SUMMER Christopher   Financial and Actuarial Mathematics
                                           Vienna University of Technology (Autriche)
Mr.   TARRES Pierre   CMLA, ENS de Cachan
Mr.   TEICHMANN Josef   Financial and Actuarial Mathematics
                                           Technische Universität Wien (Autriche)
Mr.   TINDEL Samy   Institut Galilée, Université PARIS XIII
Mr.   TUDOR Ciprian   Département de Mathématiques
                                           Université de la Rochelle
Mlle UGOLINI Stefania   Facoltà di Scienze
                                           Università degli studi di Verona (Italie)
Mr.   VACCARO David   Mathematical Institute
                                           St Giles, Oxford (Royaume Uni)
Mr.   VERSCHUERE Michel   Institute for Theoretical Physics
                                           Catholic University of Leuven, Heverlee
                                           (Belgique)
Mr.   VILLA MORALES Jose   Centro de Investigacion en Matematicas
                                           Guanajuato Gto (Mexique)
Mr.   WINKEL Matthias   Probabilités et Modèles Aléatoires
                                           Université PARIS VI
Mr.   WU Liming   LMA, Université Blaise Pascal, Clermont-Ferrand
Mme ZAGORKA Lozanov-Crvenkovic Institute of Mathematics
                                           University of Novi Sad (Yougoslavie)
Mr.   ZAMBOTTI Lorenzo   Scuola Normale, Pise (Italie)
Mr.   ZERNER Martin P.W.   Department of Electrical Engineering
                                           Technion, Haifa (Israël)

LIST OF PREVIOUS VOLUMES OF THE
"Ecole d'Eté de Probabilités"

Druck: Strauss Offsetdruck, Mörlenbach
Verarbeitung: Schäffer, Grünstadt

4. Manuscripts should in general be submitted in English. Final manuscripts should contain at least 100 pages of mathematical text and should include
   - a general table of contents;
   - an informative introduction, with adequate motivation and perhaps some historical remarks: it should be accessible to a reader not intimately familiar with the topic treated;
   - a global subject index: as a rule this is genuinely helpful for the reader.

5. Lecture Notes are printed by photo-offset from the master-copy delivered in camera-ready form by the authors. Springer-Verlag provides technical instructions for the preparation of manuscripts. We strongly recommend that all contributions in a volume be written in the same LaTeX version, preferably LaTeX2e. Macro-packages in LaTeX2e are available from Springer's web-pages at

   http://www.springer.de/math/authors/index.html .

   Careful preparation of manuscripts will help keep production time short and ensure satisfactory appearance of the finished book. After acceptance of the manuscript authors/volume editors will be asked to prepare the final LaTeX source files (and also the corresponding dvi- or pdf-file) together with the final printout made from these files. The LaTeX source files are essential for producing a unified full-text online version of the book

   (http://www.springerlink.com/link/service/series/0304/tocs.htm).

   The actual production of a Lecture Notes volume takes approximately 12 weeks.

6. Volume editors receive a total of 50 free copies of their volume to be shared with the authors, but no royalties. They and the authors are entitled to a discount of 33.3 % on the price of Springer books purchased for their personal use, if ordering directly from Springer-Verlag.

Commitment to publish is made by letter of intent rather than by signing a formal contract. Springer-Verlag secures the copyright for each volume. Authors are free to reuse material contained in their LNM volumes in later publications: A brief written (or e-mail) request for formal permission is sufficient.

Addresses:

Professor J.-M. Morel, CMLA,
Ecole Normale Supérieure de Cachan,
61 Avenue du Président Wilson, 94235 Cachan Cedex, France
E-mail: Jean-Michel.Morel@cmla.ens-cachan.fr

Professor F. Takens, Mathematisch Instituut,
Rijksuniversiteit Groningen, Postbus 800,
9700 AV Groningen, The Netherlands
E-mail: F.Takens@math.rug.nl

Professor B. Teissier, Université Paris 7
Institut Mathématique de Jussieu, UMR 7586 du CNRS
Equipe "Géométrie et Dynamique", 175 rue du Chevaleret
75013 Paris, France
E-mail: teissier@math.jussieu.fr

Springer-Verlag, Mathematics Editorial, Tiergartenstr. 17,
69121 Heidelberg, Germany,
Tel.: +49 (6221) 487-8410
Fax: +49 (6221) 487-8355
E-mail: lnm@Springer.de